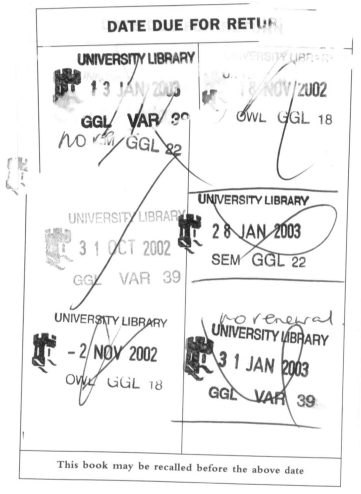

For Adélia

Plant Pathology and Plant Pathogens

JOHN A. LUCAS

IACR-LONG ASHTON RESEARCH STATION
UNIVERSITY OF BRISTOL

THIRD EDITION

 **Blackwell** Publishing

© 1977, 1982, 1998 by Blackwell Science Ltd
a Blackwell Publishing company

350 Main Street, Malden, MA 02148-5018, USA
108 Cowley Road, Oxford OX4 1JF, UK
550 Swanston Street, Carlton South, Melbourne, Victoria 3053, Australia
Kurfürstendamm 57, 10707 Berlin, Germany

First published 1977 by Blackwell Science Ltd
Second edition 1982
Third edition 1998
Reprinted 2002

Library of Congress Cataloging-in-Publication Data

Lucas, John Alexander.
  Plant pathology and plant pathogens / John A. Lucas.—3rd ed.
    p. cm.
  Rev ed. of: Plant pathology and plant pathogens / C. H. Dickinson, J. A. Lucas.
  Includes bibliographical references (p.   ).
  ISBN 0-632-03046-1
  1. Plant diseases. 2. Phytopathogenic microorganisms—Control. 3. Plant–
pathogen relationships. I. Dickinson, C. H. Plant pathology and plant pathogens.
II. Title.
SB731.L84   1998
632'.3—DC21   97-48463
                                                      CIP

A catalogue record for this title is available from the British Library.

Set by Excel Typesetters Co., Hong Kong
Printed and bound in Great Britain
by TJ International Ltd, Padstow, Cornwall

For further information on
Blackwell Publishing, visit our website:
http://www.blackwellpublishing.com

Contents

Preface

Plant pathology is an applied science concerned primarily with practical solutions to disease problems in agriculture, horticulture and forestry. The study of plant disease, and the development of methods for its control, continue to be vital elements in the drive to improve crop productivity. Projected world population growth makes this a priority. In recent years, however, experimental analysis of the interactions between plants and pathogens has become fertile territory for scientists interested in fundamental aspects of plant recognition and response systems, signal pathways, and stress physiology. The use of molecular genetic techniques has provided new insights into how pathogens cause disease, and how plants defend themselves against attack. In turn this new understanding is suggesting novel ways in which plant disease might be controlled.

This book is intended to provide an introduction to the main elements of plant pathology, including both the practical aspects of disease identification, assessment and control, and fundamental aspects of host–pathogen interaction. The aim is to place the subject not only in the context of agricultural science, but also in the wider spheres of ecology, population biology, cell biology and genetics. The recent impact of molecular biology on the diagnosis and analysis of plant disease is a recurrent theme.

A large number of people have helped in the production of this new edition. First I should acknowledge the invaluable help of many of my colleagues at IACR-Long Ashton Research Station, especially John Hargreaves, Paul Bowyer, Richard O'Connell and John Bailey in the Molecular Pathology group. David Royle, Derek Hollomon, Sheila Kendall and Eric Hislop kindly provided figures or data for inclusion. Mike Beale not only checked but also redrew many of the chemical structures. Philip Brain and Donna Hudson advised on mathematical modelling, and Keith Edwards on molecular markers. Ken Williams assisted with printing photographs, while Steve Smith provided excellent library support. Other IACR colleagues at Rothamsted or Brooms Barn who supplied information, advice or figures include Jon Antoniw, Mike Asher, Alan Dewar, Bruce Fitt, Richard Harrington, Phil Jones, Roger Plumb and Paul Verrier. Brian Case and Mike Davey at the University of Nottingham contributed to the symptom pictures. Among the many other scientists both in the UK and abroad who have helped me are Willem Broekaert, Tim Carver, Bill Clark, Mike Coffey, Andrew Entwistle, Stuart Falk, Robert Goodman, Steve Gregory, Martin Hartley, Rosemarie Honnegger, Richard Howard, Ning Huang, Chris Lamb, Morris Levy, John Mansfield, Kurt Mendgen, Tim Murray, Rupert Osborn, Nick Read, Adrian Spiers, Wijnand Swart, Philip Talboys, Bev Walpole and Martin Wolfe. Special thanks must go to Alison Daniels for her help with micrographs, and Olu Latunde-Dada for lively discussions and providing the back cover photo. Finally, a very big thank-you to Ellen Cooke, whose patient support and practical help with reading lists, figures and the appendix, as well as computer trouble-shooting, was a vital factor in bringing the project to a conclusion.

John Lucas

Acknowledgements

Permission to reproduce data and figures from the following publications has kindly been given by authors, editors and publishers. References to this list are made at the appropriate point in the text.

Ayres, P.G. & Jones, P. (1975) Increased transpiration and the accumulation of root absorbed ^{86}Rb in barley leaves infected by *Rhynchosporium secalis* (leaf blotch). *Physiological Plant Pathology* 7, 49–58.

Bailey, J.A. & Deverall, B.J. (1971) Formation and activity of phaseollin in the interaction between bean hypocotyls (*Phaseolus vulgaris*) and physiological races of *Colletotrichum lindemuthianum*. *Physiological Plant Pathology* 1, 435–449.

Basallote-Ureba, M.J. & Melero-Vara, J.M. (1993) Control of garlic white rot by soil solarization. *Crop Protection* 12, 219–223.

Basham, H.G. & Bateman, D.F. (1975) Killing of plant cells by pectic enzymes: the lack of direct injurious interaction between pectic enzymes or their soluble reaction products and plant cells. *Phytopathology* 65, 141–153.

Berg, D., Büchel, K.-H., Holmwood, G., Krämer, W. & Pontzen, R. (1990) Sterol biosynthesis inhibitors. Model studies with respect to modes of action and resistance. In: *Managing Resistance to Agrochemicals. From Fundamental Research to Practical Strategies* (eds M.B. Green, H.M. LeBaron & W.K. Moberg). ACS Symposium Series 421, pp. 184–198. American Chemical Society, Washington, D.C.

Bonas, U., Schulte, R., Fenselau, S., Minsavage, G.V., Staskawicz, B.J. & Stall, R.E. (1991) Isolation of a gene cluster from *Xanthomonas campestris* pv. *vesicatoria* that determines pathogenicity and the hypersensitive response on pepper and tomato. *Molecular Plant–Microbe Interactions* 4, 81–88.

Bracker, C.E. (1968) Ultrastructure of the haustorial apparatus of *Erysiphe graminis* and its relationship to the epidermal cell of barley. *Phytopathology* 58, 12–30.

Braun, E.J. & Howard, R.J. (1994) Adhesion of fungal spores and germlings to host plant surfaces. *Protoplasma* 181, 202–212.

Brown, I.R. & Mansfield, J.W. (1988) An ultrastructural study, including cytochemistry and quantitative analyses of the interactions between pseudomonads and leaves of *Phaseolus vulgaris* L. *Physiological and Molecular Plant Pathology* 33, 351–376.

Bunders, J. (1988) Appropriate biotechnology for sustainable agriculture in developing countries. *Trends in Biotechnology* 6, 173–180.

Café-Filho, A.C, Duniway, J.M. & Davis, R.M. (1995) Effects of the frequency of furrow irrigation on root and fruit rots of squash caused by *Phytophthora capsici*. *Plant Disease* 79, 44–48.

Campbell, W.P. & Griffiths, D.A. (1974) Development of endoconidial chlamydospores in *Fusarium culmorum*. *Transactions of the British Mycological Society* 63, 221–228.

Coffey, M.D. (1975) Ultrastructural features of the haustorial apparatus of the white blister fungus *Albugo candida*. *Canadian Journal of Botany* 53, 1285–1299.

Coffey, M.D. (1976) Flax rust resistance involving the K gene: an ultrastructural survey. *Canadian Journal of Botany* 54, 1443–1457.

Coffey, M.D. (1987) Phytophthora root rot of avocado. An integrated approach to control in California. *Plant Disease* 71, 1046–1052.

Coley-Smith, J.R., Mitchell, C.M. & Sansford, C.E. (1990) Long-term survival of sclerotia of *Sclerotium cepivorum* and *Stromatinia gladioli*. *Plant Pathology* 39, 58–69.

Cramer, H.H. (1967) *Plant Protection and World Crop Production*. Bayer, Leverkusen.

Daniels, A. & Lucas, J.A. (1995) Mode of action of the anilino-pyrimidine fungicide Pyrimethanil. 1. *In-vivo* activity against *Botrytis fabae* on broad bean (*Vicia faba*) leaves. *Pesticide Science* 45, 33–41.

Daniels, A., Lucas, J.A. & Peberdy, J.F. (1991) Morphology and ultrastructure of W and R pathotypes of *Pseudocercosporella herpotrichoides* on wheat seedlings. *Mycological Research* 95, 385–397.

Dixon, R.A. & Harrison, M.J. (1990) Activation and structure of organization of genes involved in microbial defense in plants. *Advances in Genetics* 28, 165–217.

Eckert, J.W. (1977) Control of postharvest diseases. In: *Antifungal Compounds*, Vol. I, *Discovery, Development and Uses* (eds M.R. Siegel & H.D. Sisler), pp. 269–352. Marcel Dekker, New York.

Esau, K. (1968) *Viruses in Plant Hosts*. University of Wisconsin Press, Milwaukee, Wis.

Eskes, A.B. (1983) Characterization of incomplete resistance to *Hemileia vastatrix* in *Coffea canephora* cv. Kouillou. *Euphytica* 32, 639–648.

Evans, E. (1977) Efficient use of systemic fungicides. In: *Systemic Fungicides* (ed. R.W. Marsh), pp. 198–212. Longman, London.

Falk, S.P., Pearson, R.C., Gadoury, D.M., Seem, R.C. & Sztejnbert, A. (1996) *Fusarium proliferatum* as a biocontrol agent against grape downy mildew. *Phytopathology* 86, 1010–1017.

Farmer, E.E. & Ryan, C.A. (1992) Octadecanoid precursors of jasmonic acid activate the synthesis of wound-inducible proteinase inhibitors. *The Plant Cell* 4, 129–134.

Fry, W.E. & Thurston, H.D. (1980) The relationship of plant pathology to integrated pest management. *BioScience* 30, 665–669.

González-Torres, R., Meléo-Vara, J.M., Gómez-Vázquez, J. & Jiménez-Diaz, R.M. (1993) The effects of soil solarization and soil fumigation on fusarium wilt of watermelon grown in plastic houses in south-eastern Spain. *Plant Pathology* 42, 858–864.

Greaves, D.A., Hooper, A.J. & Walpole, B.J. (1983) Identification of barley yellow dwarf virus and cereal aphid infestations in winter wheat by aerial photography. *Plant Pathology* 32, 159–172.

Green, C.F. & Ivins, J.D. (1984) Late infestations of take-all (*Gaeumannomyces graminis* var. *tritici*) on winter wheat (*Triticum aestivum* cv. Virtue): yield, yield components and photosynthetic potential. *Field Crops Research* 8, 199–206.

Hall, R. & Phillips, L.G. (1992) Effects of crop sequence and rainfall on population dynamics of *Fusarium solani* f.sp. *phaseoli* in soil. *Canadian Journal of Botany* 70, 2005–2008.

Hansen, J.G. (1993) The use of meteorological data for potato late blight forecasting in Denmark. In: *Proceedings of the Workshop on Computer-Based DSS on Crop Protection* (eds B.J.M. Secher, V. Rossi & P. Battilani), pp. 183–192. SP-Report Vol. 7, Danish Institute of Plant and Soil Science.

Hargreaves, J.A., Mansfield, J.W. & Rossall, S. (1977) Changes in phytoalexin concentrations in tissues of the broad bean plant (*Vicia faba* L.) following inoculation with species of *Botrytis*. *Physiological Plant Pathology* 11, 227–242.

Hayes, J.D. & Johnson, T.D. (1971) Breeding for disease resistance. In: *Diseases of Crop Plants* (ed. J.H. Western), pp. 62–88. Macmillan, London.

Hedrick, S.A., Bell, J.N., Boller, T. & Lamb, C.J. (1988) Chitinase cDNA cloning and mRNA induction by fungal elicitor, wounding and infection. *Plant Physiology* 86, 182–286.

Hewitt, H.G. & Ayres, P.G. (1975) Changes in CO_2 and water vapour exchange rates in leaves of *Quercus robur* infected by *Microsphaera alphitoides* (powdery mildew). *Physiological Plant Pathology* 7, 127–137.

Hickey, E.L. & Coffey, M.D. (1977) A fine structural study of the pea downy mildew fungus *Peronospora pisi* in its host *Pisum sativum*. *Canadian Journal of Botany* 55, 2845–2858.

Hirano, S.S., Rouse, D.I., Clayton, M.K. & Upper, C.D. (1995) *Pseudomonas syringae* pv. *syringae* and bacterial brown spot of snap bean: A study of epiphytic bacteria and associated disease. *Plant Disease* 79, 1085–1093.

Hirumi, H. & Maramorosch, K. (1973) Ultrastructure of the aster yellows agents: Mycoplasma-like bodies in sieve tube elements of *Nicotiana rustica*. *Annals of the New York Academy of Sciences* 225, 201–222.

Hittalmani, S., Foolad, M.R., Mew, T., Rodriguez, R.L. & Huang, N. (1995) Development of a PCR-based marker to identify rice blast resistance gene, *Pi-2 (t)*, in a segregating population. *Theoretical and Applied Genetics* 91, 9–14.

Holtz, B.A. & Weinhold, A.R. (1994) *Thielaviopsis basicola* in San Joaquin valley soils and the relationship between inoculum density and disease severity of cotton seedlings. *Plant Disease* 78, 986–990.

Honegger, R. (1985) Scanning electron-microscopy of the fungus–plant cell interface — A simple preparative technique. *Transactions of the British Mycological Society* 84, 530–533.

Hooker, W.J. (1956) Foliage fungicides for potatoes in Iowa. *American Potato Journal* 33, 47–52.

Howard, R.J. (1994) Cell biology of pathogenesis. In: *Rice Blast Disease* (eds R.S. Zeigler, S.A. Leong & P.S. Teng), pp. 3–22. CAB International, Wallingford.

Hyakumachi, M., Kanzawa, K. & Ui, T. (1990) Rhizoctonia root rot decline in sugarbeet monoculture. In: *Biological Control of Soil-Borne Plant Pathogens* (ed. D. Hornby), pp. 227–247. CAB International, Wallingford.

Jeffree, C.E., Baker, E.A. & Holloway, P.A. (1976) Origins of the fine structure of plant epicuticular waxes. In: *Microbiology of Aerial Plant Surfaces* (eds C.H. Dickinson & T.F. Preece), pp. 119–158. Academic Press, London.

Jones, D.A., Thomas, C.M., Hammond-Kosack, K.E., Balint-Kurti, P.J. & Jones, J.D.G. (1994) Isolation of the tomato *Cf-9* gene for resistance to *Cladosporium fulvum* by transposon tagging. *Science* 266, 789–793.

Jørgensen, L.N., Secher, B.J.M. & Nielsen, G.C. (1996) Monitoring diseases of winter wheat on both a field and a national level in Denmark. *Crop Protection* 15, 383–390.

Kendall, S., Hollomon, D.W., Cooke, L.R. & Jones, D.R. (1993) Changes in sensitivity to DMI fungicides in *Rhynchosporium secalis*. *Crop Protection* **12**, 357–362.

Kistler, H.C. & VanEtten, H.D. (1984) Regulation of pisatin demethylation in *Nectria haematococca* and its influence on pisatin tolerance and virulence. *Journal of General Microbiology* **130**, 2605–2613.

Kranz, J. & Royle, D.J. (1978) Perspectives in mathematical modelling of plant disease epidemics. In: *Plant Disease Epidemiology*, (eds P.R. Scott & A. Bainbridge), pp. 111–120. Blackwell Scientific Publications, Oxford.

Large, E.C. (1952) The interpretation of progress curves for potato blight and other plant diseases. *Plant Pathology* **1**, 109–117.

Latunde-Dada, A.O., Bailey, J.A. & Lucas, J.A. (1997) Infection process of *Colletotrichum destructivum* O'Gara from lucerne (*Medicago sativa* L.). *European Journal of Plant Pathology* **103**, 35–41.

Levy, M., Romao, J., Marchetti, M.A. & Hamer, J.E. (1991) DNA fingerprinting with a dispersed repeated sequence resolves pathotype diversity in the rice blast fungus. *The Plant Cell* **3**, 95–102.

Liu, L., Kloepper, J.W. & Tuzun, S. (1995) Induction of systemic resistance in cucumber against bacterial angular leaf spot by plant growth-promoting rhizobacteria. *Phytopathology* **85**, 843–847.

Livine, A. & Daly, J.M. (1966) Translocation in healthy and rust-affected beans. *Phytopathology* **56**, 170–175.

Lyr, H. (1995) Selectivity in modern fungicides and its basis. In: *Modern Selective Fungicides—Properties, Applications, Mechanisms of Action*. (ed. H. Lyr), pp. 13–22. Gustav Fischer Verlag, Jena.

MacKenzie, D.R. (1981) Scheduling fungicide applications for potato late blight with BLITECAST. *Plant Disease* **65**, 394–399.

Magyarosy, A.C., Schürmann, P. & Buchanan, B.B. (1976) Effect of powdery mildew infection on photosynthesis by leaves and chloroplasts of sugar beets. *Plant Physiology* **57**, 486–489.

Melchers, L.S. & Hooykaas, P.J.J. (1987) Virulence of *Agrobacterium*. In: *Oxford Surveys of Plant Molecular and Cell Biology* (ed. B.J. Miflin), pp. 167–220. Oxford University Press, Oxford.

Mohn, G., Koehl, P., Budzikiewicz, H. & Lefèvre, J.-F. (1994) Solution structure of pyoverdin GM-II. *Biochemistry* **33**, 2843–2851.

Mundt, C.C. & Leonard, K.J. (1985) A modification of Gregory's model for describing plant disease gradients. *Phytopathology* **75**, 930–935.

Oerke, E.-C., Dehne, H.-W., Schonbeck, F. & Weber, A. (1994) *Crop Production and Crop Protection*. Elsevier, Amsterdam.

Owera, S.A.P., Farrar, J.F. & Whitbread, R. (1983) Translocation from leaves of barley infected with brown rust. *New Phytologist* **94**, 111–123.

Paul, N.D. & Ayres, P.G. (1986a) The impact of a pathogen, *Puccinia lagenophorae*, on populations of groundsel, *Senecio vulgaris*, overwintering in the field. 1. Mortality, vegetative growth and the development of size hierarchies. *Journal of Ecology* **74**, 1069–1084.

Paul, N.D. & Ayres, P.G. (1986b) The impact of a pathogen, *Puccinia lagenophorae*, on populations of groundsel, *Senecio vulgaris*, overwintering in the field. 2. Reproduction. *Journal of Ecology* **74**, 1085–1094.

Plumb, R.T. (1986) A rational approach to the control of barley yellow dwarf virus. *Journal of the Royal Agricultural Society of England* **147**, 162–171.

Plumb, R.T. (1995) The epidemiology of barley yellow dwarf in Europe. In: *Barley Yellow Dwarf: 40 Years of Progress* (eds C.J. D'Arcy & P.A. Burnett), pp. 107–127. APS Press, St Paul, Minn.

Priestley, R.H. (1978) Detection of increased virulence in populations of wheat yellow rust. In: *Plant Disease Epidemiology* (eds P.R. Scott & A. Bainbridge), pp. 63–70. Blackwell Scientific Publications, Oxford.

Priestley, R.H. & Bayles, R.A. (1982) *Identification and Control of Cereal Diseases*. National Institute of Agricultural Botany, Cambridge.

Puchta, H., Herold, T., Verhoeven, K. *et al.* (1990) A new strain of potato spindle tuber viroid (PSTVd-N) exhibits major sequence differences as compared to all other PSTVd strains sequenced so far. *Plant Molecular Biology* **15**, 509–511.

Read, N.D., Kellock, L.J., Knight, H. & Trewavas, A.J. (1992) Contact sensing during infection by fungal pathogens. In: *Perspectives in Plant Cell Recognition* (eds J.A. Callow & J.R Green), pp. 137–192. Cambridge University Press, Cambridge.

Rogers-Lewis, D. (1985) Dried peas — A review of trials 1980–84. In: *Terrington Experimental Husbandry Farm Annual Review*. Ministry of Agriculture, Fisheries and Food, London.

Royle, D.J. (1973) Quantitative relationships between infection by the hop downy mildew pathogen, *Pseudoperonospora humuli*, and weather and inoculum factors. *Annals of Applied Biology* **73**, 19–30.

Royle, D.J., Shaw, M.W. & Cook, R.J. (1986) Patterns of development of *Septoria nodorum* and *S. tritici* in some winter wheat crops in Western Europe, 1981–83. *Plant Pathology* **35**, 466–476.

Saile, E., McGarvey, J.A., Schell, M.A. & Denny, T.P. (1997) Role of extracellular polysaccharide and endoglucanase in root invasion and colonization of tomato plants by *Ralstonia solanacearum*. *Phytopathology* **87**, 1264–1271.

Scheffer, R.P. & Yoder, O.C. (1972) Host-specific toxins and selective toxicity. In: *Phytotoxins in Plant Diseases* (eds R.K.S. Wood, A. Ballio & A. Graniti), pp. 251–272. Academic Press, London.

Schieber, E. & Zentmyer, G.A. (1984) Coffee rust in the Western Hemisphere. *Plant Disease* **68**, 89–93.

Scholes, J.D., Lee, P.J., Horton, P. & Lewis, D.H. (1994) Invertase—Understanding changes in the photosynthetic and carbohydrate-metabolism of barley leaves infected with powdery mildew. *New Phytologist* **126**, 213–222.

Scholthof, K.B.G., Scholthof, H.B. & Jackson, A.O. (1993) Control of plant-virus diseases by pathogen-derived resistance in transgenic plants. *Plant Physiology* **102**, 7–12.

Seck, M., Roelfs, A.P. & Teng, P.S. (1988) Effect of leaf rust (*Puccinia recondita tritici*) on yield of four isogenic wheat lines. *Crop Protection* **7**, 39–42.

Shaw, P.D. (1987) Plasmid ecology. In: *Plant–Microbe Interactions*, Vol. 2 (eds T. Kosuge. & E.W. Nester), pp. 3–39. Macmillan Publishing Company, New York.

Showalter, A.M., Bell, J.N., Cramer, C.L., Bailey, J.A., Varner, J.E. & Lamb, C.J. (1985) Accumulation of hydroxyproline-rich glycoprotein mRNAs, in response to fungal elicitor and infection. *Proceedings of the National Academy of Sciences of the USA* **82**, 6551–6556.

Slusarenko, A.J., Croft, K.P. & Voisey, C.R. (1991) Biochemical and molecular events in the hypersensitive response of beans to *Pseudomonas syringae* pv. *phaseolicola*. In: *Biochemistry and Molecular Biology of Plant–Pathogen Interactions. Proceedings of the Phytochemical Society of Europe* (ed. C.J. Smith), pp. 127–143. Clarendon Press, Oxford.

Smedegaard-Petersen, V. (1984) The role of respiration and energy generation in diseased and disease-resistant plants. In: *Plant Diseases: Infection, Damage and Loss.* (eds R.K.S. Wood & G.J. Jellis), pp. 73–85. Blackwell Scientific Publications, Oxford.

Smith, V.L., Campbell, C.L., Jenkins, S.F. & Benson, D.M. (1988) Effects of host density and number of disease foci on epidemics of southern blight of processing carrot. *Phytopathology* **78**, 595–600.

Song, W.-Y., Wang, G.-L., Chen, L.-L. *et al.* (1995) A receptor kinase-like protein encoded by the rice disease resistance gene, *Xa21*. *Science* **270**, 1804–1806.

Spiers, A.G. & Hopcroft, D.H. (1996) Morphological and host range studies of *Melampsora* rusts attacking *Salix* species in New Zealand. *Mycological Research* **100**, 1163–1175.

Staskawicz, B.J., Ausubel, F.M., Baker, B.J., Ellis, J.G. & Jones, J.D.G. (1995) Molecular genetics of plant disease resistance. *Science* **268**, 661–667.

Staub, T. (1991) Fungicide resistance: practical experience with antiresistance strategies and the role of integrated use. *Annual Review of Phytopathology* **29**, 421–442.

Stover, R.H. (1973) Method 106. In: *Crop Loss Assessment Methods. FAO Manual on the Evaluation and Prevention of Losses by Pests, Diseases and Weeds.* FAO and Commonwealth Agricultural Bureau, Oxford.

Sweetmore, A., Simons, S.A. & Kenward, M. (1994) Comparison of disease progress curves for yam anthracnose (*Colletotrichum gloeosporioides*). *Plant Pathology* **43**, 206–215.

Talboys, P.W. (1987) Verticillium wilt in English hops: retrospect and prospect. *Canadian Journal of Plant Pathology* **9**, 68–77.

Taylor, S., Massiah, A., Lomonossoff, G., Roberts, L.M., Lord, J.M. & Hartley, M. (1994) Correlation between the activities of five ribosome-inactivating proteins in depurination of tobacco ribosomes and inhibition of tobacco mosaic virus infection. *The Plant Journal* **5**, 827–835.

Tomiyama, K. (1971) Cytological and biochemical studies of the hypersensitive reaction of potato cells to *Phytophthora infestans*. In: *Morphological and Biochemical Events in Plant–Parasite Interaction* (eds S. Akai & S. Ouchi), pp. 387–401. Ednki Shoin, Tokyo.

Tottman, D.R. & Broad, H. (1987) The decimal code for the growth stages of cereals, with illustrations. *Annals of Applied Biology* **110**, 441–454.

Turner, J.G. (1984) Role of toxins in plant disease. In: *Plant Diseases: Infection, Damage and Loss* (eds R.K.S. Wood & G.J. Jellis), pp. 3–12. Blackwell Scientific Publications, Oxford.

Vanderplank, J.E. (1963) *Plant Diseases: Epidemics and Control.* Academic Press, New York.

Vernooij, B., Friedrich, L., Ahl Goy, P., Staub, T. & Kessmann, H. (1995) 2,6-Dichloroisonicotinic acid-induced resistance to pathogens without the accumulation of salicylic acid. *Molecular Plant–Microbe Interactions* **8**, 228–234.

Wellings, C.R. & McIntosh, R.A. (1990) *Puccinia striiformis* f.sp. *tritici* in Australasia: pathogenic changes during the first 10 years. *Plant Pathology* **39**, 316–325.

Williams, P.H., Aist, J.R. & Bhattacharya, P.K. (1973) Host–parasite relations in Cabbage Clubroot. In: *Fungal Pathogenicity and the Plant's Response* (eds R.J.W. Byrde & C.V. Cutting), pp. 141–148. Academic Press, London.

Wolfe, M.S. (1993) Can the strategic use of disease resistant hosts protect their inherent durability? In: *Durability of Disease Resistance* (eds Th. Jacobs & J.E. Parlevliet), pp. 83–96. Kluwer Academic Publishers, Dordrecht.

Zadoks, J.C. & Schein, R.D. (1979) *Epidemiology and Plant Disease Management.* Oxford University Press, Oxford.

Zhu, Q., Maher, E.A., Masoud, S., Dixon, R.A. & Lamb, C.J. (1994) Enhanced protection against fungal attack by constitutive co-expression of chitinase and glucanase genes in transgenic tobacco. *BioTechnology* **12**, 807–812.

Part 1
Plant Disease

'We see our cattle fall and our plants wither without being able to render them assistance, lacking as we do understanding of their condition.' [J.C. Fabricius, 1745–1808]

The health of green plants is of vital importance to everyone, although few people may realize it. As the primary producers in the ecosystem, green plants provide the energy and carbon skeletons upon which almost all other organisms depend. The growth and productivity of plants determines the food supply of animal populations, including the human population. Factors affecting plant productivity, including disease, therefore affect the quantity, quality and availability of staple foods throughout the world. Nowadays crop failure, due to adverse climate, pests, weeds or diseases, is rare in developed agriculture, and instead there are surpluses of some foods. Nevertheless disease still takes a toll, and much time, effort and money is spent on protecting crops from harmful agents. In developing countries the consequences of plant disease may be more serious, and crop failure can damage local or national economies, and lead directly to famine and hardship. Improvements in the diagnosis and management of plant disease are a priority in such instances. Furthermore, the pressures on plant productivity are increasing. Currently there are 0.3 ha of land per person on our planet, and as the population continues to multiply this figure will decline.

As well as supplying staple foods, plants provide many other vital commodities such as timber, fibres, oils, spices and drugs. The use of plants as alternative renewable sources of energy and chemical feedstocks is becoming more and more important, as other resources are depleted. Finally, the quality of the natural environment, from wilderness areas to urban parks, sports fields, and gardens, also depends to a large extent on healthy plants.

The science of **plant pathology** is the study of all aspects of disease in plants, including causal agents, their diagnosis, physiological effects, population dynamics and control. It is a science of synthesis, using data and techniques from fields as diverse as agriculture, microbiology, meteorology, genetics and biochemistry. But first and foremost plant pathology is an applied science, concerned with practical solutions to the problem of plant disease. Part of the appeal of the subject is to be found in this mixture of pure and applied aspects of biology.

The scope of plant pathology is difficult to define. On a practical level any shortcoming in the performance of a crop is a problem for the plant pathologist. In the field, he or she may well be regarded in the same way as the family doctor—expected to provide advice on all aspects of plant health! A distinction is often drawn between **disease** caused by infectious agents, and **disorders** due to non-infectious agents such as mineral deficiency, chemical pollutants, or adverse climatic factors. The main emphasis of this book is on disease caused by plant pathogenic micro-organisms such as fungi, bacteria and viruses. Under favourable conditions these pathogens can multiply and spread rapidly through plant populations to cause destructive disease **epidemics**. Many of the principles discussed apply equally well, however, to other damaging agents such as insect pests and nematodes.

A fundamental concept in plant pathology is the **disease triangle** (Fig. 1a), which shows that disease results from an interaction between the host, the pathogen and the environment. This can be enlarged to include a further component, the host–pathogen complex (Fig. 1b), which is not simply the sum of the two partners, as the properties of each are changed by the presence of the other.

A comprehensive analysis of plant disease must take all four components into account. Obviously one needs to be familiar with the characteristics of the host and the pathogen in isolation. The successful establishment of a pathogen in its host gives rise to the host–pathogen complex. Unravelling the dynamic sequence of events during infection, the molecular

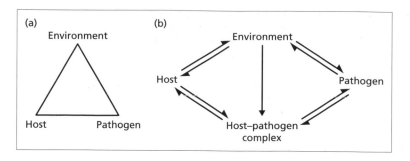

Fig. 1 Host–pathogen interactions.

'cross-talk' taking place between the partners, is one of the most challenging problems in experimental biology. In addition, the effects of the environment on each of the other components must be understood. These include not only physical and chemical factors but also macro- and microbiological agents. The two-way arrow between the host and the environment in Fig. 1b should not be overlooked, as populations of crop plants often have important effects on their surrounding microclimate. For example, the relative humidity within a crop canopy is higher than that outside and this will favour the development of some microorganisms. Effects of pathogens on the environment are more subtle but may be significant; some fungi, for instance, produce the volatile hormone ethylene, which can in turn affect the development of adjacent host plants.

This book is intended to provide an outline of the main elements of modern pathology. The approach is designed to achieve a balance between laboratory and field aspects of the subject, and to place the phenomenon of plant disease in a wider biological context. Research in plant pathology can be broadly divided into **tactical** and **strategic** aspects. The former is concerned with providing solutions to disease problems by identifying causal agents, and evaluating the most cost-effective options for their control. The latter is longer-term and aims to understand fundamental aspects of plant disease such as pathogen ecology, population biology, host–pathogen interactions, and plant defence systems. This knowledge can then be applied to devise improved methods of disease control.

The book is divided into three parts. Part 1 focuses on the problem of plant disease, the causal agents and their significance, disease diagnosis, and the development of epidemics in plant populations. This account highlights the influence of environmental factors on

the multiplication and spread of pathogens, and the use of climate data in disease prediction.

Part 2 deals with host–pathogen interactions; how pathogens gain entry to the host, how their growth and development in the plant leads to disease symptoms, and how the plant responds. The outcome of any host–pathogen confrontation depends on a dynamic interplay between factors determining microbial pathogenicity, and the various mechanisms of plant defence. This interaction ultimately determines host–pathogen specificity, whereby any pathogen is able to cause disease in only a restricted range of host plants. Increasingly, biologists and biochemists are studying host–pathogen systems as experimental models to probe mechanisms of gene expression and regulation, using the powerful new tools of molecular biology.

Part 3 deals with the practical business of disease control, often described as **crop protection**. This covers the management of disease by means of chemicals, resistance breeding, and alternative biological approaches. Finally, the combined use of cultural practices and all these other measures to provide sustainable, integrated systems for disease control is described.

A comprehensive treatment of individual diseases and the methods used in their control is beyond the scope of a text of this length. For the sake of brevity, specific pathogens or the diseases they cause are often mentioned without further explanation. This approach may be likened to that adopted in many ecology texts, in which the reader is expected to be familiar with most of the higher plants or animals discussed therein. There is, however, an appendix listing the pathogens and diseases mentioned in the book, together with brief details which will enable the reader to obtain further information about particular diseases. More detail concerning specific aspects of pathology may be obtained by consulting the recommended further reading.

Further reading

Books

Agrios, G. (1997) *Plant Pathology,* 4th edn. Academic Press, New York.

Gareth-Jones, D. (1987) *Plant Pathology. Principles and Practice*. Open University Press, Milton Keynes.

Parry, D.W. (1990) *Plant Pathology in Agriculture.* Cambridge University Press, Cambridge.

Reviews and papers

Many scientific journals contain reviews and research papers relevant to plant pathology. One especially useful source is the *Annual Review of Phytopathology.* Others include:

Advances in Botanical Research
Annals of Applied Biology
Crop Protection
European Journal of Plant Pathology
Fungal Genetics and Biology
Journal of Phytopathology
Molecular Plant–Microbe Interactions
Mycological Research
Pesticide Science
Physiological and Molecular Plant Pathology
Phytopathology
The Plant Cell
Plant Disease
The Plant Journal
Plant Pathology
Review of Plant Pathology

Databases and electronic information

Increasingly, information on plant pathology and crop protection can be accessed from CD-ROM databases, and through the Internet. Useful sources are the American Phytopathological Society (website at http://www.scisoc.org/) and CAB International (who have recently launched Plant-PathCD, a database of plant pathology literature—website at http://www.cabi.org).

The British Society for Plant Pathology runs an electronic journal, *Molecular Plant Pathology On-Line,* which can be accessed at: http://www.bspp.org.uk/mppol.

1   The Diseased Plant

'Since it is not known whether plants feel pain or discomfort, and since, in any case, plants do not speak or otherwise communicate to us, it is difficult to pinpoint exactly when a plant is diseased.' [G.N. Agrios, 1997]

The significance of disease in plants varies depending upon both biological and socio-economic factors. At one extreme, disease may be so severe that the need for control measures is immediately obvious. Such outbreaks may devastate a crop and cause complete failure, with serious consequences for the grower and others dependent upon the crop for food or income. There are, however, many other diseases where it is difficult to define the symptoms, where the cause of the problem is far from obvious and where any benefits obtained from control measures are not immediately apparent. This chapter discusses the nature of disease, and surveys the range of pathogens, pests and other agents which adversely affect plants. The impact of disease, both in natural plant communities, and in agriculture, forestry and horticulture, is then considered.

Concepts of disease

To fully understand the nature of disease one must first identify the processes occurring during the growth and development of the healthy plant. Such an analysis may be conducted at three levels:

1 the sequence of events comprising the normal plant life cycle;
2 the physiological processes involved in plant growth and development;
3 the molecular reactions underlying these processes.

Seed germination, maturation of vegetative structures, the initiation of reproduction and the formation and dispersal of fruits and seeds are all critical phases of the life cycle at which disease may occur (Fig. 1.1). At each stage in this developmental sequence, the integration of several physiological processes is essential for the continued development of the plant. Cell division and differentiation, the fixation and utilization of energy (photosynthesis and biosynthesis), transport of water and food materials (transpiration and translocation), and storage of reserve materials are all necessities for growth. Each of these functions involves a complex series of molecular events which comprise the overall metabolism of the plant. The nature and regulation of metabolism is itself determined by the genetic makeup of the plant.

Disease may disrupt the activities of the plant at one or more of these levels. Some disorders involve subtle alterations in metabolism which do not affect the successful completion of the life cycle. Many diseases caused by viruses have only slight effects on the growth of the plant; in such cases it may be difficult even to recognize the existence of a disease problem. For instance, potato virus X was known as potato healthy virus until virus-free seed potatoes became widely available. Comparisons with infected plants then showed the virus to be capable of causing a 5–10% loss in yield. Other more destructive diseases may interfere with numerous molecular, cellular and physiological processes and lead to premature death of the plant.

While everyone is familiar with the idea of disease, in practice there may be difficulties in drawing a precise distinction between healthy and diseased plants. No single definition of disease has found universal acceptance; the most widely used ones involve some reference to the 'normal' plant, for instance 'a condition where the normal functions are disturbed and harmed' (see Holliday (1989) *A Dictionary of Plant Pathology*). However, there is no consensus as to the exact extent of deviations from this norm which may constitute the diseased state.

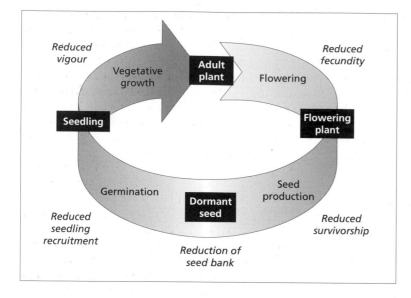

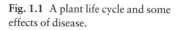

Fig. 1.1 A plant life cycle and some effects of disease.

The problem of defining normality, in terms of the processes outlined above, is further complicated by the variation inherent in all plant populations. Such variation is particularly common in natural populations, especially where hybrids occur, but even within apparently uniform populations of crop plants there may be differences between individuals. Such differences either have a genetic basis or are due to environmental factors operating during the growth of the crop. If, for instance, one sows seed of an old cereal variety alongside that of a modern, improved cultivar, one will observe major differences between the two crops. In particular, the modern cultivar will be shorter, form much larger seed heads and heavier grains, and the final yield will be greater. The difference in this case is due to intensive selection and genetic improvement, rather than to any disease in the old variety, but this example highlights the importance of understanding the initial potential of the plant before accurate estimates of disease can be obtained.

Damage or disease?

It can be argued that short-term harmful effects on plants, such as injury due to grazing, do not constitute disease. Indeed, some plants, such as the grasses, are well-adapted to regular grazing and respond with increased growth if so affected. In cases where damage is sustained over a longer period of time, for

example progressive destruction of roots by migratory nematodes, or distortion of aerial shoots by exposure to persistent herbicides, the outcome is clearly within the scope of pathology. However, these fine distinctions are of limited use in arriving at a working definition of disease. Such a definition will depend in part on the situation in which it is intended to be used. For example, the biochemist may well be concerned with a malfunction involving a single enzyme and hence view disease as a specific metabolic lesion, whereas the farmer is normally only interested in changes which affect the overall performance of the crop and reduce its value.

Although at present definitions of disease lack precision, it may ultimately be possible to describe all malfunctions in terms of biochemical changes. To date, this has been achieved in only a few exceptional cases, notably in diseases caused by fungi which produce host-selective toxins, where all of the symptoms are due to a single toxic compound acting at a specific target site (see Chapter 8).

Symptoms of disease

A doctor diagnoses illness in a patient by looking for visible or measurable signs that the body is not functioning normally. Such signs are known as **symptoms** and they may occur singly or in characteristic combinations and sequences. For example, someone suffering from influenza may have a sore throat, a fever, and muscular aches and pains. Such a group of symptoms occurring together and in a regular sequence is termed a disease **syndrome**. For many diseases the occurrence of a particular combination of symptoms is sufficient to arrive at an accurate diagnosis. Alternatively, symptoms may be common to a wide variety of diseases (for instance, fever is a generalized response to both infection and certain types of injury). In such cases detailed microbiological and biochemical analyses will be necessary to detect other diagnostic symptoms.

Similar considerations apply to the diagnosis of disease in plants. Just as with doctors and human disease, plant pathologists must be aware of the range of disease symptoms (Plates 1–9, facing p. 12) and what these suggest as the cause of the problem. The major symptoms of disease in plants are listed in Table 1.1 on the basis of the functions affected. This approach is used because it directs attention to the underlying nature of the disorder. For instance, the presence of galls or other cancerous growths immediately suggests some malfunction in the control of cell division; this in turn implicates a hormonal imbalance and/or genetic change in host cells. It should be real-ized that this classification of symptoms is to some extent arbitrary and non-specific. Permanent wilting provides a useful example. Although this symptom suggests that something is interfering with the uptake and transport of water, the symptom itself tells us little about the actual site or cause of the interference. The problem could be due either to a blockage in the vascular system, as in vascular wilt diseases, or to a general destruction of root tissues. It is also possible that the problem has little to do with water uptake or transport; in some diseases, for example infections of leaves by rust fungi, wilting is a sign of excessive water loss due to increased transpiration.

The symptoms listed in Table 1.1 will interact in numerous ways. In club root of cabbage the basic symptoms are hypertrophy and hyperplasia in root tissue (Fig. 1.2), but the first visible symptom is often wilting of the aerial parts of the plant. Any disruption of normal root development inevitably affects other functions such as water and nutrient transport. In view of the highly integrated nature of life processes, it is hardly surprising that attempts to define symptoms lack precision.

The relative importance of any symptom will vary,

Table 1.1 Symptoms of plant disease.

Symptom	Function affected	Examples
Stunting	General development	*Pseudomonas* wilt of tobacco, barley yellow dwarf virus, citrus exocortis viroid
Necrosis (cell death)	General	Whole plant—damping-off Leaf tissues—potato late blight Storage tissues—*Erwinia* soft rots of carrot and potato Woody tissues—apple canker, fireblight
Chlorosis	Photosynthesis	Black stem rust, beet mild yellowing virus, halo blight of bean
Permanent wilting	Water relations	Panama disease of bananas, *Verticillium* wilt of tomatoes, bacterial wilt of cucurbits
Hypertrophy (abnormal cell enlargement)	Growth regulation	Club root of cabbage, maize smut, peach leaf curl
Hyperplasia (uncontrolled cell division)	Growth regulation	Crown gall, peach leaf curl
Leaf abscission	Growth regulation	Leaf cast of rubber, coffee rust
Epinasty (downward growth of petioles)	Growth regulation	*Verticillium* wilt of tomatoes
Etiolation	Growth regulation	Bakanae disease of rice
Inhibition of flowering	Reproduction	Choke of grasses
Inhibition of fruit formation	Reproduction	Ergot on grasses, barley loose smut
Changes in pigmentation	Secondary metabolism	Peach leaf curl, tulip-breaking virus, grapevine leaf roll virus

(a)

(b)

Fig. 1.2 Club root disease of crucifers. (a) Primary infection causes distortion of root hairs, which contain plasmodia of the fungal pathogen *Plasmodiophora brassicae*. (b) Later the main root becomes swollen due to repeated division and enlargement of root cortical cells.

depending not only upon its duration and severity but also on the habit or life form of the plant affected. Thus, necrosis in the stem of a herbaceous seedling will probably lead to the death of the whole plant, while necrotic lesions in the stem of a woody perennial may only result in the loss of a twig or branch. If, however, such a lesion girdles the trunk of a tree, then translocation will be disrupted to the extent that the plant will die. Pathogens which actually kill plants are the exception. More commonly, disease symptoms indicate an impairment of the efficiency of plant physiological and biochemical processes (Table 1.2). Some symptoms, such as local changes in pigmentation, may be trivial in terms of overall plant performance. Often the most important consideration is the stage in the life cycle at which symptoms first appear. Severe chlorosis of the first-formed leaves of a cereal may have little effect upon the final yield, as most of the photosynthetic products required for grain filling are provided by the flag leaf and ear tissues. Accelerated abscission of leaves is unlikely to be a problem in annual herbaceous plants, but in perennials it may exert a severe drain on the food reserves of the plant. For instance, Fig. 1.3 shows defoliation of some willow clones due to infection by the rust fungus *Melampsora*. Loss of photosynthetic tissue reduces the biomass produced by the crop. A similar symptom can be seen in coffee bushes affected by another rust, *Hemileia*, or in rubber trees affected by the leaf blight fungus *Dothidella*. In both these evergreen crops, early leaf fall is often followed by the production of a second flush of leaves. If these are also prematurely lost due to further infection then the plant may die.

Causes of disease

Any agent capable of adversely affecting green plants may be regarded as lying within the scope of plant pathology. The principal agents involved in plant disease are shown in Fig. 1.4. Partial or total crop failure may be due to one or more agents. Where more than one agent is responsible, each may act independently or they may interact. In the latter instance there may be **synergism**, that is, two or more agents act in combination to produce changes which either is unable to produce alone.

Synergism has been demonstrated to occur with several combinations of viruses. For example, tobacco mosaic virus and potato virus X each cause

Table 1.2 Disease symptoms in relation to function in higher plants.

	Vegetative organs			Reproductive organs	
	Roots	Stems	Leaves	Flowers, fruit	Seeds, seedlings
Functions	Uptake Transport Anchorage	Support Transport	Photosynthesis Gas exchange Transpiration	Fertilization Development	Survival Germination
Symptoms	Necrosis Hypertrophy Hyperplasia Excessive branching	Necrosis Etiolation Gall formation Excessive branching Lodging	Chlorosis Pigment changes Necrosis Wilting Epinasty Hypertrophy Abscission	Inhibition Substitution Necrosis	Necrosis
Disease examples	Root rots Club root Rhizomania	Heart rots Foot rots Cankers Crown gall Witch's broom Bakanae disease Cereal eyespot	Mosaic Leaf spots Blight Leaf roll/leaf curl Vascular wilts Leaf cast Coffee rust	Choke Ergot Anther smut Storage rots	Seed decay Damping-off

relatively mild mottling symptoms in tomato. If by chance they occur together in the same host, then severe necrosis develops and this may even result in the death of the plant.

A useful distinction can be drawn between animate (**biotic**) and inanimate (**abiotic**) causes of disease (Fig. 1.4). Many of the animate agents, including the microbial **pathogens**, the parasitic angiosperms and some of the animal pests, are infectious. Due to their capacity for growth, reproduction and dispersal, these agents spread from one host plant to another. Under particularly favourable circumstances, they may be dispersed rapidly over wide areas and even entire continents.

Pests

Amongst the animals exploiting plants are many pests which cause damage to roots, leaves, shoots, flowers and fruits. Usually these pests, which include insects such as aphids and leaf hoppers, and some nematodes, spend relatively brief periods on individual plants before moving on to explore new food supplies. Other pests, such as leaf miners, gall-forming sawflies and endoparasitic nematodes, spend their entire life cycle, or a major part of it, on one plant. Pest attack may result simply in a drain on host nutrients or, alternatively, in extensive destruction of tissues. Aphids on leaves and stems extract sap from the phloem with an almost clinical efficiency. Many caterpillars are simply small herbivores; nevertheless they can consume large areas of the leaf lamina. Other pests cause more complex host responses or symptoms. Developing cynipid wasp larvae induce the formation of cherry or spangle galls on oak leaves, while nematodes such as *Meloidogyne* spp. cause swellings, termed 'knots', on the roots of tomatoes and potatoes. When these root-knot nematodes and the related endoparasitic cyst nematodes penetrate root tissues, host cells adjacent to the vascular system become enlarged and provide a specialized feeding site where nutrients are transferred to the sedentary worm. Such pests often show highly specialized adaptations to their respective hosts; conversely, the plants are able to produce defence reactions in response to attack which may be similar to those induced by pathogenic microorganisms.

Fig. 1.3 Rust of willows (*Salix*) caused by *Melampsora* species. (a) Aerial view of experimental trial of willow clones in summer. Plots which appear empty have lost their leaves due to severe rust infection. (Courtesy of D.J. Royle.)

(b) Scanning electron micrograph of rust pustule on willow leaf showing spiny uredospores of the fungus. Scale bar = 20 μm. (From Spiers & Hopcroft 1996.)

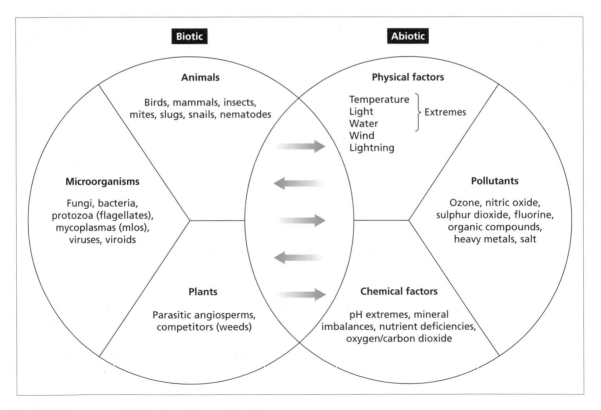

Fig. 1.4 Agents responsible for plant diseases, disorders and damage.

Larger animals such as birds or mammals can also be destructive pests. Winter grazing by rabbits can seriously reduce the final yield of autumn-sown crops such as wheat and oilseed rape. In Europe pigeons also cause damage to oilseed rape, while in parts of Africa flocks of seed-eating finches, such as *Quelea*, are a major threat to crops of sorghum and millet.

Parasitic plants and weeds

Higher plants may cause disorders of or damage to other plants, either by acting directly as parasites in diverting nutrients and water, or as vigorous competitors or antagonists within mixed populations. Parasitic angiosperms are rare enough to be curiosities in many cool temperate countries, but elsewhere they are nuisances or economically important parasites (Table 1.3). The dwarf mistletoes, *Arceuthobium*, can kill or deform pines and other conifers, and even minor attacks reduce the quality of timber by causing the production of numerous large knots and irregularly grained, spongy wood. These parasites spread their sticky seeds by an explosive dispersal mechanism, leading to patches or foci of infestation within a plantation. By contrast, root parasites such as

Orobanche and witchweed (*Striga*) produce numerous tiny seeds which lie dormant in the soil. The seeds are triggered to germinate by a stimulant from host roots. The parasite then attaches itself to the root by means of specialized organs known as haustoria, which divert water and nutrients, leading to wilting, chlorosis and stunting of the host. These parasites are difficult to control due to the large number of seeds they produce (in the case of *Striga* as many as 2×10^5 per plant) and the long periods over which they remain viable. On a world scale the most important angiosperm parasite is *S. hermonthica*, which attacks cereals such as maize, sorghum, millet and rice. In many of the agricultural areas where it is most prevalent, for example sub-Saharan Africa, there are insufficient resources to support expensive control measures, and infested land may eventually be abandoned. *Orobanche* is a significant problem in sunflower, tobacco, tomato and especially faba bean, with a substantial proportion of the crop area in the Mediterranean region affected.

The deleterious effects of other higher plants are due to competition for space, light, water and nutrients. Species which are vigorous competitors with crop plants are usually described as **weeds**. As well as

Table 1.3 Angiosperms parasitic on other higher plants.

Family, common name, genus	Geographic area	Crops attacked
Convolvulaceae Dodder (*Cuscuta*)	Europe, North America	Alfalfa, clover, potatoes, sugarbeet
Lauraceae Dodder (*Cassytha*)	Tropics and subtropics	Citrus trees
Loranthaceae Dwarf mistletoe (*Arceuthobium*)	Worldwide	Gymnosperms
American true mistletoe (*Phoradendron*)	North America	Angiosperm trees
European true mistletoe (*Viscum*)	Europe	Angiosperm trees, especially apple
Orobanchaceae Broomrape (*Orobanche*)	Europe	Tobacco, sunflower, beans
Scrophulariaceae Witchweed (*Striga*)	Africa, Asia, Australia, North America	Maize, sorghum, rice, cowpea

affecting crop development, weeds may interfere with harvesting and their seeds can contaminate grain samples. They may also be important as alternative hosts for pests or pathogens which can subsequently spread to crops. In addition to simple competitive effects, some plants produce chemicals which inhibit the growth of neighbouring plants. This phenomenon, analogous to microbial antibiosis, is known as **allelopathy**. Plant roots release a diverse range of chemicals which can act as potential inhibitors or defence compounds, but it is difficult to determine the extent to which these interactions operate in nature. Allelopathy is believed to influence plant succession and pattern in natural communities, and may also have significant effects in agricultural systems. The chemicals involved are of interest, both as potential herbicides or as signal molecules affecting the growth and behaviour of other organisms.

Abiotic agents

Green plants, in common with all other organisms, only flourish within a relatively narrow range of environmental conditions. Inside the plant, individual cells are able to exert control over their internal environment and thereby maintain conditions suitable for normal metabolism. However, the extent to which living cells can withstand alterations in the external environment is limited. Fluctuations in environmental conditions outside an acceptable range are therefore harmful, and may result in irreversible damage. Green plants, unlike animals, are particularly susceptible to the effects of inanimate agents because they are sedentary and so are unable to escape from local changes in the environment. Plants also lack the sophisticated homeostatic mechanisms possessed by higher animals.

Many abiotic disease agents are, under other circumstances, normal components of the environment. The harmful effects of physical factors are associated with the incidence of extreme conditions. Light, for instance, while essential for green plants, may in excess cause a type of necrosis termed scorch on susceptible aerial parts of the plant. Low temperatures often result in frost damage. Plants differ greatly in their sensitivity to frost and typical symptoms include morphological deformation or death of part or all of the plant. Some subtropical plantation crops, such as citrus and coffee, are especially vulnerable. For instance, frosts in southern Brazil in 1975 affected most of the coffee-growing area, with production losses estimated at more than 60%. Many of these physical effects are relatively unsubtle and the symptoms associated with them are non-specific. Drought is also an important cause of loss, due to reduced yield or even complete crop failure. Recent dry summers in the USA and Europe have highlighted this threat to crop productivity, and raised concerns about global warming. While the current models of global climate change predict widely different future scenarios, they all agree that greater fluctuations in temperature and rainfall are likely to occur, with consequent annual variations in crop yield. High winds can also have catastrophic effects on plants; hurricanes have in recent years destroyed entire banana plantations in the West Indies.

There are other, less common, environmental hazards. The massive eruption of Mount St Helens in Washington State in 1980 deposited volcanic ash over a wide area with a variety of effects on agriculture. Ash on plant leaves reduced photosynthesis by as much as 90% — some crops such as alfalfa actually lodged under the weight of ash — but eventual crop losses were less than expected at around 7% of the total.

Chemical deficiencies or imbalances often result in distinctive symptoms, which may be diagnostic in the case of deficiencies of essential cations. For example, magnesium deficiency in swedes is associated with an abnormal purplish pigmentation in interveinal leaf areas, whereas boron deficiency in the same crop causes brown-heart symptoms in the storage root. Such deficiency diseases are commonplace, especially in the intensive cropping systems of present-day agriculture. In the UK, recent reductions in atmospheric sulphur, and subsequent deposition by rainfall, are now leading to deficiencies occurring in sulphur-demanding crops such as oilseed rape. Sulphur deficiency can affect crop quality as well as quantity, for instance by reducing the bread-making quality of wheat flour.

Excess amounts of certain mineral ions may be equally harmful, due to their effects on the availability or uptake of other essential ions. When insufficient iron is taken up, plants become chlorotic. Such a shortage may be due to inhibition of iron uptake by high levels of calcium or manganese in the soil, rather than any absolute shortage. Imbalances of soil nitrogen, phosphorus and potassium result in the development of plant tissues which are particularly

(a)

(b)

Plate 1 (a) Potato late blight, caused by *Phytophthora infestans*, showing necrotic lesions on leaflets. (b) Late blight in a potato crop. Under favourable conditions the entire crop canopy can be destroyed.

(a)

(b)

Plate 2 (a) Foot and collar rot of citrus, caused by soil-borne *Phytophthora* species such as *P. parasitica* and *P. citrophthora*. Early symptoms are chlorosis of leaves, followed by defoliation and eventual death of the tree. (b) A typical lesion, or canker, caused by infection at the base of the trunk.

[*Facing p. 12*]

(a)

(b)

Plate 3 (a) Tobacco mosaic virus (TMV) disease showing characteristic systemic symptoms with variegation and blistering of infected leaves. (b) Tobacco mosaic virus in a resistant tobacco variety, showing formation of necrotic local lesions with dark margins.

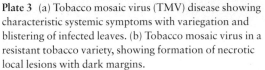

Plate 4 Rust disease, *Uromyces fabae*, on broad bean, showing abundant production of rust-coloured urediospores from pustules formed on leaves, which remain green and photosynthetically active.

Plate 5 Black spot of rose, caused by the fungus *Diplocarpon rosae*. Dark lesions and some chlorosis on infected leaves, which are later shed prematurely.

(a)

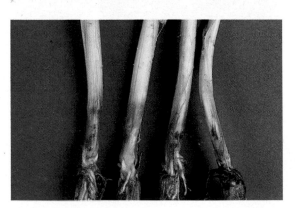

(b)

Plate 6 (a) Eyespot disease of cereals, caused by *Pseudocercosporella* (*Tapesia*) *herpotrichoides*, showing typical necrotic lesions with dark centres at base of stem. (b) Eyespot damage in a wheat crop, Washington State, USA.

Weakening of the stem base has caused the crop to collapse (lodge), particularly in areas at the base of slopes where the wetter conditions favour infection. (Courtesy of Dr Tim Murray.)

Plate 7 Blossom end rot of courgette, caused by the grey mould fungus *Botrytis cinerea*. The pathogen has colonized the remains of the senescing flower, and is now spreading into healthy tissues.

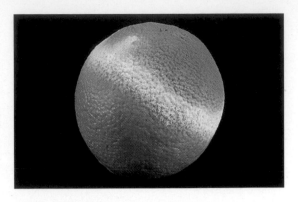

Plate 8 Green rot of citrus, caused by the fungus *Penicillium digitatum*. The post-harvest pathogen gains entry through wounds.

Plate 9 Panama disease of banana, a wilt disease caused by the soil-borne fungus *Fusarium oxysporum* f.sp. *cubense*. (Courtesy of AgrEvo UK Ltd.)

prone to infection by microorganisms or damage by other agents.

A common difficulty in diagnosing disorders caused by chemical agents is the similarity between the symptoms they produce and those due to infection by microorganisms. Foliar symptoms in barley resulting from a deficiency of manganese resemble those caused by the leaf blotch fungus *Rhynchosporium*. Symptoms of other deficiency diseases bear a striking resemblance to those caused by viruses. Recently, molecular work on the cellular systems responsible for the uptake of specific nutrient ions has identified some of the transporter proteins involved. The genes encoding these transporters are in many cases regulated by levels of the appropriate nutrients. In the long term it may therefore be possible to engineer components of such transport pathways to recognize a deficiency or excess of ions by producing a reporter chemical which is visible in the plant. These so-called 'smart plants' would be sown at intervals within a crop and hence provide an early warning of nutrient imbalance.

Pollutants are substances which are either unnatural components of the environment, for example polychlorinated biphenyls (PCBs) and dichlorodiphenyltrichloroethane (DDT), or naturally occurring substances present in abnormal concentrations, for example ozone, sodium chloride and the acid mist which forms when oxides of sulphur and nitrogen dissolve in atmospheric moisture. Photochemical smog is increasingly common in urban areas and results from the interaction of waste gases—especially from automobiles — particulates and sunlight. High concentrations of certain chemicals may be a normal feature of some habitats, but problems arise when human activities redistribute these substances. High salt concentrations, which are non-toxic to salt-marsh plants, can severely injure inland species, as in the case of roadside communities which are damaged by splash following the application of sodium chloride to roads to prevent ice formation. Irrigation of desert soils can also lead to an accumulation of salts.

The influence of many of these compounds on higher plants is now well known and the symptoms induced include abnormal growth due to meristem damage, chlorosis and necrosis. Some species of plants are especially sensitive to particular pollutants. However, within a species the response of different cultivars may vary considerably. Plant species or genotypes which are tolerant of high levels of pollutants, and which can accumulate them from soil, are of value in reclaiming contaminated sites, a process known as bioremediation.

This discussion has only considered the direct effects of biological, physical and chemical agents acting independently. In reality, all these agents interact with each other in a more or less complex manner. For instance, infection of the stem base of many crop plants predisposes them to collapse in wind or heavy rain; such lodging then results in problems at harvest (Plate 6b, facing p. 12). It is difficult to distinguish between the damaging effects of each of these factors. Other interactions are even more complex. The widespread defoliation and death of trees observed in some industrialized countries, a condition known as forest decline, is believed to be due to aerial pollution. However, there is dispute over the relative importance of different atmospheric pollutants and acid rain as contributory factors. It has even been claimed that the premature death of some trees is a normal part of the forest cycle. Most likely, several factors interact to affect tree health, including direct toxic effects, soil acidity, release of toxic ions such as aluminium, and indirect effects on root function, including inhibition of beneficial mycorrhizal fungi, and enhanced activity of minor root pathogens. Forest decline provides a good example of a complex disease syndrome, and also illustrates the difficulty of reaching a conclusive diagnosis when several interacting factors are involved.

Significance of disease

It is often assumed that disease outbreaks are less frequent and less severe in wild plant populations, than in crops. This is because there are several important differences between natural and agricultural plant

Table 1.4 Natural and agricultural plant communities compared.

	Natural	Agricultural
Genotypes	Diverse	Uniform
Age structure	Mixed	Uniform
Distribution	Dispersed	Crowded
Nutrient status	Often low	High
Co-evolution span	Long	Short

communities (Table 1.4). Wild species are more diverse, both genetically and in the age structure of the population. Hence individual plants will differ in their relative susceptibility to infection. Natural populations also tend to be spatially dispersed as part of a mixed plant community, thereby reducing opportunities for the spread of disease. Often, nutrients are added to crops as fertilizers, and in some cases this can lead to increased susceptibility to infection. Finally, it is likely that plants in natural communities have co-evolved with their pathogens over long periods of time, leading to some kind of host–pathogen equilibrium, while plant breeding and the deployment of crops in new areas have disturbed this balance.

Surveys of wild plants show, however, that disease is commonplace in natural communities, with a proportion of the plant species present showing signs of pathogen attack each season. In natural ecosystems,

disease is one of the many factors which regulate populations (see Fig. 1.1) and hence determine the spectrum of species which are successful in any habitat. Pathogens affect the reproduction and longevity of plants and are therefore agents of natural selection. Figure 1.5 shows the effect of rust infection on the survival and reproductive capacity of the annual weed groundsel: infected plants produced fewer mature flowers, and hence seeds, and died earlier than uninfected plants. Repeated over several generations, such disease would therefore have a significant effect on population size. (There is a practical spin-off from this observation as some pathogenic fungi, known as **mycoherbicides** (see p. 237), may be useful as agents to control weeds.) In particular, disease will tend to limit the spread of species to less favourable geographic regions or habitats, as the impact of pathogens will be greater on plants growing under suboptimal conditions. Disease may also accelerate change within established plant communities. For example, forest trees may be killed, thereby opening the canopy and allowing regeneration to proceed. The recent global epidemic of Dutch elm disease completely altered the structure and species composition of woodlands where elm was one of the dominant trees. An even more dramatic example of a disease outbreak in natural vegetation, however, has

Fig. 1.5 Effect of infection by the rust fungus *Puccinia lagenophorae* on survival and reproduction of the annual weed groundsel (*Senecio vulgaris*). (a) Percentage of the original population surviving in control and rust-infected populations. (b) Number of mature inflorescences formed by control and rust-infected populations. (After Paul & Ayres 1986a,b.)

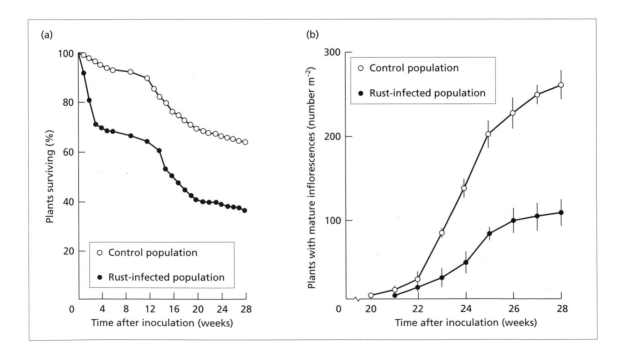

occurred in Australia where the root rot fungus *Phytophthora cinnamomi* has killed large areas of native forest (see Table 1.5). The unusual feature of this epidemic was that many different species of trees and shrubs were affected by the same pathogen. One explanation for the severity of the outbreak is that the fungus responsible was almost undoubtedly introduced from another continent, and hence spread through a host population not previously exposed to infection. The term **new-encounter disease** has been proposed to describe such an epidemic arising from contact between a previously separate host and pathogen.

In agricultural ecosystems, disease is one of the factors influencing crop yield and quality. Farmers, foresters and horticulturalists are, by and large, interested only in those changes in crop performance which influence cash return per hectare. The ideal situation, in which pathogens are avoided, excluded or eliminated, is therefore a theoretical rather than practical goal. On the farm other priorities may prevail; choice of crop or cultivar is usually based on likely profitability, rather than resistance to pests or pathogens. Unless a financial return is guaranteed, control measures may be ignored or reduced in scale. As a consequence, the significance of disease, as perceived by the grower, will depend to a large extent on the market value of the crop. Inputs of chemicals, or other actions designed to reduce disease, are only justified when the likely impact on yield or quality will outweigh the cost of the measure. Even a possible bonus, such as restricted carry-over of the pathogen until the following season, may not provide sufficient incentive for any financial outlay.

Other parties with an interest in crop diseases are government advisory or extension pathologists, consultants, and representatives of the agrochemical industry. The relative resistance of new crop varieties to pathogens, and the efficacy of commercial formulations of pesticides, are assessed by government scientists under field conditions; recommendations for use may be based on these field trials. Independent consultants offer growers an overall package of advice for crop management, part of which concerns disease. Agrochemical companies also influence decision-taking by promoting sales of their products. While advertising pressure encourages farmers to utilize chemicals, the approach now adopted is more cautious than in the past, and chemical use is discouraged in situations where environmental side-effects, or problems of resistance to pesticides, might occur. In recent years pressure has been applied, both through legislation and voluntary schemes, for more stringent assessment of the toxicity of the new compounds. This has inevitably created extra difficulties in the development and registration of new agrochemicals and so has added to the costs incurred prior to market release (see Chapter 11).

The relationship between the amount of disease and loss of income is complex due to the many possible interactions between symptoms of disease and the final determinants of crop yield and quality. Figure 1.6 defines a number of yield levels for a hypothetical crop. The theoretical maximum yield is a value based on predictions from crop physiology; under field conditions this yield level is not a practical possibility and hence there is some unavoidable loss. Attainable yield indicates the maximum level to be expected under optimum conditions in the field. With appropriate inputs of fertilizer and pesticides this is the best yield the farmer can realistically hope for. The difference between this value and the actual yield obtained from the crop can be defined as an avoidable loss. In practice, attainable yield is not a realistic goal for most crops for simple economic reasons. To increase yield to this level requires so many inputs that the cost is greater than any return at the end of the season.

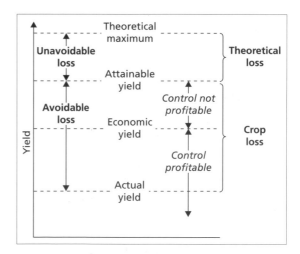

Fig. 1.6 Relationship between yield levels and crop loss, indicating economic benefits of control. (From *Epidemiology and Plant Disease Management* by J.C. Zadoks & R.D. Schein. Copyright © 1979 by Oxford University Press, Inc. Used by permission of Oxford University Press, Inc.)

Instead we can define a slightly lower threshold, the economic yield, which represents the break-even point at which the cost of the inputs is balanced by the extra productivity of the crop. Any shortfall below this level is an avoidable loss which justifies the expense of a control measure.

For some crops, especially high-value fruits, vegetables or ornamental plants, the quality of the product is as important as the yield. Under these circumstances very little disease is tolerated, as any damage or blemish may have a disproportionate effect on crop value. Not surprisingly the most intensive disease and pest control regimes available are used for such crops.

The impact of disease

There is a tendency to portray plant disease as a constant war between farmers and agents that threaten to destroy their crops. This is misleading as in most instances disease is just one of several factors affecting the overall productivity of plants. Nevertheless, one can cite cases where the impact of disease has been, quite literally, a matter of life and death, or where the course of history has been changed (Table 1.5). No one would seek to underestimate the tragic effects of potato late blight in Ireland during the 1840s, or the devastation wrought by coffee rust in Sri Lanka in the late 19th century. Diseases such as

these have taken a dreadful toll in terms of human suffering. Coffee rust has now spread to most major coffee-growing areas (see Fig. 5.8) with serious consequences for the economies of some of the world's major producers. Attempts to eradicate the destructive cocoa swollen shoot virus (CSSV) from Ghana by means of statutory removal of infected and surrounding trees entailed not only the most costly eradication campaign ever attempted, but also led to political unrest. In the USA and South America, spread of the bacterial disease citrus canker has only been contained by burning huge numbers of infected citrus trees and nursery stock, combined with vigilant quarantine measures. Other diseases have precipitated major changes in agriculture or natural landscapes over vast areas. Nor are such devastating disease epidemics a thing of the past. The major epidemic of southern corn leaf blight in the USA in 1970 (see Fig. 5.1) raised doubts about the wisdom of achieving genetic uniformity in modern cereal crops and forced a reassessment of the breeding methods employed in the production of new cultivars.

Even in areas of high agricultural efficiency, losses due to pathogens, pests and weeds make constant inroads into production, and hence profits (Table 1.6). These estimates are notoriously difficult to compile but suggest that more than one-third of total production is being lost. Furthermore, comparison with data from earlier surveys (Fig. 1.7) shows that

Table 1.5 Some examples of the impact of plant diseases.

Type of impact	Disease	Agent	Country/region affected
Famine	Late blight of potato	*Phytophthora infestans*	Europe 1845–1846
	Brown spot of rice	*Helminthosporium oryzae*	India 1942–1943
	Cassava bacterial blight	*Xanthomonas campestris* pv. *manihotis*	Zaire 1970–1975
	Failure of maize crop	Maize mosaic virus?	Guatemala, 9th-century Mayan civilization
Economic	Coffee rust	*Hemiliea vastatrix*	Sri Lanka 1870; now worldwide
	Cocoa swollen shoot	CSSV (virus)	Ghana/Nigeria 1930–
	Panama disease of banana	*Fusarium oxysporum* f.sp. *cubense*	Central America 1930–1955
Agricultural	Southern corn leaf blight	*Bipolaris maydis*	USA 1970
	Citrus canker	*Xanthomonas campestris* pv. *citri*	USA 1910– ; South America
Ecological	Dutch elm disease	*Ophiostoma novo-ulmi*	Northern hemisphere 1930, 1970–
	Jarrah die-back	*Phytophthora cinnamomi*	Western Australia 1920–

Table 1.6 Crop losses due to pathogens, pests and weeds in the years 1988–1990, for (a) the eight most important food and cash crops and (b) for the same crops analysed by continent. (From Oerke *et al.* 1994.)

(a)

Crop	Crop losses (%) due to			
	Pathogens	Pests	Weeds	Total
Rice	15.1	20.7	15.6	51.4
Wheat	12.4	9.3	12.3	34.0
Barley	10.1	8.8	10.6	29.4
Maize	10.8	14.5	13.1	38.3
Potatoes	16.4	16.1	8.9	41.4
Soybeans	9.0	10.4	13.0	32.4
Cotton	10.5	15.4	11.8	37.7
Coffee	14.9	14.9	10.3	40.0

(b)

Continent	Loss (%) of production due to			
	Pathogens	Pests	Weeds	Loss overall
Africa	15.6	16.7	16.6	48.9
North America	9.6	10.2	11.4	31.2
South America	13.5	14.4	13.4	41.3
Asia	14.2	18.7	14.2	47.1
Europe	9.8	10.2	8.3	28.2
USSR	15.1	12.9	12.9	40.9
Australasia	15.2	10.7	10.3	36.2

the situation has not improved in recent years. Some of the changes taking place in modern agriculture may in fact have increased vulnerability to disease. The demand for improved agricultural productivity has led to large areas being planted with high-yielding, genetically identical cultivars. Similarly, the increasing cost of labour and the trend towards mechanization which involves major capital expenditure, have also contributed to a reduction in the diversity of crop types planted. As a result, long-established systems of crop rotation have been abandoned in many areas.

A second aspect of modern agriculture which has undoubtedly aggravated disease problems is increased world trade in crop plants and plant products. Following the gradual shift from self-sufficiency, at a community and in many instances also at a national level, and the increasing relative affluence of some countries, large-scale transport of plant material and food produce has become commonplace. The consequences of such transportation over long distances are well known (graphically demonstrated by 'new-encounter diseases'; see above). One need only cite the introduction of Dutch elm disease into North America in the 1930s, and the reintroduction of a virulent strain into Europe on imported elm logs in 1969–1970, to emphasize the hazards of exposing host plants to novel forms of the pathogen. The catastrophic epidemic of potato late blight (Plate 1, facing p. 12) in 1845–1846 was itself most likely due to the introduction of the fungus from America. In addition, there is increasing long-term storage of produce, which, in turn, brings further pathological problems. Losses due to post-harvest diseases have tended to be underrated. With some crops, spoilage losses while in transit or in store are actually greater than those which occur in the field. It has been estimated that one-third of all tropical produce is destroyed, by a variety of agents, before it reaches the consumer. Even in highly developed countries, the scale of potential losses during long-term storage, transport and marketing can be daunting (Table 1.7). Post-harvest pathology is now recognized as an important aspect of crop protection, and

	Commodity	Country of origin	Potential loss (%)
Loss during low-temperature storage	Apples	England, USA	2–50
	Carrots	England, USA	6–38
	Citrus fruits	Italy, USA	3–52
Loss during transport and marketing	Apples	USA	29
	Citrus fruits	USA	0–25
	Lettuce	USA	10–15
	Peaches	USA	15–24
	Strawberries	USA	25–35

Table 1.7 Estimates of post-harvest losses likely to occur in the absence of effective disease-control measures. (Data from Eckert 1977.)

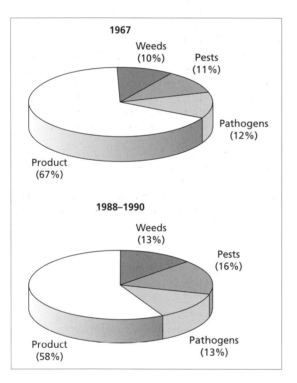

Fig. 1.7 Comparison of proportions of total production of major food and cash crops lost due to pathogens, pests and weeds estimated in 1967 (data from Cramer 1967) and 1988–1990 (data from Oerke *et al.* 1994). Note: Crops included in the calculation are not identical for the two surveys.

various methods have been developed to reduce these losses, including pesticides designed for use in store, careful control of the storage environment, and rapid-processing systems such as freezing or canning which prevent microbial spoilage.

Disease is a dynamic phenomenon

All in all, the intensification of agriculture through plant breeding, by widespread use of fertilizers and plant growth regulators, with larger fields and shorter rotations, has brought in its wake new and sometimes severe disease problems. Changes in the types of crops grown, or in crop management practices, almost invariably lead to new challenges. Irrigation has opened up whole regions to agriculture, but the same water which brings life to the crop can also nourish and spread its microbial enemies. The extension of crops into new geographical areas has exposed them to novel disease agents; for instance, tropical plant species such as cocoa and cassava, which originated in South America, are grown extensively in Africa where they have succumbed to virus diseases spreading from native species. While it is true that breeding programmes and chemical control measures have won notable victories in the campaign against plant disease, the situation is never static. As in medical microbiology, rapid pathogen evolution can confound attempts to control disease, and there is a need for constant vigilance to ensure that the plants we grow remain healthy and productive.

Further reading

Books

Ainsworth, G.C. (1981) *An Introduction to the History of Plant Pathology.* Cambridge University Press, Cambridge.

Burdon, J.J. & Leather, S.R. (1990) *Pests, Pathogens and Plant Communities.* Blackwell Scientific Publications, Oxford.

Cramer, H.H. (1967) *Plant Protection and World Crop Production.* Bayer, Leverkusen.

Crawford, R.M.M. (1989) *Studies in Plant Survival*. Blackwell Scientific Publications, Oxford.

Davidson, R.H. & Lyon, W.F. (1987) *Insect Pests of Farm, Garden and Orchard*. John Wiley & Sons, New York.

Evans, K., Trudgill, D.L. & Webster, J.M. (eds) (1993) *Plant Parasitic Nematodes in Temperate Agriculture*. CAB International, Wallingford.

Fowden, L., Mansfield, T. & Stoddart, J. (eds) (1993) *Plant Adaptation to Environmental Stress*. Chapman & Hall, London.

Gratwick, M. (ed.) (1992) *Crop Pests in the UK*. Chapman & Hall, London.

Holliday, P. (1989) *A Dictionary of Plant Pathology*. Cambridge University Press, Cambridge.

Jones, F.G.W. & Jones, M.G. (1984) *Pests of Field Crops*. Edward Arnold, London.

Mukhopadhyay, A.N., Singh, U.S., Kumar, J. & Chaube, H.S. (1992) *Plant Diseases of International Importance*, Vols 1–4. Prentice Hall, New Jersey.

Oerke, E.-C., Dehne, H.-W., Schönbeck, F. & Weber, A. (1994) *Crop Production and Crop Protection*. Elsevier, Amsterdam.

Press, M.C. & Graves, J.D. (eds) (1995) *Parasitic Plants*. Chapman & Hall, London.

Putnam, R.J. (ed.) (1984) *Mammals as Pests*. Chapman & Hall, London.

Robinson, J.B.D. (ed.) (1987) *Diagnosis of Mineral Disorders in Plants*, Vols 1–3. HMSO, London.

Schumann, G.L. (1991) *Plant Diseases: Their Biology and Social Impact*. APS Press, St Paul, Minn.

Smith, I.M., Dunez, J., Phillips, D.H., Lelliott, R.A. & Archer, S.A. (eds) (1988) *European Handbook of Plant Diseases*. Blackwell Scientific Publications, Oxford.

Williams, M.A.J. (ed.) (1994) *Plant Galls*. Clarendon Press, Oxford.

Zimdahl, R.L. (1993) *Fundamentals of Weed Science*. Academic Press, San Diego, Calif.

Reviews and papers

Andrivon, D. (1996) The origin of *Phytophthora infestans* populations present in Europe in the 1840s: a critical review of historical and scientific evidence. *Plant Pathology* **45**, 1027–1035.

Bos, L. & Parlevliet, J.E. (1995) Concepts and terminology on plant/pest relationships: towards consensus in plant pathology and crop protection. *Annual Review of Phytopathology* **33**, 69–102.

Jarosz, A.M. & Davelos, A.L. (1995). Effects of disease in wild plant populations and the evolution of pathogen aggressiveness. *New Phytologist* **129**, 371–387.

Oerke, E.-C. & Dehne, H.-W. (1997) Global crop production and the efficacy of crop protection—current situation and future trends. *European Journal of Plant Pathology* **103**, 203–215.

Skelly, J.M. (1994) Waldsterben in the forests of Central Europe and Eastern North America: fantasy or reality? *Plant Disease* **78**, 1021–1032.

Zadoks, J.C. (1985) On the conceptual basis of crop loss assessment: the threshold theory. *Annual Review of Phytopathology* **23**, 455–473.

2   The Microbial Pathogens

'*It was first necessary to determine if characteristic elements occurred in diseased parts of the body, which do not belong to the characteristics of the body, and which have not arisen from body characteristics.*' [Robert Koch, 1843–1910]

Heterotrophic microorganisms, unlike autotrophs, are entirely dependent upon an external supply of organic carbon compounds. The ultimate source of most carbon compounds is green plants, but there are a variety of routes by which microbes can obtain these nutrients.

A large number of microorganisms are decomposers. These organisms utilize substrates in dead tissues and their activities eventually lead to the disappearance of plant and animal remains. Such decomposers play a key role in the ecosystem by releasing nutrients which would otherwise remain locked up in plant litter. Some microbes have, in addition, an ability to parasitize living plants; if, during invasion of the plant, they kill their host this ensures a supply of dead tissues on which they can continue to grow. Other microorganisms are only able to obtain nutrients from living host plants, and establish more balanced relationships which may be of mutual benefit. The effects of microbes on plants therefore vary from severe damage and even death, to diversion of nutrients, to associations in which both partners gain some advantage. Hence, heterotrophic microorganisms are involved in a variety of ways in the movement of fixed carbon between different trophic levels in the ecosystem.

A comprehensive analysis of plant disease caused by microorganisms requires several different types of information. Firstly, the causal agents must be identified; however, the usual criteria employed for distinguishing between microbial species are of limited value when dealing with microorganisms isolated from plants. Different isolates of the same species may vary widely in their ability to cause disease. It is important to understand the genetic basis of such variation, and the corresponding variations in the plant's response. Secondly, the nature of the host–parasite relationship needs to be considered; the biology of infection, sources of nutrients, the basis of damage to the host, and the effects of the environment. The diversity of relationships is enormous, but identifying some common features is helpful in providing basic guidelines for the control of contrasting types of pathogens.

Pathogens and pathogenesis

Considerable confusion surrounds the terms pathogen and parasite. While they are generally used to describe microbial disease agents, in particular the fungi, bacteria and viruses, the distinction between the two terms has often been overlooked. They are not synonymous; a parasite is an organism having a particular type of nutritional relationship with a host, while the term pathogen refers to the ability of an organism to cause disease. They may be defined as follows.

• **Parasite:** an organism or virus living in intimate association with another living organism (host) from which it derives some or all of its nutrients, while conferring no benefit in return.

• **Pathogen:** an organism or virus able to cause disease in a particular host.

The allied term **pathogenesis** describes the complete process of disease development in the host, from initial infection to production of symptoms.

At first sight the distinction between a parasite and a pathogen might appear subtle; indeed in many cases the parasitic activities of an organism automatically lead to its being a pathogen as well. The diversion of nutrients from the host will cause some metabolic stress, which will normally be expressed as disease.

However, in other host–microorganism associations this stress may be offset by the microbe contributing nutrients in return. This is the case with root nodules of legumes, where the bacterium *Rhizobium* obtains carbohydrates from the host but also fixes atmospheric nitrogen, some of which the host subsequently utilizes. Mycorrhizal fungi infect plant roots but actually stimulate growth by assisting the uptake of scarce nutrients, especially phosphates, from the soil. The definition of a parasite given above takes account of situations such as these.

Where the invading microbe confers some beneficial effect the term **symbiosis** has been used. As originally conceived, symbiosis (literally 'living together') referred to any intimate or close association between organisms, irrespective of benefit or harm, and was subdivided as shown in Fig. 2.1. The advantage of this scheme is that it can accommodate relationships where the balance may shift from mutual benefit, termed **mutualism**, to injurious effects on one partner.

If one considers the terms parasite and pathogen from the reverse viewpoint, in other words the ability to cause disease, the difference becomes more obvious. While all parasites are potentially pathogenic due to their redirection of host nutrients, many of the characteristic symptoms of disease cannot be explained on the basis of nutritional stress alone. The growth and development of a pathogen in its host, along with the response of the host to the presence of an alien organism, involves other interactions which have little to do with nutrition. Many of the more injurious effects of pathogens may be traced to toxic chemicals whose production is apparently incidental to their parasitic way of life (see Chapter 8). Looked at in this way, the statement 'a good parasite is a poor pathogen' may appear to be justified. Any organism which is dependent upon another organism for its

supply of nutrients might be expected to restrict its pathogenic effects to a minimum.

Biotrophs and necrotrophs

Although there is an enormous variety of pathogens, an important distinction can be made between those which rapidly kill all or part of their host, and others which coexist with host tissues for an extended period without inflicting severe damage. The former category, referred to as **necrotrophs**, are often opportunist pathogens which invade wounds and juvenile or debilitated plant tissues. They grow intercellularly producing cytolytic factors and then utilize the dead host tissues as a resource. The ability to attack a living host distinguishes these organisms from the **saprotrophs**, which subsist exclusively on organic debris. In contrast, **biotrophs** do not kill their host immediately. They are, in fact, dependent upon viable host tissue to complete their development. Extreme biotrophy resembles mutualism in that it is difficult to discern any marked pathogenic effects. The contrasting features of necrotrophs and biotrophs are summarized in Table 2.1.

Biotrophs, in keeping with their more specialized parasitism, usually attack only a limited range of hosts. Biotrophic fungi frequently form differentiated infection structures including modified intracellular hyphae termed haustoria (see Chapter 6). Generally speaking, these fungi do not produce large quantities of extracellular enzymes or toxins; during their co-evolution with the host, synthesis of degradative enzymes may have been repressed, or limited to localized sites where host cells are penetrated. Eventually the ability to elaborate such enzymes may have been lost altogether. An alternative view of these different lifestyles is that necrotrophs may have evolved from biotrophs through an increasing ability to produce enzymes capable of degrading complex substrates. This theory proposes that the first fungi were dependent on living plants, but gradually evolved independence by developing enzyme systems able to deal with polymeric carbon sources in plant litter. Such schemes can therefore be extended to include free-living saprotrophs, but in the absence of any adequate fossil record both versions are speculative. The advent of molecular techniques for analysing genome structure may, however, provide fresh evidence to support or refute such evolutionary models.

A human analogy for these contrasting types of

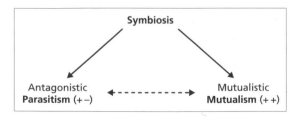

Fig. 2.1 Symbiotic relationships: +, positive effects on partner; –, negative effects on partner.

Table 2.1 Major characteristics of necrotrophic and biotrophic pathogens.

Necrotrophs	Biotrophs
Biochemical and morphological features	
Host cells rapidly killed	Host cells not rapidly killed
Toxins and cytolytic enzymes produced	Few or no toxins or cytolytic enzymes produced
No special parasitic structures formed	Special parasitic structures, e.g. haustoria, typically formed
Host penetration via wounds or natural openings	Host penetration direct or via natural openings
Ecological features	
Wide host range	Narrow host range
Able to grow saprophytically away from the host	Unable to grow away from the host
Attack juvenile, debilitated or senescing tissues	Attack healthy hosts at all stages of development

parasite has been proposed as follows; necrotrophs are 'thugs', while biotrophs are 'con-artists', reflecting their more devious way of obtaining resources from the host plant!

The impression may have been given that biotrophy and necrotrophy represent absolute categories; in reality there is a continuous gradation between the two types of pathogen. At one extreme are the viruses, which can replicate only within living cells, and fungal biotrophs, such as the downy and powdery mildews. At the other extreme are necrotrophs, such as the damping-off fungi and soft rot bacteria. In between, one encounters pathogens with intermediate characteristics. For instance, the potato late blight fungus *Phytophthora infestans* (Plate 1, facing p. 12) exhibits a high degree of host specificity and other biotrophic features such as haustoria, but it also causes relatively rapid necrosis of invaded tissues. Some pathogens pass through both a biotrophic and a necrotrophic phase during their life cycle. Plant pathogenic bacteria such as *Pseudomonas syringae* initially proliferate in intercellular spaces in leaves or fruits without apparent damage to host cells, but water-soaked lesions which become necrotic then appear. The apple scab fungus, *Venturia inaequalis*, grows beneath the cuticle of host leaves for several days without causing obvious necrosis, but as the lesions age the host tissues are eventually killed and the typical scabs develop (Fig. 2.2). Some species of the anthracnose fungus, *Colletotrichum*, penetrate directly into host cells, which remain alive for several days (see Fig. 6.7a); subsequently, necrotic, spreading lesions are formed. The term **hemibiotroph** has been used to describe

such behaviour. The factors responsible for this switch from a balanced mode of parasitism to rapid killing of host cells have in most cases not been identified.

In nature, necrotrophs may grow on both living and dead host tissues. Pathogens such as *Pythium* and *Rhizoctonia* may be found growing actively in soil, or on subterranean or aerial plant surfaces in competition with the natural microflora. In the absence of a suitable host they may successfully complete their life cycle by utilizing dead organic resources. The ability of biotrophs to compete for dead organic matter is very limited or even non-existent. These differences in patterns of natural occurrence of pathogens are reflected in their growth on laboratory culture media. Most necrotrophs are nutritionally undemanding; they grow well on a wide range of simple media. Biotrophs, on the other hand, have traditionally been regarded as fastidious organisms and in extreme cases cannot be grown on any known culture media.

Distinctions based on the criterion of culturability are used to divide pathogens into two nutritional types, **facultative** and **obligate** parasites. A further refinement of this scheme distinguishes pathogens which are able to grow relatively well in pure culture, but which in nature are unable to compete with non-parasitic microbes. Such parasites are termed **ecologically obligate**, in contrast to **biochemically obligate** organisms, which are unable to grow apart from the living host either *in vivo* or *in vitro*. The basis of obligate parasitism remains unresolved; such microorganisms may have metabolic blocks or incomplete nutrient uptake mechanisms that

(a)

(b)

Fig. 2.2 Apple scab caused by the fungus *Venturia inaequalis*. (a) Lesions on apple fruit. (b) Scanning electron micrograph of fracture through infected leaf showing sporulation of the pathogen on the leaf surface, but intact, uncolonized host tissue beneath. Arrows indicate subcuticular hyphae above epidermal cells (E). P, palisade and; S, spongy mesophyll. Scale bar = 10 μm. (Courtesy of Alison Daniels.)

require integration with host cells to function effectively. Figure 2.3 summarizes these different relationships and modes of nutrition in heterotrophic microorganisms.

Pathogen classification

The classification of pathogenic microorganisms is based initially on the same morphological and physiological criteria as other groups. However, conventional taxonomy does not accommodate all the characteristics of importance in pathology. Thus, different isolates of a pathogen which may appear identical in morphology and cultural characters can differ in pathogenicity and in the range of host species attacked. The same problem also occurs in medical microbiology. For instance, the common gut bacterium *Escherichia coli* is normally a harmless species resident in the human intestine, but certain isolates of this species can infect the gut, causing gastroenteritis. The differences between the pathogenic isolates and normal *E. coli* are relatively minor and are coded for by a few genes, often carried on extrachromosomal plasmids. Similar subtleties are common with plant

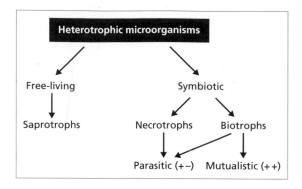

Fig. 2.3 Nutritional modes among heterotrophic microorganisms.

pathogens. In some cases, differences in pathogenic behaviour may be due to only a single gene (see p. 130). Differences in host range may be sufficient to define particular groups, or **pathotypes**, adapted to particular host species. In fungi, where such host specialization is clear, it may be possible to recognize **form species**. For instance, the black stem rust fungus, *Puccinia graminis*, occurs on various grasses including wheat (*P. graminis* f.sp. *tritici*) and barley (*P. graminis* f.sp. *hordei*). With plant pathogenic bacteria, particular **pathovars** adapted to different host plants may also be distinguished (see p. 34).

The classification of plant pathogenic viruses presents particular problems as many have very wide host ranges infecting different plant species, genera and even families. Nevertheless different strains occur which vary in important properties such as the relative severity of disease they cause, or frequency of transmission by different insect vectors. Such variation needs to be accommodated in any scheme for classifying viruses responsible for disease in plants.

Koch's postulates

To determine with certainty that a particular microorganism is the cause of a disease rather than some incidental contaminant it is necessary to examine critically its relationship with the host plant. This dilemma was first recognized in studies of pathogens of man and other animals. In 1876, Robert Koch provided the first experimental proof of disease causation by applying a set of rules which have since

come to be known as 'Koch's postulates'. Koch considered that these rules must be satisfied before any microorganism can be regarded as a pathogen. The rules involve five steps outlined below.

1 The suspected pathogen must be consistently associated with the same symptoms.

2 The organism should be isolated into culture, away from the host. This precludes the possibility that the disease may be due to malignant tissues or other disorders of the host itself.

3 The organism should then be re-inoculated into a healthy host.

4 Symptoms should then develop which are identical to those observed in the original outbreak of disease.

5 The causal agent should be re-isolated from the test host into pure culture and should be shown to be identical with the microorganism initially isolated.

An actual example of the use of Koch's rules is shown in Fig. 2.4. An apparently new disease of orange trees, with symptoms of chlorosis, stunting, and die-back of branches, was recently reported in South America. Leaves from affected trees were surface sterilized and plated onto a nutrient medium. Colonies of a small Gram-negative bacterium were obtained. Suspensions of the bacterium were then injected into healthy citrus saplings, and after a period of incubation, some of these artificially inoculated trees developed symptoms very similar to those seen in the original infected tree. The same small bacterium was re-isolated from these trees.

This procedure satisfied Koch's postulates and showed that the new disease, named citrus variegated chlorosis, was due to a bacterium. In reality, a lot more work, including light and electron microscopy, and the use of specific antisera, was required to actually identify the agent as a new strain of the xylem-inhabiting pathogen *Xylella fastidiosa*. Procedures for the detection and diagnosis of specific pathogens are described in more detail in Chapter 4.

This example shows that even today Koch's rules are still relevant, although they cannot be rigidly applied in their original form to all pathogens. The most important exceptions in plant pathology are when the pathogen cannot be grown in artificial culture, for example the viruses and some biotrophic fungi. The problem of isolating viruses from their host plants is generally overcome by using indicator plants. These are alternative hosts which develop symptoms which are specific for a particular virus. Healthy specimens of the original host may then be

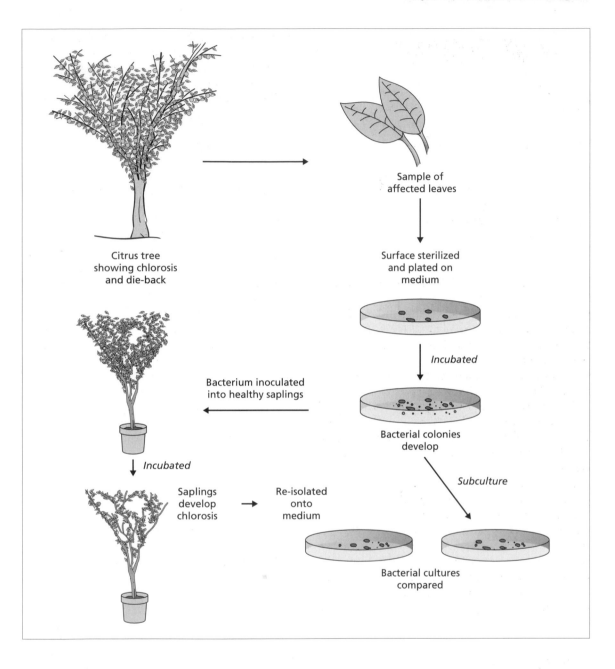

Fig. 2.4 Application of Koch's postulates to establish the aetiology of a new disease of citrus. (Based on Hartung *et al.* 1994.)

re-inoculated. In addition, electron microscopy of plant sap or of purified crystalline samples of the virus, coupled with serological techniques, may be employed to investigate the type(s) of virus present at each step of the procedure. There are also a number of new and powerful methods for detecting nucleic acid sequences specific to particular pathogens (see p. 55).

The application of Koch's rules to non-culturable fungal pathogens presents fewer problems because these agents produce spores. Such propagules can be removed from the host and then be used in reinoculation experiments. In many instances, spore morphology is also a valuable aid to identification of the inoculated and re-isolated pathogens.

Further difficulties in satisfying these postulates may be experienced in cases where symptoms result from mixed infections or when dealing with previously undescribed disease agents. For instance, few pathologists would have predicted the existence of the viroids, which scarcely conform to our preconceptions of a successful parasite (see Chapter 3).

Host resistance and pathogen virulence

All crops are exposed to a wide variety of potentially pathogenic microorganisms present in soil, water and the surrounding atmosphere. Yet most plants remain healthy most of the time. Consequently the majority of pathogens are unable to infect the plants with which they come into contact. Even where a specific pathogen can attack a particular host species there are marked variations in the extent to which individual plants succumb to disease. These differences are paralleled by variations in the pathogen population, reflecting differences in the genetic constitution of both the host and the pathogen. The ability of the pathogen to cause disease, and of the host to respond to invasion, have been shown to be determined by specific genes. This discovery has important implications both in the analysis of disease and in its control.

Resistance and susceptibility

When a microorganism makes contact with a plant it may be able to penetrate the potential host or it may be completely excluded. Following penetration, development of the pathogen may be halted by a host response or, alternatively, growth continues within the host tissues.

Describing the interaction between a microbial pathogen and its host presents problems as the outcome needs to be defined in terms of both partners (Fig. 2.5). At one extreme is the situation where a microorganism is incapable of causing disease in the host under any conditions, and so is described as a non-pathogen of that host. Likewise, plants able to completely prevent penetration by a microbial agent are non-hosts, and are considered to be immune to that organism. The majority of interactions between microbes and green plants are likely to be of this type. It may, however, be difficult to establish whether a plant is immune to a particular pathogen, as the absence of visible symptoms does not automatically imply a failure to penetrate. Hence the term 'immunity' is best avoided, or should only be used in cases where precise descriptions of the microbe–plant interactions are available.

In some instances (see Chapter 9) pathogens penetrate their host only to be immediately prevented from further colonization by the death of the first living cells they enter. Such restricted development indicates that the host is **resistant** to the pathogen. Resistance, unlike immunity, is not an all-or-nothing property of the host. In practice, there may be a whole range of responses. These vary from high resistance, where no visible symptoms are manifest, to low resistance, where the host succumbs completely to disease. Between these two extremes, resistance is described by a number of adjectives which though imprecise are of practical use in distinguishing between host reaction types.

Alternatively, differing degrees of pathogen development may be described in terms of host **susceptibility** (Fig. 2.5). For each degree of resistance there is a corresponding level of susceptibility. For instance, high resistance is equivalent to low susceptibility. Complete susceptibility is of considerable biological interest, as it appears to constitute the exception

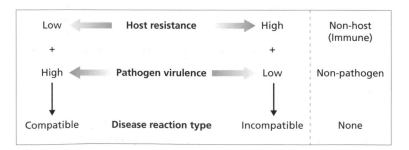

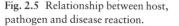
Fig. 2.5 Relationship between host, pathogen and disease reaction.

to the general rule that plants exhibit a degree of resistance. However, in this book we will normally consider these complementary descriptions of host responses only in terms of resistance. This approach has significant practical advantages in that it emphasizes the character which is selected for by plant breeders (see Chapter 12).

The terms 'resistance' and 'susceptibility' describe conditions of the host. However, just as the host may vary in its ability to resist infection so pathogens differ in their ability to invade and cause disease. Those microorganisms which cannot under normal circumstances induce disease in a host are regarded as non-pathogenic with respect to that host. Others which are able to penetrate but which have insignificant effects on the host may be termed **avirulent**; where the effects are more drastic they are described as possessing some degree of **virulence**.

It should be noted that there are still some problems with this terminology. In clinical microbiology a distinction was originally drawn between non-pathogenic and pathogenic microorganisms. Pathogenicity was considered to be an absolute property. Virulence was used to describe differences in the extent to which different strains of a pathogen caused disease. These clear-cut distinctions are valid for some pathogenic species, but it is now appreciated that the situation is more complex. Many microorganisms have the ability to acquire or lose traits which have a major effect on disease reaction type. Ultimately the phenomenon of pathogenicity needs to be analysed in genetic terms, and the differences between species, strains and pathotypes described in terms of the molecular interactions determining the disease phenotype. Avirulence, for instance, has a more precise meaning associated with the presence in the pathogen of a specific gene encoding a product which can be recognized by the plant, thereby triggering resistance (see below and Chapter 9). Hence some authors prefer to use the term **aggressiveness** to describe differences between pathogen strains in the amount of disease they cause in the host.

Genetic control of resistance and virulence

In common with all other biological characteristics, host resistance and pathogen virulence are genetically determined. However, these two properties can only

be assessed in the presence of the other partner. In the majority of cases, host resistance or pathogen virulence are not obviously correlated with other phenotypic characters. Features of the pathogen, such as rapid and extensive growth or the production of cell-wall-degrading enzymes, may or may not be related to virulence. Assessments of resistance and virulence are therefore based on disease reaction types. An interaction where symptoms are clearly expressed is described as a **compatible** disease reaction as opposed to an **incompatible** reaction where symptoms do not develop and the effect on the plant is minimal (Fig. 2.5).

Host resistance is controlled by one or a few genes whose individual effects may be easily detected, or by a multiplicity of genes, each of which contributes only a small fraction of the property as a whole. The practical implications of this are described in more detail in Chapter 12. In a few instances, disease reaction type has been shown to be controlled by factors inherited through the host's cytoplasm. The best-known example of such cytoplasmic inheritance involves the reaction of maize to the leaf blight fungus *Bipolaris maydis*. In the past the production of hybrid maize has involved the laborious task of detasselling by hand to avoid self-pollination occurring. The discovery of a cytoplasmically inherited factor for male sterility (*Cms*), which meant that cross-pollination was essential, removed the need for this operation. Because of this, cultivars possessing *Cms* came to predominate throughout the USA. Unfortunately, *Cms* was also correlated with susceptibility to a particular strain (race T) of *B. maydis*. As a result, the occurrence in 1970 of favourable conditions for the development of the pathogen resulted in a disastrous epidemic (see Fig. 5.1). In any breeding programme, the possibility that the cytoplasm may be important in disease resistance must therefore be considered.

Although pathogen virulence, host resistance and disease reaction types are genetically determined, the environment may modify the expression of any of these characters. For example, the *Sr6* gene conferring resistance in wheat to black stem rust is effective at 20°C, but is inoperative at 25°C.

Gene-for-gene theory

From an evolutionary viewpoint it is predictable that genetic systems determining virulence in the pathogen will be paralleled by genes conferring resistance in the host. This is because any mutation to virulence in a pathogen population will be countered by the selection of hosts able to resist this more aggressive pathogen. Population geneticists describe such a dynamic process of complementary changes as an 'arms race'. Thus, in an ideal world we might envisage a perpetual stalemate, with host and pathogen populations being closely matched in resistance and virulence. Hence over a period of several years disease would be neither completely absent nor would it become rampant.

Support for these ideas has been obtained by field observations and experimental studies on a number of host–pathogen combinations. The planting of crop cultivars containing a limited number of specific resistance genes will tend to select for pathogen strains possessing complementary virulence. Strains of a pathogen which differ in specific virulence genes are usually called **races**. The races present in any one season or in a particular locality may be identified by their behaviour on a selection of cultivars carrying different combinations of genes conferring resistance. A hypothetical interaction between several races of a pathogen and a differential set of host cultivars is shown in Table 2.2.

Further analysis of such differential interactions has shown that there is a precise genetic relationship between the two partners. The pioneering study was conducted by Harold Flor, who analysed the genetics of host resistance and pathogen virulence using flax rust, *Melampsora lini*, as a model. Flor showed that for each host gene conferring resistance there is a complementary gene in the pathogen determining virulence. This finding has become widely known as the **gene-for-gene** theory of host–pathogen interactions. On the basis of Flor's gene-for-gene theory, the possible interactions between a pair of alleles governing resistance in a plant and the corresponding pair determining virulence in the pathogen can be shown as a quadratic check (Fig. 2.6). In this scheme a resistant reaction occurs only where an allele for resistance in the host plant interacts with an allele for avirulence in the pathogen.

The gene-for-gene theory has important practical implications. The sequential introduction of new host cultivars which differ in resistance genes has been accompanied by corresponding changes in pathogen populations, whereby new races have successively come to predominate. The implications of this will be

Table 2.2 Hypothetical interaction between four host cultivars and four pathogen races.

Host	Pathogen			
	Race 1	Race 2	Race 3	Race 4
Cultivar 1	+	–	–	–
Cultivar 2	–	+	–	–
Cultivar 3	–	–	+	–
Cultivar 4	–	–	–	+

+, Compatible disease reaction (host susceptible; pathogen virulent).

–, Incompatible disease reaction (host resistant; pathogen avirulent).

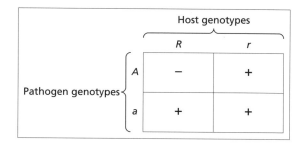

Fig. 2.6 The quadratic check, showing interactions between alleles of a host resistance gene and a pathogen gene for avirulence. Resistance (*R*) and avirulence (*A*) are normally dominant.

discussed further in Chapter 12. The gene-for-gene theory is also an important starting point for molecular models of host–pathogen specificity. Where single genes determine the outcome of a particular interaction, identification of the gene products involved should clarify how host–pathogen recognition occurs. Progress towards this goal is described in Chapter 10.

Further reading

Books

Isaac, S. (1992) *Fungal–Plant Interactions*. Chapman & Hall, London.

Mims, C.A. (1995) *The Pathogenesis of Infectious Disease*, 5th edn. Academic Press, London.

Smith, D.C. & Douglas, A.E. (1987) *The Biology of Symbiosis*. Edward Arnold, London.

Reviews and papers

Andrivon, D. (1993) Nomenclature for pathogenicity and virulence: The need for precision. *Phytopathology* **83**, 889–890. [And subsequent correspondence; see *Phytopathology* (1995) **85**, 518–519.]

Cooke, R.C. & Whipps, J.M. (1980) The evolution of modes of nutrition in fungi parasitic on terrestrial plants. *Biological Reviews* **55**, 341–362.

Crute, I.R. (1994) Gene-for-gene recognition in plant–pathogen interactions. *Royal Society of London. Philosophical Transactions. Series B. Biological Sciences* **346**, 345–349.

Hartung, J.S., Beretta, J., Brinasky, R.H., Spisso, J. & Lee, R.F. (1994) Citrus variegated chlorosis bacterium: axenic culture, pathogenicity, and serological relationships with other strains of *Xylella fastidiosa*. *Phytopathology* **84**, 591–597.

Mansfield, J.W. & Brown, I.R. (1986) The biology of interactions between plants and bacteria. In: *Biology and Molecular Biology of Plant–Pathogen Interactions* (ed. J.A. Bailey), pp. 71–98. Springer-Verlag, Berlin.

3   Pathogen Structure and Function

'Scarcely any part of the organised world is free from the attack of parasites, a provision which is clearly one amongst many ordered by the Creator to maintain that balance amongst living beings.' [Rev. M.J. Berkeley, 1803–1889]

By virtue of their lifestyle parasites are faced with a number of problems which, though not unique, are nevertheless more acute than those confronting free-living organisms. Dependence on a host which is often dispersed in space and time means that parasites must possess effective mechanisms for transmission and survival between hosts. Furthermore, their habitat is a living organism. Each host has the ability to mount a response to infection in order to ward off potential invaders. The host population is variable and can undergo changes from one generation to the next, so that the habitat is never static. The genetic flexibility of microorganisms, combined with their capacity for prolific reproduction, ensures that they maintain their status as successful parasites. Hence microbial parasites are able to exploit many of the niches afforded by the diversity of green plants in natural communities.

The groups of microorganisms with which this book is concerned, the fungi, bacteria, mycoplasma-like organisms (MLOs), viruses and viroids, have certain features in common as plant pathogens. For instance, they are all capable of rapid reproduction or replication which results in the formation of numerous infective spores, cells or particles. Collectively these may be described as propagules. Each group has, however, a distinctive and characteristic vegetative morphology (Fig. 3.1). Such differences in vegetative form have important implications as regards the behaviour of these different groups as plant pathogens.

What is the relative significance of these groups of microorganisms in pathology? Most notorious dis-eases of humans and other animals are caused by bacteria and viruses. By contrast, plants are more commonly affected by fungi and viruses. This difference in the importance of particular groups of microorganisms can be explained partly on the basis of several differences between plants and animals as habitats for microbial growth (Table 3.1). Bacteria generally prefer warm, alkaline conditions with high nitrogen levels. Because bacteria are unicellular their spread within the host is enhanced by a circulatory system. Filamentous fungi are more effective parasites of higher plants as their requirements are generally in direct contrast to those of bacteria. While these generalizations appear to hold good today, the extent to which such preferences are due to progressive adaptation to animal or plant hosts is not known.

Fungi as plant pathogens

Vegetative growth of fungi

Most plant-pathogenic fungi form **hyphae** (singular = hypha), i.e. filamentous cells which extend by apical growth and an ordered system of branching. The network of hyphae which results from such growth is called a **mycelium**, and the interconnected hyphal network derived from one propagule is termed a **colony**.

The apical mode of growth of most fungi is the key to the success of these organisms both as saprotrophs and parasites. Unlike unicellular organisms, filamentous fungi are able to extend through soil, plant litter or living tissues. As a hypha grows through the substrate it secretes extracellular enzymes which digest complex molecules. The products of this process are then absorbed by the hypha. As the nutrients become exhausted the hypha simply grows on to explore a new area. Not all filamentous fungi have the same

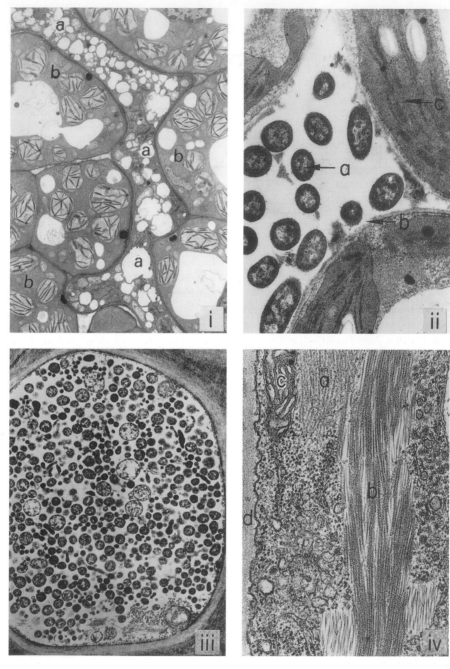

Fig. 3.1 (i) Intercellular hypha of *Peronospora viciae* (*P. pisi*) in submeristematic shoot tissue of pea (×3000) showing fungal hyphae (a) growing between host cells (b) which contain developing chloroplasts. The dark deposits in host cells adjacent to hyphae mark sites at which intracellular haustoria will develop. (ii) *Pseudomonas phaseolicola* cells in an intercellular space in a *Phaseolus* bean leaf (×12 700) showing bacterial cells (a), host cell walls (b), and host cytoplasm containing chloroplasts (c).

(iii) Mycoplasma-like bodies in sieve tube of *Nicotiana rustica* (×11 050) showing the variable morphology of these organisms. (iv) Tobacco mosaic virus in phloem parenchyma cell of tobacco (×67 500) showing striate crystalline aggregates of the virus in various orientations (a, b), a host cell mitochondrion (c), and the host cell wall (d). (Photographs from (i) Hickey & Coffey 1977; (ii) H.A.S. Epton; (iii) Hirumi & Maramorosch 1973; (iv) Esau 1968.)

Table 3.1 Higher plants and animals compared as hosts.

Plants	Animals
Anatomical features	
Rigid cell wall	No cell wall
No circulatory system	Circulatory system
Internal environment	
Acid pH	Alkaline pH
High C:N ratio	Low C:N ratio
No temperature regulation	Temperature regulation

enzymatic capabilities. For example, many of the necrotrophic pathogens which attack herbaceous tissues are noted for their production of pectolytic enzymes. In contrast, heart rot pathogens of trees can secrete cellulase and ligninases which enable them to utilize the complex constituents of wood.

The relationship between growing hyphal tips and the older, first-formed parts of the colony varies at different stages in the life cycle of a fungus. Three different patterns of behaviour may be identified. In some fungi, as hyphae extend at their apices the older portions of the colony become moribund and may autolyse, or be destroyed by bacteria and grazing animals. Amongst pathogenic fungi this type of growth is found in *Pythium* and *Rhizopus*. Sporulation in these fungi occurs at or near the advancing margin of the colony. Following disintegration of the first-formed hyphae the older parts of the lesion are quickly invaded by secondary organisms.

The hyphae of other fungi are longer-lived. They can act as transport systems from the older parts of the colony to the hyphal tips. Such transport systems are probably not important while the pathogen is still colonizing the original host, but they are essential when fungi attempt to grow from one host to another. Unless the fungus is then able to obtain nutrients by competing with saprotrophic microorganisms, it must fuel both its growth across inhospitable terrain and the subsequent infection processes needed to establish itself on a new host, by transporting nutrients from its established food base. Hyphae growing on the plant surface or away from the host are in an adverse environment and face problems such as desiccation and the destructive activities of the associated microfauna. Hyphae may therefore exhibit adaptations which improve their efficiency and success in

this role. The **runner hyphae** of the take-all fungus *Gaeumannomyces graminis*, which facilitate rapid and extensive growth along the surface of roots, have thicker, darker-coloured walls than hyphae involved in penetration and colonization of the root tissues. Other fungi produce cordlike structures consisting of a number of hyphae aggregated together. At their simplest, these **mycelial strands** involve relatively little differentiation. In more massive and elaborate structures, known as **rhizomorphs**, the outermost hyphae, which have thickened and pigmented walls, form a resistant rind while the internal hyphae are differentiated into a highly efficient conducting system for water and nutrients. Rhizomorphs can grow five to six times faster than normal vegetative hyphae. They are formed by many tree pathogens, such as *Armillaria mellea* and *Fomes lignosus*.

The third type of colonial behaviour is exhibited by the rust fungi and other biotrophic pathogens. In these the whole colony remains functional for a relatively long period and transport of nutrients takes place from the colony margin to the centre, where there is usually some continuing activity such as sporulation. Rust fungi develop these integrated colonies within photosynthetic tissues and the continued activity of their colonies depends on these tissues remaining viable.

Individual hyphae are frequently modified in form to accomplish particular functions, and this is especially common amongst pathogenic species. These fungi have in many cases evolved specialized hyphal structures to aid adhesion, penetration and colonization of the host. Appressoria, infection hyphae and haustoria (Fig. 3.2) are all examples of such hyphal adaptations (see Chapter 6).

Reproduction in fungi

The fungi exhibit a very wide range of reproductive mechanisms. Asexual reproduction seems to be of prime importance in increasing the number of individuals. Epidemic spread of pathogens usually depends on the production of countless asexual spores. The Mastigomycotina and the Zygomycotina form asexual spores in **sporangia**. In some species these are motile, when they are called zoospores, but in others they are non-motile. The asexual spores of the Ascomycotina and Basidiomycotina, which are all non-motile, are called **conidia**.

(a)

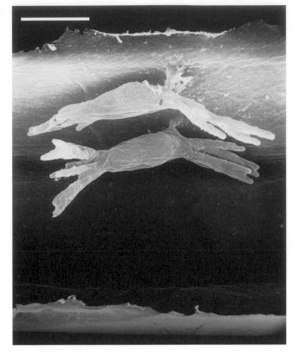

(b)

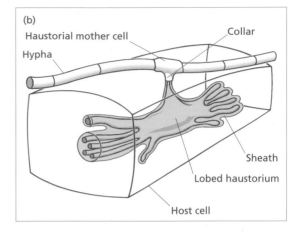

Fig. 3.2 Specialized parasitic structures, known as haustoria, formed by the powdery mildew fungus, *Erysiphe graminis*. (a) Scanning electron micrograph of a fractured epidermal cell showing two branched haustoria inside the host cell. Scale bar = 10 μm. (From Honegger 1985.) (b) Diagrammatic interpretation showing the fungal hypha on the leaf surface and a haustorium within the epidermal cell. (After Bracker 1968.)

Sexual reproduction fulfils the dual role of increasing variation in the population and assisting survival through unfavourable periods. It occurs regularly in the life histories of many fungal pathogens, such as *Armillaria mellea, Claviceps purpurea* and *Peronospora parasitica*. In some species, however, it is apparently an exceptional or very rare event. In fact, one group of fungi, the Fungi Imperfecti (or Deuteromycotina), is defined on the basis of the absence or infrequency of sexual reproduction. *Penicillium* on citrus fruits (Plate 8, facing p. 12), *Botrytis* on strawberries and *Cladosporium fulvum* on tomatoes are all examples of pathogens which seem to have relegated sexual reproduction to a minor role in their life cycles. With some imperfect fungi, however, the existence of a sexual stage may have been overlooked. The cereal eyespot pathogen, *Pseudocercosporella herpotrichoides*, was thought to reproduce only by conidia, but recently a sexual stage (*Tapesia*) has been discovered on straw in several different countries. Even when sexual reproduction is a regular feature of a life cycle, it usually only occurs under a more limited range of host and environmental conditions than permit asexual reproduction.

The nature of sexual reproduction varies considerably among the major groups of fungi. Four types of sexual spore can be identified, with **oospores, zygospores, ascospores** and **basidiospores** being formed by the Mastigomycotina, the Zygomycotina, the Ascomycotina and the Basidiomycotina, respectively. Some sexual spores germinate soon after dispersal, as in the case of *Puccinia* basidiospores on barberry leaves, and *Claviceps* ascospores on grass stigmas. In these fungi, the survival function ascribed to sexual reproduction has usually been accomplished prior to spore formation, by the perennation of structures which allow the fungus to overwinter or oversummer. In contrast, the sexual spores formed by the lower fungi, and some ascospores and basidiospores, are themselves able to withstand environmental stresses and thus facilitate survival between periods of pathological activity.

Some economically important pathogens, notably the rusts, have extremely complex life cycles in which as many as five different types of spore are formed in regular sequence. The most complex rust life cycles involve urediospores, teliospores, basidiospores, pycniospores and aeciospores. Epidemic spread is essentially by urediospores, which are asexual conidia, whereas variation and survival are ensured by the other four types of spore.

Bacteria as plant pathogens

Given the large number of bacterial genera, relatively few have been recorded as plant pathogens (Table 3.2). Some of the most important plant pathogenic species, for example *Pseudomonas syringae* and *Xanthomonas campestris*, are, however, subdivided into numerous pathovars distinguished by specialization on different hosts. The taxonomy of some of these is under review as studies of DNA homology have revealed large variations in relatedness between pathovars. The identity of other bacteria has been resolved only recently; small rickettsia-like organisms were first discovered in the vascular tissues of grapevines in 1973, but were later successfully cultured and shown to be a distinct species *Xylella fastidiosa*. The host range of these xylem-limited species appears to be very wide.

The majority of plant-pathogenic bacteria are unicellular (Fig. 3.3), with cell division by binary fission; the subsequent activities of the separated cells are not coordinated in any way. The major exceptions to this generalization are the bacterial pathogens classified as Actinomycetes. *Streptomyces*, for example, forms a rudimentary branching mycelium composed of relatively narrow, septate filaments. Sometimes individual bacterial cells aggregate to form substantial colonies, as in crown gall or in the cankers resulting from fireblight (*Erwinia amylovora*) infection. In other cases cells spread throughout an entire organ or physiological system. Soft rot bacteria in potatoes spread indiscriminately through tuber tissues, while vascular wilt bacteria in cucurbits are widely dispersed within the xylem. Growth of single-celled bacteria, like fungi, also involves the production of extracellular enzymes. However, most plant-pathogenic bacteria have only a limited ability to degrade cell-wall polymers, such as cellulose and lignin. Instead, pectolytic enzymes digest substances in the middle lamella which provides the bacteria with nutrients and allows them to spread between the newly separated cells. There is less integration in the development of a colony (which may be defined as the sum total of cells originating from one individual) than is possible in fungi. Neither the filamentous forms nor the more common single-celled bacteria are able to produce modified cells comparable to those formed by fungi which facilitate penetration into or extraction of nutrients from host tissues. However, many plant-pathogenic bacteria possess flagella (Fig. 3.3) and are therefore motile and capable of moving along nutrient gradients and towards host signal molecules. This may be important in habitats such as the soil surrounding plant roots, although to date there is little firm evidence that motility contributes directly to host infection.

Bacteria also reproduce sexually, but this mode of reproduction is less frequent than in fungi. Owing to their normally haploid state and the absence of a clearly defined nucleus, this process is more irregular

Table 3.2 Main genera of plant pathogenic bacteria and some example species.

Genus	Species	Subspecies or pathovar	Disease
Gram-negative			
Agrobacterium	*A. tumefaciens*		Crown gall
Erwinia	*E. carotovora*	ssp. *atroseptica*	Blackleg of potato
Pseudomonas	*P. syringae*	pv. *glycinea*	Bacterial blight of soybean
Ralstonia	*R. solanacearum*		Bacterial wilt
			Moko disease of banana
Xanthomonas	*X. campestris*	pv. *citri*	Citrus canker
Xylella	*X. fastidiosa*		Pierce's disease of grapevine
Gram-positive			
Clavibacter	*C. michiganense*	ssp. *insidiosus*	Bacterial wilt of alfalfa
Streptomyces	*S. scabies*		Potato common scab

than is found elsewhere in the plant and animal kingdoms. DNA passes from one cell, the donor, to another, the recipient, by transduction, transformation or conjugation. Transduction, which is mediated by bacteriophages, is possibly the most common method occurring in nature.

Mycoplasmas as plant pathogens

A number of diseases once thought to be of viral aetiology are in fact caused by a group of very small prokaryotic agents known as mycoplasma-like organisms (MLOs). In particular, several yellows diseases, characterized by extensive chlorosis and a gradual decline in the host, are associated with the presence of large numbers of these organisms in the phloem. Despite their superficial similarity in electron micrographs of diseased tissue, it is now clear that this group includes at least two types of agent, the **phytoplasmas** and the **spiroplasmas**. Both types may be regarded as bacteria which have lost the ability to

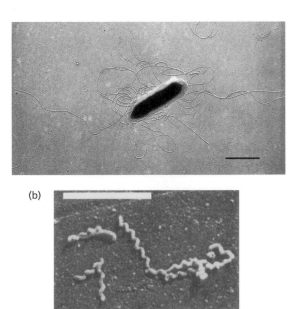

Fig. 3.3 (a) Electron micrograph of *Erwinia amylovora*, the bacterium causing fireblight disease, showing rod-shaped cell with flagella. Scale bar = 1 μm. (Courtesy of R.N. Goodman.) (b) Scanning electron micrograph of *Spiroplasma* cells from a pure culture. Scale bar = 5 μm. (Courtesy of P.G. Markham, R. Townsend & J. Burgess, John Innes Institute.)

form a rigid cell wall; they are highly pleomorphic, and it seems probable that under natural conditions they cannot survive apart from their hosts or vectors. MLOs are similar to viruses in their inability to multiply outside host cells, and in being transmitted by grafting, or insect vectors. They are sensitive to tetracyclines and treatment with these antibiotics leads to remission of symptoms. To date, no phytoplasma has been cultured away from its host or vector, so little is known about their physiology and other properties. The spiroplasmas can be cultured on artificial media, on which they typically assume a motile, helical form (Fig. 3.3).

More than 100 diseases in a variety of crops are now known to be caused by phytoplasmas. These include coconut lethal yellowing (see Fig. 4.2), which has virtually destroyed the coconut industry in the Caribbean, and rice yellow dwarf. Spiroplasmas have been associated with only two diseases, corn stunt and citrus stubborn.

Protozoa as plant pathogens

Parasitic protozoa are mostly thought of in the context of devastating human diseases such as malaria and sleeping sickness, but in fact a number of ailments of tropical perennial crops are associated with the presence of flagellate protozoa, often in phloem tissues. These include a wilt disease of coffee, heart rot of coconuts, and sudden wilt of oil palm. The flagellates, classified as *Phytomonas,* can be transmitted from plant to plant by insect vectors. A disease of cassava occurring in northern Brazil, typified by chlorosis and poor root development, may also be due to such pathogens, as large numbers of flagellates occur in the latex ducts of affected plants. No natural vector has been identified, but the disease can be transmitted by grafting.

Viruses as plant pathogens

Viruses are much simpler in structure than cellular microbial pathogens, consisting only of a nucleic acid core surrounded by a protein coat or capsid; within infected cells they occur either as individual particles or as crystalline aggregates (see Fig. 3.1). At first sight viruses might appear ill-equipped to act as pathogens, due to their extreme dependence on living cells. However, their diverse and effective methods of transmission between hosts, coupled with efficient replication once established within living cells, ensures that they are potentially devastating plant pathogens.

Viral parasitism is unique, in that the parasite is incorporated into the metabolism of the host cell. After gaining entry into a living cell, the nucleic acid component of the virus is released from its protein coat. The viral genome is then translated and replicated, and numerous new virus particles are assembled from the newly synthesized nucleic acid and protein (see Fig. 10.8). A virus can thus be visualized as a set of instructions for making more virus, packaged in a protective coat. In contrast to fungi and bacteria, viruses do not attack the structural integrity of their host tissues, but instead they subvert the synthetic machinery of the host cell, acting as 'molecular pirates'.

While viruses are simple parasites, they exhibit great diversity in their structure, morphology and mode of replication. Five classes of plant virus have been described, containing a DNA or RNA genome in either single- or double-stranded form; where the nucleic acid is single-stranded, the polarity of the strand may be the same as messenger RNA or complementary to it (Table 3.3). The genome may be a single molecule (monopartite) or divided into two or more pieces (bipartite or multipartite). The precise sequence of events during replication within the host cell is determined by the type of nucleic acid present (see Dimmock & Primrose (1994) *Introduction to Modern Virology*). To date, no examples of plant viruses belonging to class VI have been described; this group is noteworthy as replication involves copying RNA to DNA via the enzyme reverse transcriptase, and several important animal viruses, including the human immunodeficiency virus (HIV), are in this class.

The majority of plant viruses so far described (Table 3.3) belong to class IV, and contain a single strand of RNA which, once freed from its coat protein, can act directly as messenger RNA in the synthesis of further virus particles. This messenger must code for at least two components, an RNA replicase enzyme and new virus-coat protein. The first plant virus discovered, tobacco mosaic virus (TMV), serves as an example; TMV is a helical rod-shaped particle containing a single strand of RNA approximately 6400 nucleotides long (Fig. 3.4). Translation of the RNA strand yields four polypeptides differing in molecular weight. The two largest (P126 and P183) are

Table 3.3 The Baltimore system for virus classification based on the type of nucleic acid present (RNA or DNA), whether it is double- (ds) or single-stranded (ss) and whether the strand is of the same (+) or opposite (−) polarity to messenger RNA.

	Genome	Number known*	Examples
Class I	ds(±)DNA	13	Cauliflower mosaic virus (CaMV)
Class II	ss(+)DNA	26	Gemini viruses, e.g. African cassava mosaic virus (ACMV) Maize streak virus (MSV)
Class III	ds(±)RNA	27	Wound tumour virus (WTV)
Class IV	ss(±)RNA	487	Tobacco mosaic virus (TMV) Cucumber mosaic virus (CMV)
Class V	ss(−)RNA	82	Rhabdoviruses, e.g. lettuce necrotic yellows virus (LNYV)
Class VI	ss(+)RNA transcribed to DNA for replication	0	No plant-infecting examples known
Class VII	ssRNA which does not contain structural genes and has no protein coat	16	Viroids

* Approximate figures as new viruses are constantly reported.

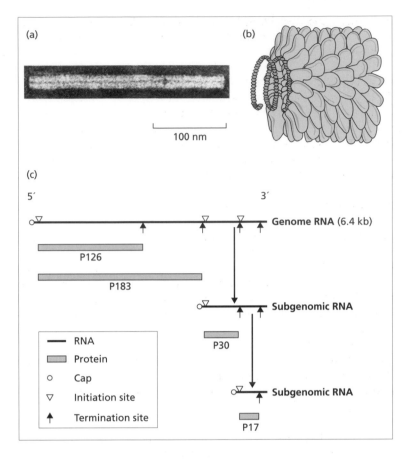

Fig. 3.4 Structure of tobacco mosaic virus (TMV). (a) Electron micrograph of an intact TMV particle. (b) Organization of particle with helical arrangement of coat protein subunits around the RNA. (c) RNA genome, subgenomic RNAs and the four proteins produced.

initiated from the same start site, with the longer polypeptide arising by readthrough of a stop signal. The two other proteins (P30 and P17) are produced via synthesis of smaller copies of parts of the TMV genome, described as subgenomic RNAs. Thus even with a relatively simple viral genome, the molecular events accompanying expression may be quite complex.

By introducing mutations at precise points in the RNA one can obtain clues as to the function of the different protein products. Changes in the region coding for the large polypeptides interfere with virus replication. Mutations in the P30 coding region inhibit the movement of virus from cell to cell. The smallest product, P17, is the coat protein, and is needed in large quantities as each virus particle contains more than 2000 capsid subunits. Production of a subgenomic RNA is a mechanism allowing rapid synthesis of multiple copies from a small template. This type of functional analysis is useful not only in

understanding the virus life cycle, but also in devising new ways of engineering plants for resistance to viruses (see p. 229).

In other plant viruses, the viral nucleic acid must first be transcribed to yield messenger RNA, from either an RNA or DNA template. The polymerase enzyme necessary to achieve this transcription may already be present in the intact virus particle.

Viroids as plant pathogens

Viroids differ from viruses in the size of their RNA genomes and in their lack of a protein coat. These circular RNA molecules comprise only about 300 nucleotides, which is 10 times smaller than a typical RNA-virus genome. This amount of genetic information is minimal, and insufficient to code for even a single enzyme. The absence of an AUG initiation codon also suggests that no protein is produced from the viroid RNA. This raises the interesting question as

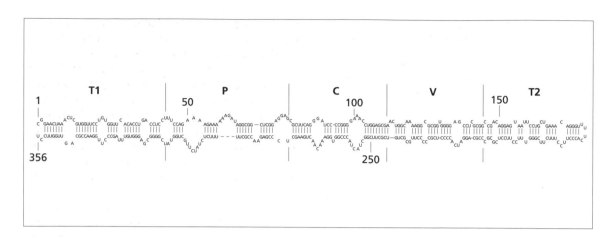

Fig. 3.5 The nucleotide sequence and secondary structure of one strain of potato spindle tuber viroid (PSTVd). Analysis of viroid sequences has identified five regions, or structural domains, in the molecule as indicated. T1 and T2 are the left and right terminal domains, P is the pathogenicity domain believed to be involved in symptom causation, C is a central conserved region and V is variable. (After Puchta *et al.* 1990, Fig. 1; with kind permission from Kluwer Academic Publishers.)

to how these agents replicate or modify host cells to cause disease.

The first disease definitely attributed to a viroid was potato spindle tuber (PSTVd), and at least 15 other diseases are now known to be caused by these infectious agents, including coconut cadang cadang (CCCVd), which to date has destroyed about 30 million trees in the Philippines. Speculation suggests that viroids may be of recent origin, favoured by intensive agriculture, and that cryptic viroids may be present in many apparently healthy plants. Analysis of viroid genome sequences (Fig. 3.5) has revealed similarities between these molecules and introns, the pieces of RNA spliced out of precursor messenger RNA during normal processing. Might viroids therefore be rogue molecules that have somehow escaped from regulated cellular functions?

Variation in microbial pathogen populations

An important property of all pathogenic micro-

organisms is their capacity for change. Variation in pathogen populations is maintained by several means (Table 3.4). Mutation, the underlying source of genetic variation, is probably of equal significance in all groups. Mechanisms for the maintenance of variation, and for the recombination of genes, differ in importance between groups. Viruses would appear to have few options other than mutation, although recombination can occur between RNA genomes where there is more than one piece of nucleic acid per particle, and hybridization is possible where more than one strain of a virus simultaneously infects a host cell. The sexual cycle is clearly important for many fungi and for some bacteria. Heterokaryosis (the long-term coexistence of two or more genetically different nuclei in a common cytoplasm) and the parasexual cycle are peculiar to the fungi. Fusion between the hyphae of two distinct individuals followed by nuclear migration can create a heterokaryon. In the parasexual cycle this may be followed by nuclear fusion and genetic recombination to produce novel variants without the normal steps involved in sexual reproduction taking place. It must be emphasized that for any species this phenomenon only occurs during the formation of a tiny percentage of its asexual spores, but as literally millions of spores are produced the parasexual cycle assumes considerable significance. Different isolates of the same fungus may also differ in ploidy, or in the number of chromosomes per nucleus. For instance, potato late blight, *Phytophthora infestans*, is usually diploid, but tetraploid and polyploid isolates occur; the rice blast fungus, *Magnaporthe*

Table 3.4 Mechanisms maintaining variation in pathogen populations. The more pluses, the more commonly the mechanism is used.

	Fungi	Bacteria	Virus
Mutation	+	+	+
Sexual reproduction	+++	+	–
Heterokaryosis	+++	–	–
Parasexual cycle	++	–	–
Cytoplasmic factors	++	+++	–
RNA recombination	–	–	++

grisea, shows wide differences in chromosome number between strains, due mainly to the occurrence of very small minichromosomes, which partly explains the highly variable nature of this pathogen in the field.

Cytoplasmic genetic elements are important in both bacteria and fungi. Most pathogenic bacteria contain extrachromosomal self-replicating circles of DNA known as **plasmids**. One or more copies of a plasmid may be present, and they can also move from cell to cell, thereby transmitting information, and increasing the genetic flexibility of the population. With many animal pathogenic bacteria, plasmids encode important virulence functions, such as adhesion to host cells, and toxin production. Less information is available for plant-pathogenic bacteria, with the notable exception of *Agrobacterium* species,

which harbour tumour-inducing (Ti) or root-inducing (Ri) plasmids (see p. 136). Genes located on plasmids may also influence the host range, nutrition and survival of plant-infecting bacteria. For instance, some genes for avirulence (*avr* genes) and antibiotic production can occur on plasmids. Variability is further enhanced by transposable elements (**transposons**), so-called 'jumping genes' which can insert at different sites in the genome, or move from plasmid to chromosome or vice versa. Recently, genetic equivalents of plasmids have also been discovered in fungi.

Together these mechanisms ensure that in nature pathogen populations encompass a range of variation which enables them to contend with changes in their hosts or in the environment. Indeed, the apparently endless capacity for variation amongst pathogens remains a major stumbling block in the search for durable, disease-resistant crops.

Dispersal of pathogens

Colonization of a particular habitat is followed sooner or later by spread, as space and nutrients become limiting. The problem of dispersal is a fundamental feature of the life cycle of all living organisms. However, because of their adaptation to a narrow ecological niche, namely a living host, most pathogens require an especially efficient means of solving this problem. Dispersal is usually thought of in terms of spatial spread (Fig. 3.6), but the time scale

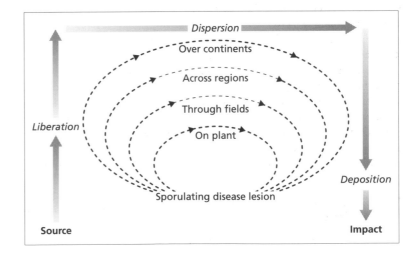

Fig. 3.6 The processes involved in pathogen dispersal and the scale on which they operate.

over which this spread takes place is also important. While dispersal may be completed in a few hours or even minutes, in other cases spread may take place over several years.

Many pathogens spread through crops in a spectacular manner. Some can cover long distances at remarkable speeds. In general the most dramatic examples involve pathogens with airborne propagules. Pathogens which infect subterranean organs often exhibit a restricted pattern of dispersal through the soil. These generalizations apply particularly to pathogens which have developed their own independent processes for dispersal. Pathogens which rely on another, vector organism (see below) for spread are at the mercy of these agents, in respect of the distances and speeds achieved.

There are almost as many dispersal mechanisms as there are pathogens (Fig. 3.7). Some mechanisms are

both elegant and highly efficient in that they are closely adapted to the biology of the host. The synchronization of spore release in certain pathogens, for instance *Venturia* and *Claviceps*, with bud-break or flowering is essential if the propagules are to find vulnerable host tissues. Other pathogens merely saturate the environment with propagules in an apparently haphazard and wasteful manner. For example, the fruit-bodies of bracket fungi causing heart rots release literally billions of spores over several months or years. Only occasionally will a spore land on the exposed wood of a newly damaged tree, where it can germinate, colonize and cause extensive tissue destruction.

Aerial environment

Dispersal through the atmosphere involves three dis-

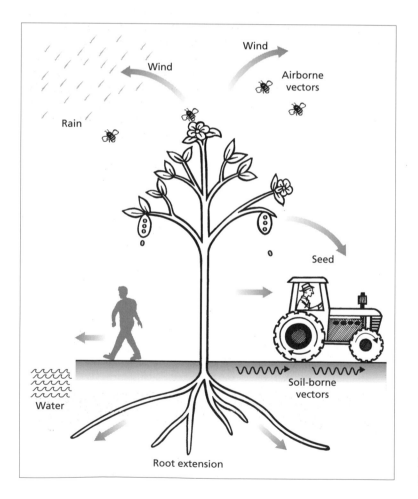

Fig. 3.7 Some mechanisms by which pathogens are dispersed.

tinct processes: liberation of the propagules; dispersion; and, finally, deposition (see Fig. 3.6). In the first phase the pathogen itself may participate actively, but during the two latter processes propagules are transported passively.

Liberation has two aspects: (i) the release of propagules from the parent colony; and (ii) their take-off from the host surface. Take-off is of special interest as it involves the boundary layer phenomenon. This is a zone of still air which surrounds all surfaces including leaves, stems and petals, as well as the crop canopy, inasmuch as this forms a definite layer. The thickness of the boundary layer varies according to a number of factors, particularly wind speed and turbulence and the size and shape of the surface. Some pathogens, notably many fungi, have evolved sophisticated methods which enable their spores to pass through the boundary layer into the turbulent air beyond (Fig. 3.8). Other fungi, and many bacteria, take advantage of external dispersal agencies. Foremost amongst these are rain drops, which splash or puff and tap spores from wet and dry surfaces, respectively (Fig. 3.9). The potential importance of rain drops as agents of dispersal can be simply illustrated. In regions with a rainfall of 100 cm year^{-1}, which is typical of northwest Europe, the northeast USA and many other regions, each square metre of ground will be hit each year by about one thousand

million rain drops. Each drop can break up into as many as 5000 splash droplets which may carry spores through the boundary layer. These droplets plus spores or cells can bounce directly onto an adjacent plant or, if the water evaporates, the propagules can become airborne. With some pathogens, for instance *Septoria tritici* on wheat, the risk of disease can be directly related to rainfall events (see p. 68).

Liberation is followed by dispersal. This may be rapid and local, as when spores travel in splash drops, or alternatively may involve transport in the upper atmosphere across continents over several weeks. Many fascinating questions concerning this stage in the propagule's journey remain to be answered. Can spores cross the oceans? Do sequential disease outbreaks across a region signify the movement of a pathogen? What factors limit the survival of spores in the upper reaches of the earth's atmosphere? Most microbial cells, for instance, are sensitive to exposure to ultraviolet (UV) light. Modelling the long-range transport of plant pathogens is a complex science intimately linked to meteorology. Nevertheless there is good evidence that some pathogens can disperse through the atmosphere on a continental scale. Urediospores of rust fungi can be trapped at high altitudes, and the first occurrence of yellow stripe rust (*Puccinia striiformis*) in New Zealand in 1980 is

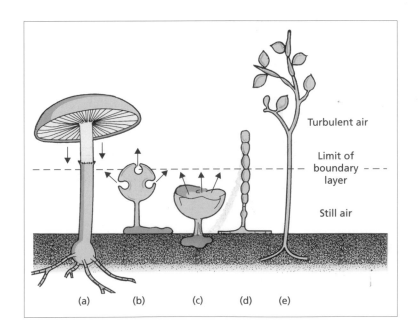

Fig. 3.8 Devices employed by fungi to escape from the boundary layer. Gravity, (a) *Armillaria*; violent discharge, (b) *Claviceps* and (c) *Sclerotinia*; chain extension, (d) *Erysiphe*; long sporangiophores, (e) *Phytophthora*.

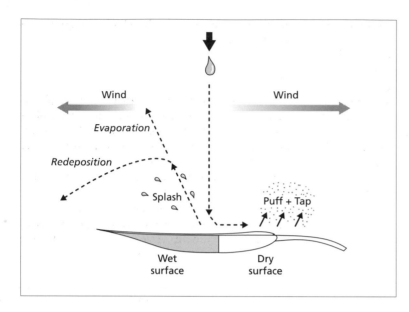

Fig. 3.9 Spore dispersal by rain drops from wet and dry leaves.

assumed to have been due to the wind transport of urediospores from eastern Australia.

Other unanswered questions concern deposition onto a plant surface. In calm conditions, diffusion brings propagules into contact with the boundary layer. On entering this layer of still air they fall onto the surface under the influence of gravity. In windy conditions propagules can be impacted onto leaf hairs and other projections. This process is not, however, very efficient and it is only of any importance for large spores which become impacted on small, narrow surfaces at high windspeeds. The number of spores deposited in this way is increased if the air currents are turbulent.

Rain is also important during the deposition phase as rain drops collect the air spora and deposit it on plant surfaces. Both large and small propagules are washed out from the atmosphere. Less is known about deposition due to electric charges or temperature gradients, but both may be important for some pathogens and/or crops.

Soil environment

Patterns of spatial dispersal of soil-borne pathogens are generally restricted, at least in the short term. Some spread only slowly through crops, forming circular patches, or **foci**, within which most plants become diseased. Other pathogens which primarily

attack subterranean tissues are not restricted to the soil and emerge briefly from below ground solely for the purpose of dispersal. The take-all fungus, *Gaeumannomyces*, spreads by mycelial growth along roots, but it can also produce an annual crop of airborne ascospores from perithecia embedded in stubble. Ascospores were almost certainly responsible for the outbreaks of take-all on newly reclaimed polders in Holland. Other soil-borne pathogens are passively if irregularly dispersed as a result of soil erosion and flash flooding. Agricultural practices such as ploughing can also aid dispersal.

Whilst soil may not offer the same opportunities for rapid, long-distance dispersal as the aerial environment, it does in part compensate by providing a better medium for microbial growth. Amongst microbial pathogens only the fungi possess the capabilities to really exploit this advantage. Two aspects are involved: the pathogen must have the capacity to obtain nutrients from soil substrates to continue growth; and it must also be able to withstand antagonism from the soil microflora and fauna. Pathogenic fungi vary widely in their potential for saprotrophic growth in soil. Some, such as *Plasmodiophora,* are unable to grow at all away from the living host, whilst others, for example *Pythium* and *Rhizoctonia*, are almost equally good saprotrophs as they are parasites. This enables

George Green Library - Issue Receipt

Customer name: Elbeyati, Selma

Title: Asking questions in biology : a guide to
hypothesis-testing, analysis and presentation in
practical
ID: 1005197824
Due: 11 Dec 2012 23:59

Title: Plant pathology and plant pathogens / John
A. Lucas.
ID: 1002819964
Due: 11 Dec 2012 23:59

Total items: 2
16/10/2012 13:26

All items must be returned before the due date
and time.
The Loan period may be shortened if the item is
requested.

WWW.nottingham.ac.uk/is

them to spread extensively through soil by hyphal growth.

Between these extremes there is one group of fungi which, though not such efficient saprotrophs as *Pythium*, are nevertheless able to spread effectively in soil. The tree-infecting pathogens *Armillaria* and *Fomes* form rhizomorphs which grow along root surfaces and through soil for considerable distances. In this instance, nutrition is being supplied by the previously parasitized host rather than by debris in the soil.

The capacity of plant-pathogenic bacteria to grow in soil is probably limited, although some can metabolize compounds typically found in plant debris, and others produce bacteriocins,

chemicals which specifically inhibit the growth of rival bacteria.

Vectors

A great diversity of animals feed on green plants. During their feeding activities, many move from plant to plant, sometimes over long distances. Not surprisingly, pathogens exploit some of these organisms as agents of dispersal. Any organism which transmits or disperses a pathogen is termed a **vector**. Some examples are shown in Fig. 3.10.

Two main types of pathogen–vector relationship may be identified. In the first the pathogen is carried externally, on the body or mouthparts, or within the

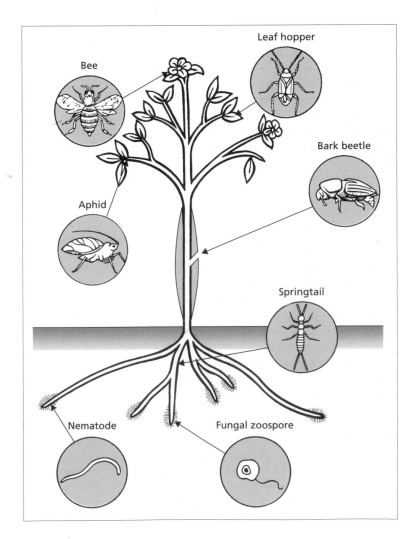

Fig. 3.10 Vectors exploited by plant pathogens.

digestive tract of the vector. The pathogen does not cross membrane barriers to enter or exit the vector, and does not multiply within the vector. In effect, the vector is contaminated with the pathogen. However, a distinction may be made among these vectors according to the extent to which their lifestyle is matched with that of the pathogen. In the least integrated relationships the pathogen is spread sporadically by a variety of vectors. Alternatively, the relationship may be quite specific, and require a close match between the partners for effective transmission to take place. In the second type of pathogen–vector relationship, the pathogen crosses membranes to enter the body of the vector and may multiply within it. In essence, the vector is infected by the pathogen and may be capable of transmitting this infection to plant hosts for the remainder of its life. Sometimes the pathogen may actually be passed on to the progeny of the vector. These remarkable dovetailed patterns of co-evolution have resulted in dispersal systems which are highly efficient.

Vectors are frequently thought of as only being involved in the transmission of viruses. However, whilst it is true that vectors are of supreme importance in the infection cycle of numerous viruses, it should be emphasized that they are also essential for the dispersal of many bacteria and fungi. For example, fireblight (*Erwinia amylovora*) and Dutch elm disease (*Ophiostoma novo-ulmi*) are spread by pollinating insects and bark beetles, respectively. Many fungi produce spores which can survive the digestive processes in animal guts. Ingestion and subsequent egestion may thus result in spores being dispersed and deposited within a nutritious faecal pellet. MLOs are another group of pathogens that are disseminated in nature by insects, usually leaf hoppers in which they can also multiply; the relationship is thus analogous to that seen with certain viruses. For instance, the agent of citrus stubborn disease, *Spiroplasma citri,* is spread by several species of leaf hopper.

Where pathogen dispersal is largely dependent on a vector organism, the disease triangle (see p. 1) becomes a tetrahedron (Fig. 3.11). To analyse disease in these cases requires an understanding of vector behaviour, and the interactions between the vector, the host plant, the pathogen, and the environment. The risk of disease may in fact be directly related to vector activity; for instance, outbreaks of barley yellow dwarf (BYDV) disease in winter cereals can be

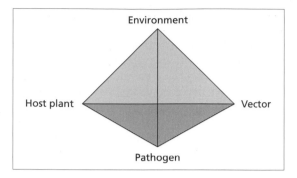

Fig. 3.11 The disease tetrahedron.

related to the movement of aphids carrying the virus in autumn (see p. 61).

Virus vectors

Due to their morphological simplicity, many viruses have come to depend entirely upon vectors for their dispersal. Most plant virus vectors are themselves pests or parasites, which ensures that the virus particles they carry are introduced into living host cells. Thus aphids, leaf hoppers, nematodes and even parasitic fungi are doubly important in that they may carry one or more viruses, as well as causing significant crop losses themselves.

Virus–vector interactions influence not only the spread of a virus, but in some cases host specificity as well. Among the repertoire of proteins coded for by the virus are some, known as **helper** proteins, which assist in transmission by vectors. For example, cauliflower mosaic virus (CaMV) encodes a small protein, P18, otherwise known as aphid transmission factor, which appears to interact with virus particles and vector mouthparts to promote binding, and hence to increase the efficiency of aphid transmission of infection. In other cases the virus coat protein appears to play a role in vector transmission.

Three important stages in virus–vector interaction can de defined: (i) **acquisition**, the time taken for the virus to be taken up by the vector; (ii) **latent period**, the time between acquisition and the vector becoming infective to other hosts; and (iii) **inoculation**, the time required for the virus to be introduced into a new host. The term **retention** describes the time during which the vector remains infective to new host plants. The diverse relationships between viruses and their vectors, with particular reference to patterns of infectivity, are indicated in Table 3.5.

Among these different vectors, most is known about the activities of sap-sucking insects which infest leaves and other photosynthetic tissues. Aphids and leaf hoppers insert their stylets directly into the phloem, and while extracting sap they also acquire virus particles. In some vectors this inoculum may be transferred immediately to the next plant that the insect probes as a possible food source. The insect may then be free from virus until it feeds on another diseased plant. Contamination of other vectors is followed by a more or less extended period during which the virus becomes established in the vector. The virus may or may not multiply in the vector, which is not infective to other hosts until the end of this establishment phase. Subsequently, the vector is able to infect plants over an extended period, which in extreme cases covers the lifetime of the vector. In some vectors infection may even be transmitted through eggs to the next generation.

A number of alternative schemes have been proposed to describe the fascinating relationships outlined in Table 3.5. One such scheme emphasizes the location of the virus in the vector, with some viruses being described as **stylet-borne**, while those that cross membranes and enter the haemolymph of the insect are termed **circulative**. These categories approximate to 'non-persistent' and 'persistent', respectively, which refers to the duration of infectivity of the vector to other hosts. As it is not known for certain where the virus particles are actually located in many vectors, the scheme outlined in Table 3.5 is preferred.

A few viruses are dispersed through the soil by nematodes and parasitic fungi. Grapevine fanleaf (GFLV) and raspberry ringspot (RRV) are two examples of viruses spread by nematodes, as the worms feed on living plant roots. Lettuce big vein (LBVV) and tobacco stunt viruses (TSV) are transmitted by *Olpidium*, a parasitic, root-infecting fungus. Beet necrotic yellow vein virus (BNYVV), the causal agent of the important rhizomania disease of sugarbeet, is transmitted by another root-invading fungus, *Polymyxa betae*, while several viruses infecting cereals, such as barley yellow mosaic (BaYMV) and barley mild mosaic (BaMMV), are spread by the related fungus *P. graminis*. Fungal zoospores move

Table 3.5 Strategies involved in virus–vector relationships.

	Persistence of virus in vector		
	Non-persistent (Stylet-borne)*	Semipersistent (Stylet-borne)	Persistent (Circulative, propagative/non-propagative)
Vectors	Aphids, chytrids, mites	Aphids, beetles, leaf hoppers, mealy bugs, whiteflies	Aphids, beetles, bugs, fungi, leaf hoppers, mites, nematodes, thrips, whiteflies
Location of virus in vector	Externally, on or near mouthparts	Near junction of stylet and tip	Internally, e.g. in haemolymph
Vector/virus specificity	Mostly low	Intermediate	Mostly high
Minimum time from acquisition of virus to inoculation of virus-free plant	Several seconds or minutes	30 min to several hours	12 hours to several days
Period during which vector remains infective	<10 hours (usually <4)	10–100 hours	>100 hours
Multiplication of virus in vector	No	No	Yes, in some instances

* Refer to those members of the fauna which have appropriate mouthparts.

most effectively in waterlogged soils, and some recent horticultural practices, in which plants are grown in a circulating liquid film of nutrients, provide favourable opportunities for the rapid dispersal of zoospore-borne viruses.

Two interesting routes of virus dispersal involve the parasitic angiosperm, dodder, and transmission via pollen. Dodder has been used experimentally to transmit viruses between different host species, but it is doubtful if it is of any practical significance under field conditions. Pollen offers a potentially ideal route for virus spread, because it becomes widely dispersed and the virus need never be exposed to the external environment. Contaminated pollen may transmit virus through the stigma to either the seed parent or the seed, or to both. From an agricultural viewpoint it is fortunate that such an efficient transmission system is uncommon. Amongst cases which are known are a number of fruit crop viruses, such as prunus necrotic ringspot (PNRV) and raspberry bushy dwarf (RBDV), which are both spread effectively as a result of cross-pollination from affected plants.

In those viruses which are seed-borne it appears that contamination of the seed from the pollen is as important as contamination from the mother plant,

with both processes occurring with equal frequency in most cases of seed-transmitted viruses.

Humans as vectors

Human activities have had a profound influence on pathogen dispersal. Human beings travel further and faster than any other potential vector, and also deliberately move plants and seed from place to place.

Humans have substantially increased the threat posed by seed-borne disease; all seeds may harbour a wide variety of pathogens, which are either present internally or as surface contaminants. For example, soybean seeds can carry a remarkable variety of pathogens as indicated in Fig. 3.12. Seeds of crop plants are no longer dispersed naturally, but instead are part of a massive international trade. In this way pathogens are often imported and become established in countries where the conditions for disease development may be more favourable than in the country of origin. In particular, newly exposed host populations are usually more susceptible to a pathogen than populations which have co-evolved over a long period (see p. 17).

The other way in which humans have, often unwit-

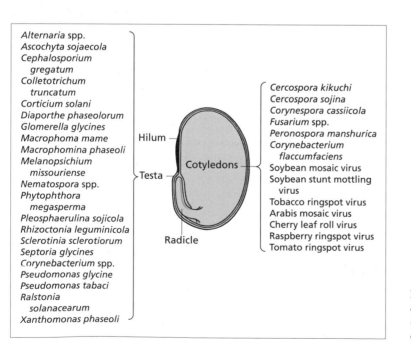

Alternaria spp.
Ascochyta sojaecola
Cephalosporium gregatum
Colletotrichum truncatum
Corticium solani
Diaporthe phaseolorum
Glomerella glycines
Macrophoma mame
Macrophomina phaseoli
Melanopsichium missouriense
Nematospora spp.
Phytophthora megasperma
Pleosphaerulina sojicola
Rhizoctonia leguminicola
Sclerotinia sclerotiorum
Septoria glycines
Corynebacterium spp.
Pseudomonas glycine
Pseudomonas tabaci
Ralstonia solanacearum
Xanthomonas phaseoli

Hilum
Testa
Cotyledons
Radicle

Cercospora kikuchi
Cercospora sojina
Corynespora cassiicola
Fusarium spp.
Peronospora manshurica
Corynebacterium flaccumfaciens
Soybean mosaic virus
Soybean stunt mottling virus
Tobacco ringspot virus
Arabis mosaic virus
Cherry leaf roll virus
Raspberry ringspot virus
Tomato ringspot virus

Fig. 3.12 Pathogens known to be dispersed on or in soybean seed. (Most of these pathogens are omitted from the Appendix.)

tingly, increased the efficiency of pathogen dispersal is by the use of vegetative propagation. Potatoes, soft fruits, plantation crops and many ornamentals are regularly propagated by vegetative methods. Viruses have been especially favoured by the use of grafts, buds, runners, cuttings and tubers. Crops such as citrus have suffered particularly through viruses being spread by such propagation methods; for instance, citrus tristeza has spread to many countries due mainly to movement of infected budwood for grafting.

Humans also act as vectors by carrying disease from crop to crop on their person, on machines, in foodstuffs or in soil or plant debris. In every respect the threat due to such movement has increased significantly in recent years. Modern air travel means that most places in the world can be reached within a day of departure, and hence spores or other propagules are more likely to be viable than when sea voyages separated continents by weeks. Outbreaks of Jarrah dieback, a devastating disease of native Australian forests caused by the soil-borne fungus *Phytophthora megasperma*, often originate along roadsides suggesting that spores have been carried by traffic. Farm machinery has become larger and more complex, and it too presents extensive opportunities for transport of propagules from field to field and farm to farm. The pattern of infection of sugarbeet crops by rhizomania disease during the first

confirmed outbreaks in the UK suggested that following introduction of the virus, contaminated soil was spread within fields by mechanical cultivation.

International trade can result in all sorts of plant material being moved from place to place. This material is seldom treated with the respect it deserves as a possible source of disease. One example of such trade is the import of timber—the aggressive strain of the Dutch elm disease pathogen, *Ophiostoma novo-ulmi*, responsible for the recent epidemic in western Europe, was almost certainly introduced in rock elm logs shipped from North America. Propagules in animal foodstuffs may pass through the gut unharmed and then re-enter the soil in manure. Corn smut, *Ustilago maydis*, is spread in this way in the USA as a result of the use of maize cobs as food for pigs. Resting spores of *Polymyxa*, the fungal vector of the virus causing rhizomania, pass intact through sheep fed on infected crop debris, thus increasing the risk of pathogen dispersal.

Pathogen survival

Whenever pathogens are not growing within a host they face problems of survival in a potentially hostile environment. The extent of this problem for any particular pathogen depends on the duration of time between available hosts, and on the relative hostility of the environment. At one extreme, dispersal during

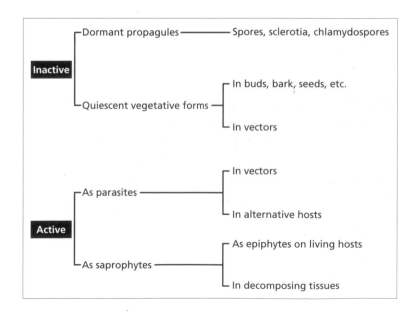

Fig. 3.13 Survival strategies adopted by plant pathogens.

epidemics may involve only brief periods when spores or other propagules are away from their hosts. Other pathogens survive between annual crops planted in successive growing seasons or in rotations when suitable hosts are available only every third, fourth or even fifth year. In the intervening periods, environmental extremes reduce the chance of pathogen survival. Drought, waterlogging and extremes of temperature can all reduce the viability of dormant pathogens. The propagules constituting pathogen inoculum are also subject to antagonism by other microorganisms, which can severely debilitate or even destroy them.

A summary of the main survival strategies adopted by microorganisms is given in Fig. 3.13. In some instances, for example when viruses become established in alternative hosts or vectors, or when fungi resort to saprotrophism, it may be argued that the term survival is inappropriate. However, plant pathologists who are concerned with potential threats to crops use the term loosely to cover all the possibilities listed in Fig. 3.13.

The fungi utilize by far the most extensive range of survival mechanisms, and many adopt several strategies which are employed in parallel to increase their chances of enduring unfavourable periods. Fungi produce a variety of specialized structures which allow long-term perennation ranging from single-celled **chlamydospores** to complex aggregations of hyphae termed **sclerotia**. Both are notable for their longevity and resistance to environmental extremes. Chlamydospores are individual cells with thick, pigmented walls and conspicuous lipid food reserves. They develop directly from hyphal compartments or from asexual conidia stranded in habitats where germination cannot immediately occur (Fig. 3.14). Chlamydospores are frequently formed by fungi living in soil, most likely in response to changes in nutrient availability, but perhaps also due to the activities of antagonistic microorganisms. Sclerotia usually consist of an outer rind of thick-walled, pigmented cells enclosing an inner medulla of thin-walled, hyaline cells. Sclerotial development appears to be more complex in that it is induced in certain species by mechanical factors, such as damage to hyphae or barriers to extension growth, and in others by nutritional factors such as changes in the carbon:nitrogen ratio and the availability of mineral ions and vitamins.

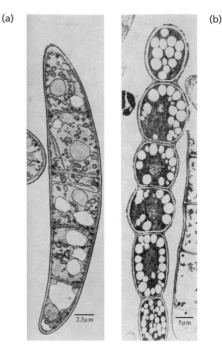

(a)   (b)

Fig. 3.14 Electron micrographs of (a) a macroconidium of *Fusarium culmorum* and (b) the development of each cell of the conidium into a swollen, resistant chlamydospore. (From Campbell & Griffiths 1974.)

Few pathogenic bacteria form resistant spores. Instead they often persist as vegetative cells on or in host tissues. Many viruses survive within the perennating organs of their principal or alternative hosts, while others persist within vectors.

Dormancy

The switch from vegetative growth to the formation of chlamydospores or other survival structures may be due to changes in the host or in the environment. In both instances the end result is the development of structures exhibiting **dormancy**, which may be defined as a rest period interrupting development. Ideally, this period should be of sufficient duration so that the reactivation of the dormant propagules coincides with renewed host activity. Precise systems of control are especially important where the pathogen has a limited ability to search for new hosts. Specific host signals may trigger the growth of dormant propagules. For example, the roots of onions and related *Allium* species exude flavour compounds known as alkyl-cysteine sulphoxides; these are in turn converted by soil microorganisms to a mixture of alkyltriols and sulphides, which specifically induce germination of dormant sclerotia of the onion white rot fungus, *Sclerotium cepivorum*. The presence of these chemicals in soil represents a biochemical 'signature', whereby the pathogen is able to detect the presence of a suitable host plant. Key stages in pathogen development may also be linked to the life history of the host plant. Ascocarp maturation in the ergot fungus *Claviceps purpurea* is controlled by environmental factors which also determine the time of flowering of its grass hosts. Hence at anthesis the air spora contains numerous ascospores which mimic pollen and infect through the exposed host stigma.

Dormancy is maintained in some propagules by **constitutive** factors. These are internal controls which must be overcome before propagules can respond to favourable external conditions. Internal checks on germination include permeability barriers and inhibitory substances. Constitutive mechanisms can be overcome by a variety of external influences such as repeated freezing and thawing, exposure to high temperatures, or alternate wetting and drying. Dormancy is also controlled by external, or **exogenous**, factors. These include temperature, water and nutrients. Thus, constitutive factors ensure that the propagule remains dormant for a minimum period

which is appropriate to the life cycle of the pathogen, whereas exogenous factors are responsible for the synchronization of renewed activity with susceptible stages in host development.

A diagrammatic representation of these processes as they affect fungal sclerotia is given in Fig. 3.15. The initial population gradually declines due to the effects of a range of lethal factors. At first, germination is prevented by constitutive factors. Once these are overcome there is a requirement for a favourable combination of exogenous factors. When sclerotia germinate they give rise either to mycelia, or to asexual or sexual spores.

Interest in germination processes has frequently centred on the possible role played by exogenous nutrients. Some propagules have a requirement for carbohydrates or similar nutrients, whereas others have sufficient endogenous reserves for pre-penetration growth.

Fungistasis

While soil is a potentially favourable habitat for microbial growth, it is a highly competitive environment. The majority of propagules deposited in soil, such as fungal spores, do not germinate. This inhibition has been termed **fungistasis**. No fully satisfactory explanation exists as to exactly how fungistasis operates but it is linked with the availability of nutrients, as dormant spores usually germinate if glucose or

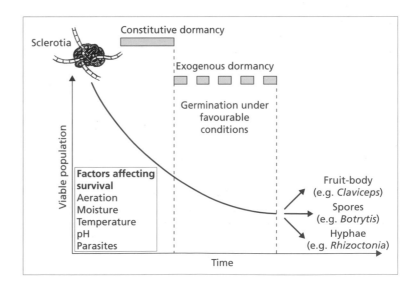

Fig. 3.15 Survival and germination of fungal sclerotia under natural conditions.

other nutrients are added to the soil. Similar inhibitory effects probably operate amongst the microbial populations on the aerial surfaces of green plants.

The pathological significance of fungistasis lies in the fact that developing roots release soluble carbohydrates and other nutrients from their apices and from the sites of emerging lateral roots. These exudates have been shown experimentally to be sufficiently concentrated to overcome fungistasis. Pathogens then commence active growth and have only a short distance to travel to the root surface. Sugars have also been shown to leak out onto aerial plant surfaces. The response to these nutrients is essentially non-specific as exudates from both host and non-host plants can stimulate germination.

Survival in hosts or vectors

Pathogens also survive in a quiescent vegetative state amongst perennating host tissues. Survival on or within the host ensures that the pathogen is ideally placed to reinfect newly formed host tissues in the subsequent growing season. The powdery mildew pathogens of perennial hosts such as apple trees or roses overwinter in dormant buds which form in the autumn. Similarly, the leaf curl fungus *Taphrina* survives in crevices in the bark of peach and almond trees. Plant-pathogenic bacteria are particularly well suited by their size and shape to exploit sheltered microhabitats provided by buds, bark and lenticels. Seeds also offer opportunities in this context as they may be contaminated superficially with propagules or infected internally with pathogens which have become dormant (see Fig. 3.12). Viruses can persist between crops within vectors which are themselves inactive. Two well-established examples of this strategy involve, firstly, tobacco rattle virus (TRV), which is able to remain infective in dormant nematodes for at least a year, and, secondly, barley yellow mosaic virus (BaYMV), which can persist for at least 10 years in the resting spores of its vector *Polymyxa graminis*. A number of viruses can remain viable in an inactive form for prolonged periods. Viable tobacco mosaic virus has been recovered from dried leaves after these had been stored for 25 years, and not surprisingly can therefore be spread through crops by workers who smoke cigarettes! Potential pathogens may be present in actively growing host tissues for an extended period without causing any visible symptoms. Factors which induce these dormant, or **latent** infections to become active include a decrease in the concentration of a toxic substance or a change in the chemistry of host tissue. Most examples involve pathogens affecting fruits, where the changes associated with ripening have been shown to stimulate a quiescent pathogen which initially infected the immature tissues. During ripening there are decreases in the level of tannins and phenols, and increases in soluble sugars, and fruits soften due to the induction of endogenous pectic enzymes. Such ripe fruits are more susceptible to attack by soft rot bacteria and fungi, an observation confirmed by recent work on non-ripening mutants of tomatoes, which remain resistant to post-harvest pathogens.

Alternative hosts

Survival can also involve parasitism in alternative hosts or vectors; these bridge the gap between two successive crops of an economically important host. A clear distinction should be drawn between such hosts and the **alternate** host species infected during the life cycle of some rust fungi. The relationships between these fungi and their alternate hosts are highly specific and essential for the completion of the pathogen life cycle. For instance, black stem rust of wheat (*Puccinia graminis*) alternates to barberry (*Berberis*), while white blister rust of pines (*Cronartium ribicola*) alternates to currants or gooseberries (*Ribes* spp.). The origin of such complex life cycles involving unrelated plant hosts is an intriguing biological mystery.

Most well-documented examples of alternative hosts involve either viruses or fungi. Some viruses in particular have extremely wide host ranges including both cultivated and wild species. Cucumber mosaic virus (CMV), for example, has been recorded in more then 700 hosts from at least 80 plant families. This virus causes severe stunting and yellowing of lettuce, and was found in one area to be present in 12 out of 14 weed species growing in the vicinity of lettuce crops. The virus was not causing any recognizable symptoms in most of these weeds, but they would undoubtedly serve as a source of infection as the main aphid vector feeds indiscriminately on numerous plants.

Wild hosts are often of particular importance when crop plants are introduced into new regions. *Cacao* is indigenous to the tropical forests of South America, yet the main areas of commercial cocoa production are in West Africa. The devastating virus disease cocoa swollen shoot (CSSV) most likely originated in the African crop through spread, by mealy bug vectors, from unrelated rainforest trees and shrubs infected with the pathogen.

Survival as saprotrophs

Some facultative parasites, including both bacteria and fungi, may be able to survive for long periods on the surfaces of their hosts without penetrating underlying tissues. These **epiphytes** are capable of continuing growth even on the limited nutritional resources present on the plant surface. Recent research has shown that bacteria such as *Pseudomonas syringae* often survive on leaf surfaces; following heavy rain, these populations multiply (Fig. 3.16) and may enter the leaf through stomata to cause disease. Most commonly, facultative parasites compete with populations of free-living decomposers. Success in this competition depends partly on the speed of germination and the subsequent growth rate. An ability to degrade a wide variety of substrates is also essential if the organism is going to be able to grow extensively in a natural habitat such as soil. During growth the organism will have to tolerate the presence of other microorganisms in its immediate environment, although it may restrict the growth of competitors by producing toxic metabolites. All these attributes are embraced in the concept of **competitive saprophytic ability**. This concept has proved useful when comparing the behaviour of pathogens which survive as saprotrophs.

Two contrasting fungal pathogens illustrate these points. *Rhizoctonia solani* attacks juvenile and senescent tissues of a wide range of hosts, and once these have been killed and decomposed the fungus is able to grow freely through the soil. *Gaeumannomyces graminis* only attacks cereals but is not restricted to any particular stage in their life cycle. This pathogen causes extensive root necrosis and persists within lesions on host debris. The extent to which *Gaeumannomyces* can survive successfully in this way is dependent upon the prevailing environmental conditions. Warm, moist, well-aerated soils promote the decomposition of roots and straw, with a consequent reduction in *Gaeumannomyces* inoculum. The nitrogen regime in the soil is also very important

as cereal tissues have a very high carbon:nitrogen ratio and this alone can limit the rate of their decomposition.

Longevity

Data concerning the longevity of particular pathogens under natural conditions are often unreliable. An outbreak of a soil-borne disease in an area which has not carried a particular crop for a known number of years has often been taken as proof of the ability of the pathogen to survive through the intervening period. Clearly, such information is of practical significance but it is usually impossible to relate it to the survival of particular propagules, such as

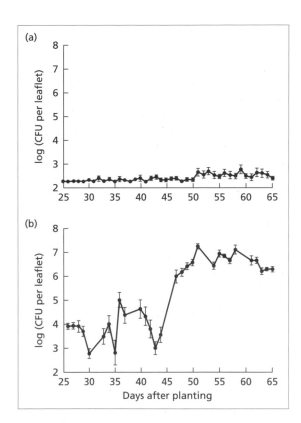

Fig. 3.16 Daily changes in populations of *Pseudomonas syringae* on bean leaves during two growing seasons. (a) During season A the population remained at a low level throughout the sampling period. (b) During season B large increases followed rainfall on days 35 and 44, and sustained wet weather after day 44. (After Hirano *et al.* 1995.)

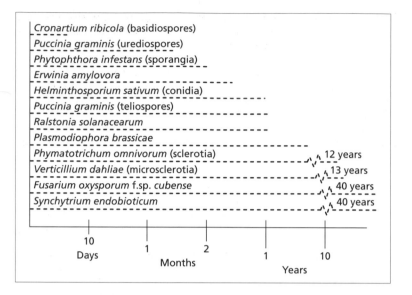

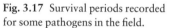

Fig. 3.17 Survival periods recorded for some pathogens in the field.

Table 3.6 Survival of sclerotia of *Sclerotium cepivorum* buried at different depths in soil. (From Coley-Smith *et al.* 1990.)

Depth of burial (cm)	Recovery of viable sclerotia (%)	
	15 years	20 years
0	42	28
7.5	90	88
15.0	96	72
22.5	88	79

spores or sclerotia. Information on the survival of such propagules may only be obtained from experiments. Given that some propagules can survive for several decades it requires little imagination to see the practical difficulties attending studies of this sort (Fig. 3.17). In addition, such experiments involve declining populations (see Fig. 3.15) and increasingly sensitive sampling techniques may be required to detect smaller and smaller numbers of viable spores. Table 3.6 shows the results of an experiment in which sclerotia of the onion white rot fungus, *Sclerotium cepivorum,* were buried for up to 20 years at different depths in field soil. Even after this extended period the majority of sclerotia from the deeper layers of soil were still viable, a finding which virtually rules out

rotation as a practical means of countering this disease.

Further reading

Books

Agarwal, V. & Sinclaire, J.B. (1997) *Principles of Seed Pathology.* Lewis Publishers/CRC Press, Boca Raton, Fla.

Andrews, J.H. & Hirano, S.S. (eds) (1991) *Microbial Ecology of Leaves.* Springer-Verlag, New York.

Blakeman, J.P. & Williamson, B. (1994) *Ecology of Plant Pathogens.* CAB International, Wallingford.

Cooke, R.C. & Whipps, J.M. (1993) *Ecophysiology of Fungi.* Blackwell Scientific Publications, Oxford.

Diener, T.O. (1979) *Viroids and Viroid Disease.* John Wiley & Sons, London.

Dimmock, N.J. & Primrose, S.B. (1994) *Introduction to Modern Virology.* Blackwell Science, Oxford.

Goto, M. (1992) *Fundamentals of Bacterial Plant Pathology.* Academic Press, San Diego, Calif.

Lynch, J.M. (ed.) (1990) *The Rhizosphere.* John Wiley & Sons, Chichester.

Matthews, R.E.F. (1992) *Fundamentals of Plant Virology.* Academic Press, San Diego, Calif.

Maude, R.B. (1996) *Seedborne Diseases and Their Control.* CAB International, Wallingford.

Mount, M.S. & Lacy, G.H. (eds) (1982) *Phytopathogenic Prokaryotes.* Academic Press, New York.

Sigee, D.C. (1993) *Bacterial Plant Pathology: Cell and Molecular Aspects.* Cambridge University Press, Cambridge.

Walkey, D.G.A. (1991) *Applied Plant Virology.* Chapman & Hall, London.

Reviews and papers

Adams, M.J. (1991) Transmission of plant viruses by fungi. *Annals of Applied Biology* **118**, 479–492.

Bonfiglioli, R.G., Webb, D.R. & Symons, R.H. (1996) Tissue and intracellular distribution of coconut cadang cadang viroid and citrus exocortis viroid determined by *in situ* hybridization and confocal laser scanning and transmission electron microscopy. *The Plant Journal* **9**, 457–465.

Broek, A.V. & Vanderleyden, J. (1995) The role of bacterial motility, chemotaxis, and attachment in bacteria–plant interactions. *Molecular Plant–Microbe Interactions* **8**, 800–810.

Coley-Smith, J.R., Mitchell, C.M. & Sansford, C.E. (1990) Long-term survival of sclerotia of *Sclerotium cepivorum* and *Stromatinia gladioli*. *Plant Pathology* **39**, 58–69.

Coplin, D.L. (1989) Plasmids and their role in the evolution of plant pathogenic bacteria. *Annual Review of Phytopathology* **27**, 187–212.

Diener, T.O. (1996) Origin and evolution of viroids and viroid-like satellite RNAs. *Virus Genes* **11**, 119–131.

Gundersen, D.E., Lee, I-M., Rehner, S.A., Davis, R.E. & Kingsbury, D.T. (1994) Phylogeny of mycoplasma-like organisms (phytoplasmas): a basis for their classification. *Journal of Bacteriology* **176**, 5244–5254.

Hirano, S.S., Rouse, D.I., Clayton, M.K & Upper, C.D. (1995) *Pseudomonas syringae* pv. *syringae* and bacterial brown spot of snap bean: A study of epiphytic phytopathogenic bacteria and associated disease. *Plant Disease* **79**, 1085–1093.

Katajima, E.W., Vainstein, M.H. & Silveira, J.S.M. (1986) Flagellate protozoa associated with poor development of the root system of cassava in the Espirito Santo State, Brazil. *Phytopathology* **76**, 638–642.

Lacey, J. (1996) Spore dispersal—its role in ecology and disease: the British contribution to fungal aerobiology. *Mycological Research* **100**, 641–660.

Leben, C. (1981) How plant pathogenic bacteria survive. *Plant Disease* **65**, 633–637.

Owens, R.A., Steger, G., Hu, Y., Fels, A., Hammond, R.W. & Riesner, D. (1996) RNA structural features responsible for potato spindle tuber viroid pathogenicity. *Virology* **222**, 144–158.

Pirone, T.P. & Blanc, S. (1996) Helper-dependent vector transmission of plant viruses. *Annual Review of Phytopathology* **34**, 227–247.

Simon, A.E. & Bujarski, J.J. (1994) RNA-RNA recombination and evolution in virus-infected plants. *Annual Review of Phytopathology* **32**, 337–362.

Symons, R.H. (1991) The intriguing viroids and virusoids: What is their information content and how did they evolve? *Molecular Plant–Microbe Interactions* **4**, 111–121.

Thresh, J.M. (1991) The ecology of tropical plant viruses. *Plant Pathology* **40**, 324–339.

Disease Assessment and Forecasting

'Plant pathogens are shifty enemies.' [E.C. Stakman, 1885–1979]

In crop husbandry the fate of the individual plant is usually irrelevant. The farmer or forester deals with millions of individuals and only when sufficient numbers succumb to disease is action deemed necessary. In the field therefore disease is a phenomenon to be considered in terms of populations.

Analysis of a disease outbreak requires accurate methods for estimating the amount or level of disease present. Ideally, this analysis should also include estimates of the future progress of the disease, so that eventual effects on the yield or quality of a crop can be predicted. This is essential if economic options for disease control are to be properly evaluated. Disease assessment therefore includes a number of interrelated activities (Fig. 4.1).

Monitoring the occurrence of a pathogen on a particular crop species over several years can provide clues as to the factors regulating its incidence and severity. This information can then be used to devise predictive systems which forecast the incidence of future outbreaks of the disease. The development of such forecasting systems is required especially for those diseases of economically important crop plants which regularly increase to epidemic proportions.

Disease detection

An essential first step in disease assessment is correct identification of the causal agent. In many cases macroscopic symptoms, often combined with microscopic examination of specimens, or isolation of an agent onto culture media, may be sufficient to detect the presence of a known pathogen. If there is uncertainty, further more rigorous tests, such as the application of Koch's rules (see Chapter 2), are necessary

to identify the disease agent. For viruses and phytoplasmas, electron microscopy, or inoculation of indicator plants may be used.

The problem with these more traditional approaches to disease diagnosis is that they are often time-consuming and labour intensive, requiring for instance a delay of days or even weeks while cultures grow, or test plants develop symptoms. Also, the sensitivity of some tests may be inadequate to detect small amounts of a pathogen. This problem is especially important when determining the level of infection in propagation material, such as seed potatoes, or in seed stocks. To detect loose smut (*Ustilago nuda*) infection in cereal seeds requires detailed microscopic examination of at least 1000 individual embryos, a laborious procedure even with recent refinements such as the use of sensitive fluorescent stains to visualize the pathogen. Relatively few contaminated seeds may be sufficient to initiate a severe disease outbreak. There is great interest therefore in developing more rapid and more sensitive methods for disease diagnosis and pathogen detection. The latest advances in this area are based on procedures first pioneered in medical microbiology and clinical pathology.

Molecular diagnostics

An ideal diagnostic test should possess the properties listed in Table 4.1. Sensitivity and accuracy are vital, but to find wide application a test should also be cheap, easy to perform, and give swift results. The most sensitive and specific methods for detecting pathogens are currently based either on antibodies, which recognize particular antigens, or nucleic acid probes, which target genomic sequences characteristic of the pathogen. An alternative approach is to look for a specific pathogen product, for instance a toxin, but while there are some examples in medical

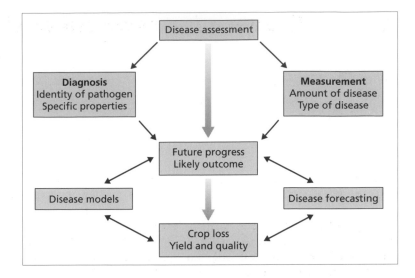

Fig. 4.1 Activities involved in disease assessment and prediction.

Table 4.1 Properties of an ideal diagnostic test.

Accurate — high specificity for target pathogen
Sensitive — able to detect small amounts of pathogen
Simple — does not require complex laboratory facilities
Rapid — gives early detection of pathogen
Cheap — affordable and disposable
Safe — non-hazardous for operator

diagnostics, this has not yet been developed to any significant extent for plant pathogens.

Serological tests utilizing the specificity of antigen–antibody reactions are by no means new, having for instance been applied for many years in the diagnosis of plant viruses. Such methods were originally based on polyclonal antisera containing a mixture of antibodies, which often reduced the reliability of the test due to non-specific cross-reactions. The development of monoclonal antibodies, which have narrow specificity against a single type of antigen, has greatly improved the accuracy of such tests. Monoclonal antibodies can, for instance, discriminate between different bacterial species, whereas polyclonal sera often react with several species within a genus. Along with improvements in the specificity of serological tests, there have been advances in sensitivity, such that tiny amounts of an antigen can now be detected. One widely used method is the enzyme-linked immunosorbent assay (ELISA), in which samples suspected of containing a

pathogen are reacted with antibody in wells on a test plate, and bound antigen is then detected by further antibody linked to an enzyme; the latter gives a colour reaction once substrate is added. As well as being very sensitive this test is also quantitative, as the intensity of colour is proportional to the amount of pathogen antigen in the original sample. It can therefore be used to estimate the level of infection in plants, or tissue samples, or seeds. Although mainly used for virus detection, ELISA tests are now available for many bacteria and phytoplasmas, and an increasing number of plant pathogenic fungi. Serological methods can also be adapted for use in microscopy, by labelling antibodies with dyes which fluoresce under ultraviolet (UV) light; bacterial cells, fungal hyphae, or virus particles can thus be specifically visualized in tissue samples.

Nucleic acid probes, which hybridize with DNA or RNA sequences characteristic of a particular pathogen, are a more recent development arising from recombinant DNA technology. Not surprisingly the first application of such probes was with viruses and viroids, which exist for part or all of their life cycle as nucleic acid strands. Hybridization with a complementary copy of part of the viral genome can therefore be used to directly detect the pathogen. It is also possible, however, to detect more complex agents such as bacteria and fungi using an appropriate DNA probe. One promising approach is to find DNA sequences that are abundant in the genome, for

(a)

(b)

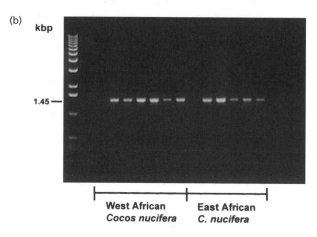

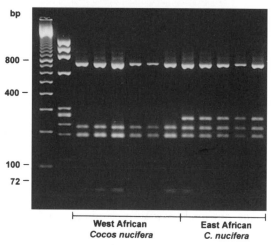

(c)

Fig. 4.2 Lethal yellowing of coconuts, caused by a phytoplasma. (a) The destructive effects of the disease in a coconut plantation in Jamaica. (Courtesy of S.J. Eden-Green.) (b) Detection of the phytoplasmas responsible for lethal yellowing by the polymerase chain reaction. The gel shows diagnostic bands obtained from West and East African samples using primers specific for 16S RNA sequences. (c) Analysis of the amplified product by digestion with a restriction enzyme reveals a polymorphism (i.e. an RFLP) between samples from the two regions, showing that the phytoplasma strains present in different areas are not identical. (b & c, Courtesy of A.M. Tymon.)

instance the repetitive DNA which codes for ribosomal RNA. Copies of such regions can then be used to specifically detect very small amounts of pathogen in a sample. The sensitivity of nucleic acid hybridization assays is comparable to that of serological tests such as ELISA, and like ELISA the results can be quantified. The limits for detection by nucleic acid probes can be greatly enhanced by using the polymerase chain reaction (PCR) to synthesize multiple copies of a particular nucleotide sequence (Fig. 4.2). Once optimized this method may be able to detect only a few cells of the target pathogen.

These new methods promise to revolutionize disease diagnosis, but some obstacles remain. Monoclonal antibodies are, for instance, expensive to produce. Initially, DNA probes relied on radioactive labels to detect hybridization, which clearly limited their use. Alternative non-radioactive markers are now being developed. PCR-based techniques require specialized equipment and controlled conditions to give reproducible results. Hence, the application of new diagnostic tests is mainly laboratory-based, for example in plant disease clinics. Provided these logistical and economic limitations can be solved, one can in the future envisage the use of rapid diagnostic kits in the field. Thus the grower might directly test a crop by processing samples with a cheap disposable kit which within minutes can detect and identify a specific pathogen. A further advantage of such simple but sensitive tests is the possibility of detecting a pathogen even before symptoms become evident in a crop, thus providing an early warning system.

Disease assessment in the field

The identification of a pathogen in a particular crop may in itself be of little interest to the grower. The likelihood of the disease increasing, of it spreading to adjacent fields, and the possible losses it might cause, are issues of far greater concern. These issues transcend the interests of individual farmers and demand the collection of data on a regional or national basis.

The potential severity of a disease outbreak is dependent upon the outcome of infection in individual plants and the capacity of the pathogen to spread through the crop. A complex series of factors are involved: the resistance of the host; the virulence of the pathogen; and the effects of the environment. Any component of the environment may influence the progress of a disease outbreak. Climatic, soil and biotic factors play a major part in determining the rate of reproduction of pathogens and their dispersal to new host plants.

Climatic conditions are of particular importance but are notoriously difficult to forecast with sufficient accuracy to predict events in the longer term. The nutrient status of the soil and hence the crop may influence disease expression in a variety of ways; for example, nitrogen deficiency may reduce the ability of the plant to repair tissue damage due to disease (e.g. the capacity to produce adventitious roots or new leaves may be restricted), while high nitrate levels can result in the development of tissues with increased susceptibility to disease. Such a change has been demonstrated in powdery mildew disease of barley, *Erysiphe graminis*, where increasing amounts of nitrogen fertilizer resulted in an increased number of colonies per leaf and a greater spore production per colony. Thus a crop cultivar which normally exhibits a high level of resistance may behave more like a susceptible host. The significance of the saprotrophic microbes which live on or in the vicinity of host plants is now appreciated. For example, the microflora present around plant roots can influence the ability of pathogens to colonize this zone.

Measuring disease

Disease assessment is concerned with the question 'How much disease?'. There are, however, several scales by which disease can be estimated. **Disease prevalence** is a broad measure of the amount of disease in a crop in a geographic region. For instance, barley yellow dwarf virus (BYDV) was prevalent in cereal crops throughout the UK during 1990, due to the survival of large numbers of virus-carrying aphid vectors through a mild winter. **Disease incidence** is a measure of the proportion of plants infected within a particular crop, often in a single field or plantation (Fig. 4.3). **Disease severity** is a measure of the degree of infection of an individual plant, usually based on the use of an appropriate assessment key. Disease assessment keys have been constructed for many pathogens, and their use for particular diseases has been standardized to ensure that assessments can be compared. Figure 4.4 shows the scheme currently used to assess the severity of banana leaf spot, or Sigatoka disease, *Mycosphaerella musicola*, on bananas.

Monitoring disease

As pathogens and pests have the potential to multiply rapidly in crop monocultures there is a need for regular crop inspection to pre-empt future problems.

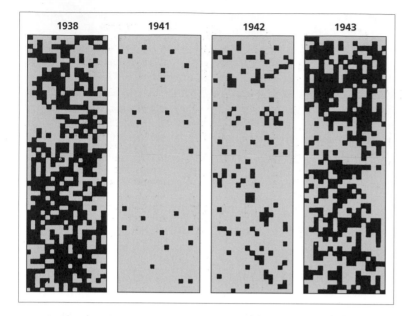

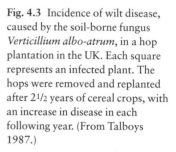

Fig. 4.3 Incidence of wilt disease, caused by the soil-borne fungus *Verticillium albo-atrum*, in a hop plantation in the UK. Each square represents an infected plant. The hops were removed and replanted after 2½ years of cereal crops, with an increase in disease in each following year. (From Talboys 1987.)

A wide variety of methods are used to gain information, including systematic surveys by government pathologists, crop consultants, or agrochemical company advisors. Outbreaks of disease in perennial crops, for example coffee, apple, or banana trees, may be monitored by regular inspection, perhaps by disease scouts whose sole job is reporting the occurrence of diseases and pests. Farmers themselves are encouraged to inspect their crops at regular intervals. Ideally the crop is examined at different growth stages (see Fig. 4.8) with samples taken over the whole area where the crop is grown. The data obtained are then collated and patterns of incidence of disease can be assessed on a local, national or international basis. For the most important diseases, data sets are kept on computers and regular reports are issued via the media to raise the level of awareness of potential problems. Comparisons can also be made between seasons to determine longer-term trends in disease or pest occurrence (Fig. 4.5).

Modern technology is providing new techniques for detecting and measuring diseases. There is increasing interest in methods for **remote sensing** of disease. Large areas of crops can be surveyed by means of aerial photography and, by using various types of film, diseased plants can be recognized as distinct patches in an otherwise uniform picture (Fig. 4.6). The technique has proved useful with diseases such as late blight of potato, where early **disease foci** can be located and their subsequent spread followed. It is also useful when sudden disaster strikes and a rapid appraisal of the situation is required. Such problems are often made worse by the large areas involved or by the inaccessibility of the terrain on which the crop is planted. In many parts of the world the wheat crop occupies large tracts of land and aerial photography can be used to determine the extent of damage by take-all or eyespot disease (Fig. 4.6 & Plate 6, facing p. 12). Similarly, the incidence of *Heterobasidion* (see Fig. 13.9) can be monitored in pine plantations covering undulating, rocky or boggy land. Developments in satellite technology, along with computerized techniques for image analysis mean that remote sensing of crops and their enemies, for instance locust swarms, may soon be possible over vast areas.

Monitoring virulence and pesticide resistance

While early detection of a pathogen in a crop is vital to ensure effective disease management, additional information on the properties of the pathogen may also be important. Two relevant features here are the virulence of the pathogen, and the relative sensitivity of the pathogen to chemicals which might be used for control.

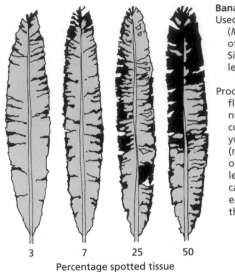

Banana leaf spot
Used for: Leaf spot
(*Mycosphaerella musicola*)
of banana (also known as
Sigatoka or black
leaf streak disease)

Procedure: Select 100 non-
flowering plants and
number the leaves on each
consecutively from the
youngest unfurled leaf
(no. 1). Record the presence
or absence of spots on each
leaf. A disease intensity score
can also be recorded for
each leaf using the key on
the left

3	7	25	50

Percentage spotted tissue

Disease intensity/loss relationship: the table below indicates the correlation
between disease level and fruit loss.

Average no. of youngest spotted leaf	Plants with spotted leaves younger than leaf no. 8 (%)	Spotted plants (%)	Fruit loss	
			Field	Transit
10–12	10	40	None	None
8–9	40–60	100	Up to 10% culled as ripe	Up to 20% ripens or colours in 7 days
5–7	60–80	100	Up to 75% culled for soft fingers or yellow pulp	Up to 50% of fruit ripens or turns colour in 7 days

Fig. 4.4 Disease assessment key for
banana leaf spot. (From Stover
1973.)

Where a pathogen occurs as races differing in virulence on particular host cultivars, identification of the
race involved in a disease outbreak is important to
predict the likelihood of spread onto other crops in
the area. For major pathogens, such as rust fungi and
powdery mildews on cereals, the race composition
of epidemics is monitored continuously to enable a
pattern to be built up over a season, as well as from
year to year. This exercise is analogous to the screening programmes conducted with human pathogens,
such as the influenza virus, to alert doctors to sudden
changes in the type of virus present. Agricultural
advisory agencies or extension services carry out
annual surveys to determine the distribution of virulence factors, with a view to pre-empting potentially
disastrous epidemics (see Chapter 12). For instance,
in the UK the yellow stripe rust (*Puccinia striiformis*)
population is analysed each year by growing sets of
differential cultivars at a number of widely spaced
sites in the wheat-growing parts of the country. Information from analyses carried out in previous seasons
permits predictions to be made as to the likely level of
disease on each cultivar. Where the amount of disease
is significantly greater than expected a sample of the

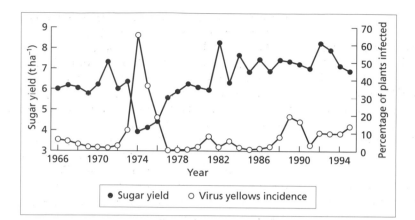

Fig. 4.5 Annual incidence of virus yellows diseases of sugarbeet in England 1966–1995, and effects on sugar yield. Note severe outbreaks in 1974–1976 correlated with reduced productivity. (Courtesy of British Sugar plc, provided by Dr A. Dewar.)

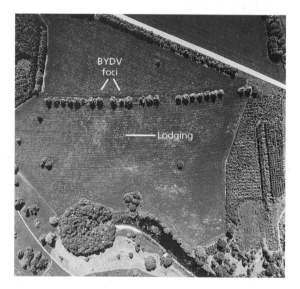

Fig. 4.6 Aerial photograph of a winter wheat crop in the UK in June, showing disease foci of barley yellow dwarf virus (BYDV), visible as lighter spots, and areas of crop collapse (lodging) caused by the eyespot fungus, *Pseudocercosporella (Tapesia)*. (From Greaves *et al.* 1983; courtesy of B.J. Walpole of the ADAS Aerial Photography Unit.)

pathogen is sent to a central laboratory for further testing. Experiments in controlled environments are then carried out to confirm the occurrence of new virulence factors. During 1988/89, yellow rust was prevalent in the UK, and especially severe on the cultivar Slejpner (see Fig. 12.5). The emergence of virulence to this cultivar led to a revised resistance

rating for Slejpner, which is now regarded as highly susceptible. Recognition of such trends alerts farmers to particular disease risks, encouraging close monitoring and early application of fungicide where necessary.

Further flexibility is possible within such a scheme by using mobile trap nurseries, comprising trays of seedlings raised under disease-free conditions. These trays are exposed within the crop, or even on the roof a travelling vehicle, and are subsequently incubated under conditions which favour disease development. This method has been used to track changes in cereal powdery mildew (*Erysiphe graminis*) populations across Europe.

The response of a pathogen to any chemicals used to control disease also needs to be evaluated, especially where problems of resistance to such chemicals (see p. 208) may have arisen. When, for instance, a fungus population occurs as a mixture of fungicide-sensitive and fungicide-resistant strains, the sensitivity of any strain responsible for a disease outbreak needs to be determined to enable choice of the most effective fungicide for control.

At present, virulence testing and evaluation of response to fungicides are both labour-intensive, involving inoculation of a range of test plants and/ or culture media containing fungicides. With the advances in diagnostic tests described above, it may eventually prove possible to directly detect virulence genes, or genes responsible for resistance to chemicals, utilizing DNA probes. Whether such powerful and sophisticated tools for defining pathogen properties will become a routine part of diagnostic pathology remains to be seen.

Monitoring vectors

Similar considerations apply to monitoring populations of vectors which may carry disease agents. In Europe, for example, aphid populations are monitored by a series of suction traps (see Fig. 4.16) which sample the atmosphere above crops and detect whether winged aphids are present (Fig. 4.7). Collating data from different regions then allows a map to be compiled showing the distribution and movement of known virus vector species.

The presence of a vector can indicate the potential for a disease outbreak, but additional information is usually required to accurately assess risk. Overall estimates of vector populations are useful, but a more valuable measure is the proportion actually carrying virus. This can be determined by feeding trapped insects on susceptible indicator plants, which develop symptoms of disease if virus is present. However, this is a relatively slow and laborious process. Once again molecular methods may be used to enhance the speed and accuracy of such tests. For example, the different viruses causing yellows diseases of sugarbeet can be detected in single aphids using an ELISA test based on monoclonal antibodies, while potato leaf roll virus (PLRV) has recently been detected by a 'reverse transcription' PCR method which first copies part of the RNA genome encoding the coat protein into a complementary DNA (cDNA) fragment, and then amplifies this specific fragment by PCR. This method is sensitive enough to detect a single virus-carrying aphid among a population of 30 aphids.

Like plant pathogens, vectors can occur as a series of strains, or biotypes, and these may differ in their host range or ability to transmit disease. The whitefly *Bemisia tabaci* is an important vector of geminiviruses, responsible for diseases such as leaf curl and cassava mosaic. The emergence of a new biotype of *B. tabaci*, which feeds on a wider range of host plants, has recently resulted in geminiviruses infecting previously unaffected crops in Europe. This new strain may also be an important factor in the currently severe epidemic of mosaic disease affecting cassava crops in West Africa. Hence information on the diversity of vector populations may be vital in assessing disease risk.

Disease severity × crop loss relationships

For disease assessments to be useful, the relationship between the amount of disease present and effects on crop yield and quality needs to be understood. This is one of the most problematical aspects of disease assessment, as in many cases the relationship is far from simple. It is important to understand the growth and physiology of the healthy plant, and how the pathogen affects these processes. The impact of disease will depend on several factors, including the growth stage of the crop, as well as overall disease severity. For instance, a wheat plant is likely

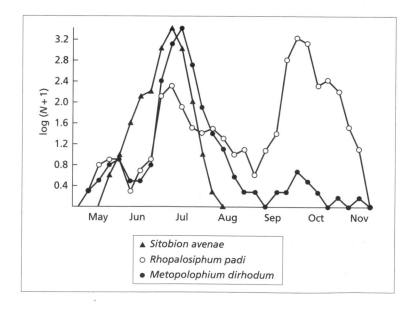

Fig. 4.7 Numbers of three aphid species trapped at Rothamsted, UK, during the 1992 season. Note autumn migration of *Rhopalosiphum padi*, a principal vector of barley yellow dwarf virus. (After Plumb 1995.)

▲ *Sitobion avenae*
○ *Rhopalosiphum padi*
● *Metopolophium dirhodum*

to be killed outright by the take-all pathogen *Gaeumannomyces graminis* if infection occurs prior to tillering (Fig. 4.8), whereas later infections result in reductions in a range of different yield components (Table 4.2).

One of the best-understood diseases in this respect is late blight (*Phytophthora infestans*) of potato (Plate 1, facing p. 12). Initially a key was created which allowed visual assessment of disease severity in the field. These assessments were then related to actual yield losses by comparing diseased plants with others kept healthy by fungicide treatment. The tuber weight losses resulting from severe blight epidemics clearly depend upon the time in the season when infection occurs (Fig. 4.9). This is predictable inasmuch as bulking of tubers is a long-term process and destruction of foliage late in the season will have little appreciable effect on yield. An interesting compari-

son can be made with cereal crops, where grain filling is a short-term and late event in the development of the crop. Direct effects on yield can in this case be related to disease affecting those photosynthetic tissues, such as the flag leaf, which are still active just prior to harvest. An example is shown in Fig. 4.10, where yield loss due to leaf rust in wheat can be related to the amount of disease present on the flag leaf, or on tillers. The amount of rust was in fact calculated by estimating the area under the disease progress curve (AUDPC); further explanation of this parameter is given in Chapter 5.

Disease losses may be purely quantitative, as when cereal mildew or rusts debilitate their hosts, or there may be additional, and overriding, qualitative effects. Apple scab (see Fig. 2.2) is now scarcely tolerated by the consumer in western countries, while potato common scab can, if severe, significantly reduce the market value of this crop. A further consideration is the effect of some pathogens on the storage properties of the produce. In the case of Sigatoka disease (see Fig. 4.4) the major effect is to cause premature ripening of the fruit. The consequences of this for an export crop which is transported to distant markets

Fig. 4.8 Growth stages in cereals. A decimal scale, from 0–100, is used to describe each stage in crop development. (After Tottman & Broad 1987.)

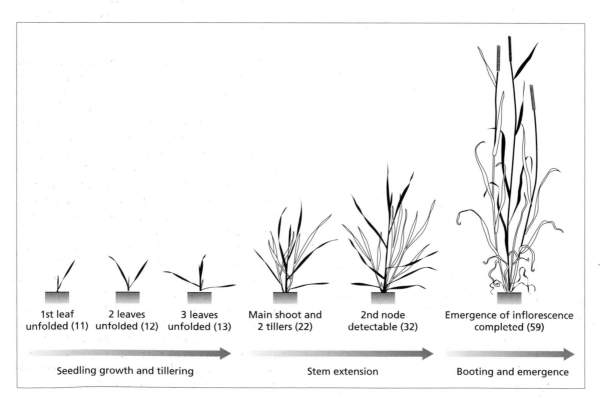

1st leaf unfolded (11) 2 leaves unfolded (12) 3 leaves unfolded (13) Main shoot and 2 tillers (22) 2nd node detectable (32) Emergence of inflorescence completed (59)

Seedling growth and tillering Stem extension Booting and emergence

Table 4.2 The effect of take-all infection on yield components of winter wheat. (Reprinted from Green & Ivins 1984; with kind permission of Elsevier Science-NL, Sara Burgerhartstraat 25, 1055 KV Amsterdam, The Netherlands.)

Yield component	Infected plants	Non-infected plants	Reduction due to take-all (%)
Ears per plant	1.60	1.80	11.1
Grains per ear	27.07	33.81	20.0
1000 grain weight (g)	29.51	46.30	36.3
Fertile spikelets per ear	14.10	15.90	11.3
Grains per fertile spikelet	1.92	2.13	9.9
Grain weight per plant (g)	1.28	2.84	54.9

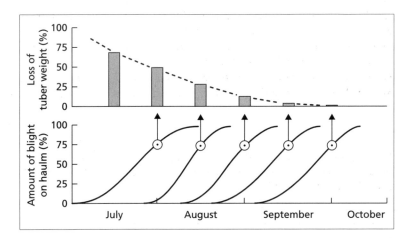

Fig. 4.9 Relation between late blight progress and yield loss of potato cultivar Majestic. (Data from Large 1952.)

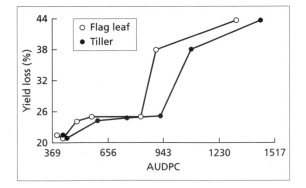

Fig. 4.10 Relationship between area under the disease progress curve (AUDPC) for leaf rust *Puccinia recondita* on the flag leaf or tillers of wheat plants and percentage yield loss. (Redrawn from Seck *et al.* 1988; copyright 1988, with kind permission from Elsevier Science Ltd, The Boulevard, Langford Lane, Kidlington OX5 1GB, UK.)

should be obvious. Even more subtle effects assume significance in products in which quality is of supreme importance. Wine made from grapes affected by grapevine leaf roll virus is of inferior quality and colour due to the low sugar and high tartaric acid levels present in diseased grapes at harvest. However, infection of grapes by the mould fungus *Botrytis cinerea* can actually improve their value; this so-called 'noble rot' imparts a distinctive characteristic to Sauternes wines. It should be noted, however, that in most circumstances *B. cinerea* is considered a serious pathogen of grapevines.

Some other fungal pathogens are important only because they produce secondary metabolites which pose a health risk to consumers. The most notorious example is ergot of cereals, *Claviceps purpurea*, which replaces infected grain with black sclerotia containing a cocktail of alkaloids. Contamination of flour by such sclerotia during milling can cause ergotism in people consuming bread made from the flour. Fortunately this problem is now very rare, although ergotism is sometimes still recorded in livestock grazing on infected grasses. Other examples of **mycotoxins** which can get into the human food chain concern species of *Fusarium* infecting cereals, especially maize. Under suitable conditions these produce potent mycotoxins such as the trichothecenes and fumonisins, which have been implicated as causing certain human cancers, for instance in parts of southern Africa. Chemical analyses of plant produce may therefore be necessary to assess potential risks to the consumer.

Disease forecasting

Because disease assessment includes a predictive element it is closely linked with disease forecasting (see Fig. 4.1). The latter, however, aims to predict whether or not disease will actually occur, as well as estimating the likely extent of its progress through a crop. As outbreaks of certain diseases show a marked correlation with particular climatic conditions, forecasting is often thought of as being synonymous with an analysis of the effects of weather on disease epidemics. Indeed, many forecasting systems are based on meteorological data, but it is important to recognize that other factors are involved. In its simplest form, forecasting merely relies on knowledge of whether a disease has occurred before in the area concerned. The prevalence of particular pathogens is so strongly influenced by regional differences in climate, that risk maps can often be drawn up to show the likely distribution of particular diseases (Fig. 4.11). Many farmers know through experience that crops grown in certain fields will be particularly prone to disease. In the light of this knowledge they avoid planting susceptible crops in those fields. Patterns of occurrence of disease in the recent past can provide an indication as to the amount of inoculum in the environment. The presence of airborne spores in the

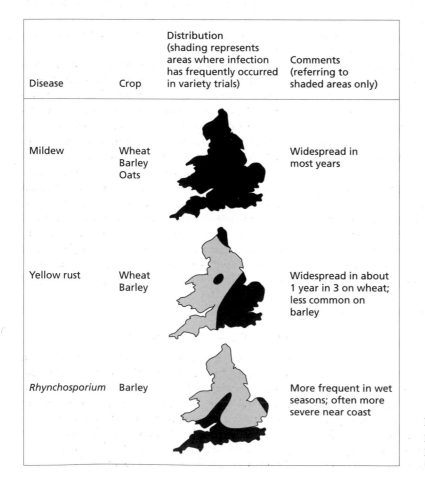

Disease	Crop	Distribution (shading represents areas where infection has frequently occurred in variety trials)	Comments (referring to shaded areas only)
Mildew	Wheat Barley Oats		Widespread in most years
Yellow rust	Wheat Barley		Widespread in about 1 year in 3 on wheat; less common on barley
Rhynchosporium	Barley		More frequent in wet seasons; often more severe near coast

Fig. 4.11 Risk maps showing prevalence of several cereal pathogens in England and Wales. (After Priestley & Bayles 1982.)

vicinity of crops can be monitored directly using spore traps which continuously sample the atmosphere (see Fig. 4.16). The presence of other host plants, such as the alternate hosts of rust fungi or alternative hosts for viruses, may also be a risk factor (see p. 50).

Diseases which are introduced in seed or on other propagules may be forecast by measuring the extent of any contamination in laboratory or greenhouse tests carried out prior to planting the crop (see p. 83). In many countries, seed testing is routinely conducted to provide certification guaranteeing the product fitness for use. A major problem with such schemes is that it is often difficult to relate levels of seed contamination to the subsequent development of epidemics, and hence to determine maximum limits above which seed must be either treated or rejected. Some such relationships have been established, for example with halo blight of *Phaseolus* beans, which are tested for the presence of *Pseudomonas* by utilizing its fluorescence characteristics and by serological tests. Field studies have shown that *Phaseolus* seed samples should have less than 0.25% infection if serious epidemics are to be avoided. This level of seed contamination will give rise to about 0.0025% primary seedling infections, and if the subsequent apparent infection rate (r, see p. 75) is 0.15 per infected plant per day, this will lead to about 4% infection in the mature crop, which is the maximum which can be tolerated before significant losses are experienced.

Other information which may have a bearing on disease forecasts includes soil type and local environmental factors, for example, microclimates created by undulating terrain or the proximity of woods and hedgerows. The intuitive approach of many farmers to disease avoidance will embrace considerations such as these. After all, they have been growing crops with reasonable success for centuries, for the most part without the advice of plant pathologists!

Climate and disease

Having made all these qualifications, it is still true that for many pathogens the dominant factors regulating disease development are those involving the weather. It is possible, in some instances, to establish direct links between climatic conditions and specific processes in the life of the pathogen, for example

spore germination or dispersal to new hosts. Furthermore, with some pathogens these effects of climate can be narrowed down to one or a few specific components, such as temperature, relative humidity or rainfall. Where the progress of an epidemic can be attributed directly to such conditions, monitoring the weather may provide accurate forecasts.

Pseudoperonospora humuli causes downy mildew disease of hops and, in common with other downy mildews, the pathogen reproduces asexually by forming sporangia on elongate sporangiophores which protrude through host stomata. The formation of sporangia is influenced by relative humidity, which must be > 90% for sporulation to occur. Sporangium release and dispersal are also affected by humidity; the former process requires fluctuating conditions and the latter is only successful if the spores are not subject to desiccation. The final step in the cycle, infection of new hosts, is similarly dependent on favourable environmental factors. Hop leaves must be wet before sporangia can produce zoospores; these swim to stomata, encyst, germinate and penetrate through the aperture.

Correlations between environmental factors and the incidence of hop downy mildew disease outbreaks are not, however, straightforward. Multiple regression analysis was employed to establish an equation which relates the incidence of disease to the amount of rainfall, the duration of leaf-surface wetness due to rain (but not dew), and airborne-sporangium concentrations (measured by trapping and counting sporangia in known volumes of air). Levels of infection predicted from the equation were compared with actual levels recorded on healthy trap plants exposed in a hop garden already affected by the disease. A very close agreement was obtained in this test (Fig. 4.12), indicating the potential of such an analytical approach as the basis for short-term forecasting of disease outbreaks.

Meteorological forecasting systems

While the relationship between environmental factors and disease may be difficult to unravel, this has not prevented the development of effective forecasting schemes. Certain diseases are more obvious candidates for forecasting systems than others, and characteristics which facilitate the development of such schemes are listed in Table 4.3. Not surprisingly, the main focus is on the most economically important

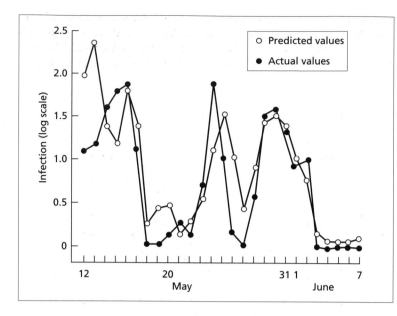

Fig. 4.12 Predicted and actual values for infection of hop plants by *Pseudoperonospora humuli*. Batches of healthy plants were exposed in a disease-affected hop garden at daily intervals and infection subsequently assessed. (Data from Royle 1973.)

Table 4.3 Factors which are important in the development of meteorological forecasting systems.

The incidence of disease varies with time due, at least in part, to the influence of climatic factors
Experimental studies have shown which climatic factors affect the pathogen and the disease
The disease is economically important
Cost-effective preventative or curative control measures are available
A system is available to issue warnings to growers

diseases of major crops. One of the most intensively studied diseases in this respect is potato late blight, caused by *Phytophthora infestans*.

In the UK, potato late blight warnings were initially based on the recognition of a 'Beaumont period', defined as a period of 48 h duration over which the temperature remains above 10°C and the relative humidity does not fall below 75%. Subsequently, a simplified system was employed in which blight warnings were broadcast if the minimum temperature remains at or above 10°C for 48 h, during which period the humidity is at 90% or more for at least 11 h. These 'Smith periods' allowed outbreaks of late blight to be forecast in maincrop potatoes in the UK

with an acceptable degree of accuracy using meteorological-station data. The combination of temperature and humidity specified allows the fungus to reproduce, spread and re-infect adjacent plants and hence continue the build-up towards an epidemic. It is obvious, however, that the significance of a Smith period will vary depending on the stage in the growth cycle of the potato plant at which it occurs and the extent to which the pathogen is already established. Such considerations have led to the recognition of zero dates, before which disease outbreaks are either rare or of little significance. Hence, in practical agricultural terms, spells of weather conducive to disease development occurring before a zero date are usually disregarded. However, the early progress of a pathogen may significantly affect later events, and it is advisable that disease development is charted throughout the entire season. Information from such a study can determine the extent of the action required when the weather favours the pathogen at a later, more critical, stage in crop development.

More sophisticated, computerized systems for providing short-term warnings of late blight attacks have since been developed in several regions, including the USA and Europe. The US system, known as BLITE-CAST, included the participation of individual farmers. Growers maintained thermohygrographs and rain-gauges in their potato fields and each week supplied weather data by telephone to a central computer programmed with data akin to the Beaumont and Smith criteria. This calculated a weekly severity value accumulated from daily risk scores based on combinations of rainfall, relative humidity and temperature (Fig. 4.13). An immediate response including an estimate of blight risk and a fungicide spray recommendation was then provided. In the German

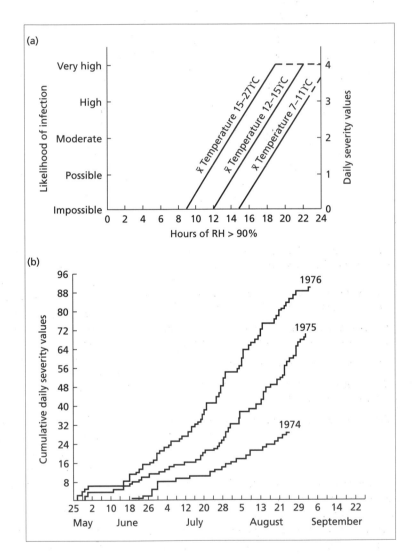

Fig. 4.13 BLITECAST parameters. (a) Relationship between relative humidity (RH), temperature and daily severity values. (b) Cumulative daily severity values for one weather station in Pennsylvania during three successive seasons. (c) Spray schedule recommendations for BLITECAST based on weekly accumulated severity values. (After MacKenzie 1981.)

(c) Spray schedule recommendations for BLITECAST based on weekly accumulated severity values.

Rainfall	Total weekly severity values				
	0–2	3	4	5–6	7 or more
Low	No spray	No spray	Alert	Moderate spray	Heavy spray
High	No spray	Alert	Moderate spray	Heavy spray	Heavy spray

scheme, Phytprog, emphasis was placed on the provision of negative forecasts, so that no action needed to be taken to control the disease for a specified period. Such negative forecasts can usefully limit the costs of control measures as well as keeping environmental damage to a minimum.

Such forecasting systems are constantly refined to improve their accuracy and usefulness for growers. Initially, forecasts were based on regional weather patterns. This gave a macroscale estimate of disease risk but did not take account of local variations due to altitude, topography, rivers and other factors. The advent of more powerful microcomputers coupled to relatively simple environmental sensors for continuous data logging (Fig. 4.14) now permits much higher resolution of risk, including on-farm forecasts of disease. As the crop develops it will influence the local microclimate; relative humidity, for example, is

usually higher within the canopy of a mature potato crop and more suitable therefore for the development of late blight than the regional weather records would suggest. Hence, it may be important to monitor conditions within the crop at different stages of development.

In recent years, leaf blotch diseases caused by the fungi *Septoria tritici* and *S. nodorum* have become more prevalent in winter wheat crops, and in many countries are now considered the most significant source of losses in this crop. The increase in incidence of *Septoria* may be due to several factors including earlier sowing of the crop, direct drilling into crop residues which may carry inoculum, and the use of dwarf varieties which are more prone to infection. During epidemic development, *Septoria* is spread by rain-splash, which carries spores from the base of the plant to the upper leaves. In this example the amount and duration of rainfall is actually less important than its quality. Short bursts of heavy rain containing large, high-energy droplets are more efficient at lifting spores onto the upper leaves than periods of lighter, continuous rainfall. In some cases disease outbreaks

Fig. 4.14 Automatic weather station providing hourly reports for several environmental parameters. Based on the Metpole used in Denmark. (After Hansen 1993.)

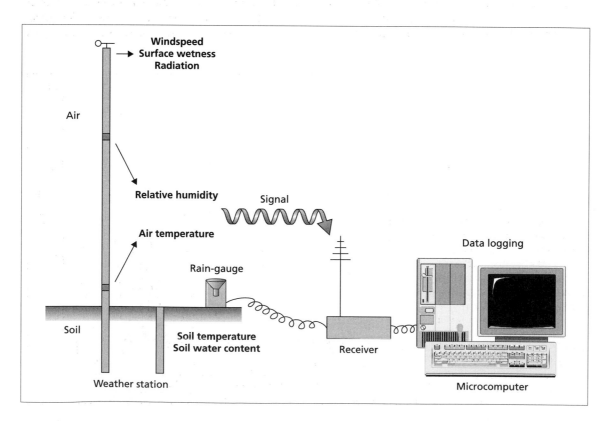

can be directly related to heavy rain events (Fig. 4.15). Conventional meteorological equipment does not monitor this factor, but the degree of rain-splash can be measured using a simple splashmeter, in which splashed droplets of coloured dye are collected on a vertical paper column (Fig. 4.16c). Outbreaks of disease have been shown to follow rainstorms in which splashes reach heights greater than 40 cm above the ground.

The way in which *Septoria* spreads upwards in a crop means that the growth pattern and structure of the host plant also influences disease risk. Recent work comparing different wheat cultivars has shown that those in which new leaves emerge below already established leaves are more prone to horizontal transfer of inoculum than cultivars in which the leaves are well spaced, one above the other. Hence crop architecture can have a significant effect on the rate of disease increase with time.

Long-range forecasts

All these forecasting systems involve relatively simple models based on actual weather observations. The progress of apple scab, downy mildew of grapevine and a number of other diseases has also been related to similar, relatively simple observations. However, in all the systems so far devised, the weather favouring a disease outbreak has already happened when forecasts are made, and control measures must be applied quickly following a disease warning. Delay will allow the pathogen to penetrate new hosts and so be less amenable to control by many pesticides. The discovery of fungicides which have a measure of curative activity has, however, to some extent reduced the need for prior warning of pathogen spread. One of the first such chemicals was dodine, which can eradicate incipient apple scab infections if applied within 48 h of spore germination. Another example is the systemic compound metalaxyl, which can destroy potato late blight infections if applied shortly after the pathogen has entered its host tissues (but see Chapter 11).

More sophisticated and commercially useful forecasts are likely to be obtained in the future by using synoptic weather charts, which indicate the likely sequence of meteorological events in succeeding days, weeks or even months. Such information is already used in some forecasting systems, such as that employed for potato late blight in Ireland. Long-term forecasts are much more valuable than short-term systems, as control measures may then be integrated

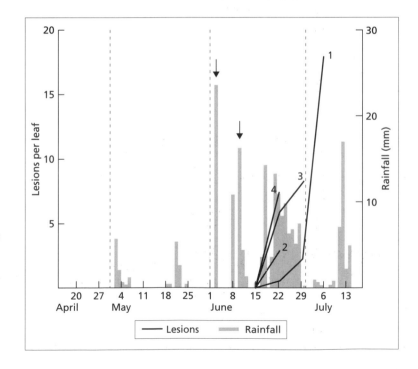

Fig. 4.15 Development of leaf blotch disease, caused by *Septoria nodorum*, on leaves of winter wheat at Long Ashton Research Station during the 1982 season. Disease progress, estimated as lesions per leaf, on the flag leaf (leaf 1) and three older, lower leaves (2–4), is shown along with rainfall records over the same period. The outbreak of disease first noted on 22 June can be correlated with heavy rainfall episodes (arrows) occurring earlier in the month. (After Royle *et al.* 1986.)

(a) (b) (c)

Fig. 4.16 Some examples of equipment used to monitor the risk of disease. (a) Aphid suction trap sampling airborne vectors 12 m above ground level. (b) Burkhardt spore trap sampling air for pathogen spores above a linseed crop. (c) A splashmeter designed to monitor rainfall episodes. Coloured dye in the ring of dishes is splashed by rain drops onto the paper column, indicating the height reached by any splash droplets formed. (c, Courtesy of David Royle.)

with other agricultural practices. They are thus more likely to be economically acceptable. Even longer-term weather predictions may indicate possible future trends in disease over several seasons, and these might influence the choice of cultivar or even of the crop itself.

The ultimate goal of all disease forecasting schemes is to provide growers with accurate predictions of disease risk on which to base control decisions. As such, disease forecasting is part of **decision support**. This will be discussed further in Chapter 14, in the context of integrated disease management.

Further reading

Books

Chiarappa, L. (ed.) (1981) *Crop Loss Assessment Methods. FAO Manual on the Evaluation and Prevention of Losses by Pests, Disease and Weeds.* CAB, Farnham.

De Boer, S.H., Andrews, J.H. & Tommerup, I.C. (eds) (1996) *Advances in Botanical Research incorporating Advances in Plant Pathology.* Vol. 23 *Pathogen Indexing Technologies.* Academic Press, London.

Duncan, J.M. & Torrance, L. (eds) (1992) *Techniques for the Rapid Detection of Plant Pathogens.* Blackwell Scientific Publications, Oxford.

Marshall, G. (Chair) (1996) *Diagnostics in Crop Production. BCPC Symposium Proceedings* **65**. British Crop Protection Council, Farnham.

Schots, A., Dewey, F.M. & Oliver, R.P. (1994) *Modern Assays for Plant Pathogenic Fungi. Identification, Detection and Quantification.* CAB International, Wallingford.

Reviews and papers

Gibson, R.W., Legg, J.P. & Otim-Nape, G.W. (1996) Unusually severe symptoms are a characteristic of the current epidemic of mosaic virus disease of cassava in Uganda. *Annals of Applied Biology* **128**, 479–490.

Hansen, J.G. (1993) The use of meteorological data for potato late blight forecasting in Denmark. In: *Proceedings of the Workshop on Computer-Based DSS on Crop Protection* (eds B.J.M. Secher, V. Rossi, & P. Battilani), pp. 183–192. SP-Report Vol. 7, Danish Institute of Plant and Soil Science.

Hatfield, J.L. (1990) Remote detection of crop stress: Application to plant pathology. *Phytopathology* 80, 37–39.

Henson, J.M. & French, R. (1993) The polymerase chain reaction and plant disease diagnosis. *Annual Review of Phytopathology* 31, 81–109.

Kendall, D.A., Brain, P. & Chinn, N.E. (1992) A simulation model of the epidemiology of barley yellow dwarf virus in winter sown cereals and its application to forecasting. *Journal of Applied Ecology* 29, 414–426.

Lovell, D.J., Parker, S.R., Hunter, R., Royle, D.J. & Coker, R.R. (1997) Influence of crop growth and structure on the risk of epidemics by *Mycosphaerella graminicola* (*Septoria tritici*) in winter wheat. *Plant Pathology* 46, 126–138.

Mackenzie, D.R. (1981) Scheduling fungicide applications for potato late blight with BLITECAST. *Plant Disease* 65, 394–399.

Markham, P.G., Bedford, I.D., Liu, S.J. & Pinner, M.S. (1994) The transmission of geminiviruses by *Bemisia tabaci*. *Pesticide Science* 42, 123–128.

Plumb, R.T., Lennon, E.A. & Gutteridge, R.A. (1986) Fore-casting barley yellow dwarf virus by monitoring vector populations and infectivity. In: *Plant Virus Epidemics: Monitoring, Modelling and Predicting Outbreaks* (eds G.D. McLean, R.G. Garrett & W.G. Ruesink), pp. 387–398. Academic Press, Sydney.

Raposo, R., Wilks, D.S. & Fry, W.E. (1993) Evaluation of potato late blight forecasts modified to include weather forecasts: a simulation analysis. *Phytopathology* 83, 103–108.

Royle, D.J., Shaw, M.W. & Cook, R.J. (1986) Patterns of development of *Septoria nodorum* and *S. tritici* in some winter wheat crops in Western Europe, 1981–83. *Plant Pathology* 35, 466–476.

Seck, M., Roelfs, A.P. & Teng, P.S. (1988) Effect of leaf rust (*Puccinia recondita tritici*) on yield of four isogenic wheat lines. *Crop Protection* 7, 39–42.

Singh, R.P., Kurz, J., Boiteau, B. & Bernard, G. (1995) Detection of potato leafroll virus in single aphids by the reverse transcription-polymerase chain-reaction and its potential epidemiologic application. *Journal of Virological Methods* 55, 133–143.

5   Plant Disease Epidemics

'Ours is a military campaign against agents that destroy our plants. We cannot wage this campaign successfully without knowing the measure of the enemy's ability to destroy.' [K. Starr Chester, 1959]

Disease outbreaks due to infectious agents are dynamic events often characterized by rapid changes in the occurrence and distribution of pathogens within a host population. The science of **epidemiology** seeks to understand the processes underlying such temporal and spatial change in disease incidence, both to identify causes and to provide a rational basis for disease control.

Early studies on the population dynamics of plant pathogens were mainly descriptive and concerned with identifying the environmental factors affecting disease. This era of epidemiology has now been superseded to a large extent by the quantitative analysis of epidemics using a mathematical approach. Simulation modelling of disease is a powerful tool, not only to predict the future progress of an outbreak, but also to evaluate the likely benefits of any control strategy. Hence, the apparently abstract world of mathematical modelling is, in this instance, closely linked with the practical business of disease management.

Epidemic development

An **epidemic** can be simply defined as an increase in disease with time. It should be added that this definition assumes a progressive increase in disease incidence within a particular population of host plants over a time scale which is relevant to the maturation of the crop. An epidemic occurring on a continental or global scale is described as a **pandemic**. There are two components in disease increase: (i) multiplication of the pathogen (i.e. increase in number of individuals); and (ii) spread of the pathogen to new hosts. Epidemiology is therefore concerned with the population dynamics of the pathogen and the factors affecting the spatial spread of disease.

A disease epidemic can only occur if three basic requirements are satisfied.
1 A large number of host plants are available at a suitable stage of development.
2 There is a source of virulent inoculum.
3 Environmental conditions are favourable for the growth and spread of the pathogen.

In natural plant communities destructive disease epidemics are rare (see p. 13). Where host and pathogen have coexisted for thousands of years it is reasonable to suggest that some form of equilibrium will have evolved between the two. The domestication of plants for human use, along with the gradual intensification of agriculture, has altered this balance. Cultivation of selected varieties of a single crop species has loaded the odds in favour of the pathogen by providing large stands of genetically uniform hosts. These monocultures provide an ideal situation for the rapid increase of any pathogen which is virulent on the crop genotype concerned. A spectacular example was the major epidemic of southern corn leaf blight, *Bipolaris maydis*, which affected around 85% of the total USA corn crop during the 1970 season. The progressive development of this disease is charted in Fig. 5.1. Fortunately the situation did not repeat itself in 1971 and an effective level of resistance has subsequently been reintroduced into the crop.

Inoculum, the portion of a pathogen that is transmitted to or grows into contact with a new host, may be present either as airborne or soil-borne propagules, or may already occur in contaminated seed or on other plant propagation material. The infection process, and the subsequent development of the pathogen to the stage at which new inoculum

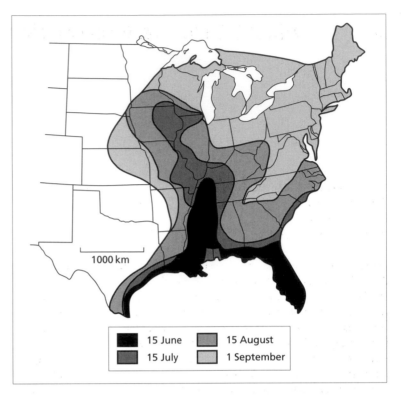

Fig. 5.1 Records of the occurrence of southern corn leaf blight (*Bipolaris maydis*) in North America during 1970. (From *Epidemiology and Plant Disease Management* by J.C. Zadoks & R.D. Schein. Copyright © 1979 by Oxford University Press, Inc. Used by permission of Oxford University Press, Inc.)

Legend:
- 15 June
- 15 July
- 15 August
- 1 September

1000 km

is produced, is dependent upon the prevailing environmental conditions.

The two latter requirements, virulent inoculum and favourable environment, are sometimes included together in the concept of **inoculum potential**. This is essentially a measure of the amount of biological energy available for the colonization of the host. A distinction can be drawn between the intensity factor, i.e. the number of infective propagules present, which is sometimes termed **inoculum density**, and the capacity factor, or **infection potential**. The latter is a measure of the ability of individual propagules to cause disease under the prevailing conditions. Environmental factors influencing the progress of an epidemic include temperature, relative humidity, dew formation, rainfall, photoperiod, windspeed and wind direction, sunshine duration and soil pH. In other words, anything which affects either the development of the pathogen or the performance of the host can influence a disease epidemic.

The disease growth curve

By plotting the amount of disease in a crop against time, one can obtain graphs which provide useful information about the dynamics of an epidemic. Two contrasting types of growth curve can be distinguished (Fig. 5.2). The first is characteristic of epidemics where all the infections occurring in a season are derived from inoculum present at the start of the season. This is typically the case with many soil-borne pathogens where an increase in diseased plants as the season advances is due to progressive contact between host roots and pathogen propagules present in the soil, rather than to reproduction of the pathogen. In the specific example shown, the fungus *Sclerotium rolfsii* attacking carrots, initial infection is from sclerotia surviving in the soil, but spread from plant to plant can occur by mycelial growth. The density of host plants therefore affects the shape of the curve. Such epidemics are termed **monocyclic**. In contrast, where the pathogen passes through several generations in one growing season, and each generation produces further inoculum, a **polycyclic** epidemic occurs. Examples of this type of epidemic are airborne foliar pathogens, such as the rusts, mildews and potato late blight, where each spore is capable of initiating a lesion in which further sporulation

73

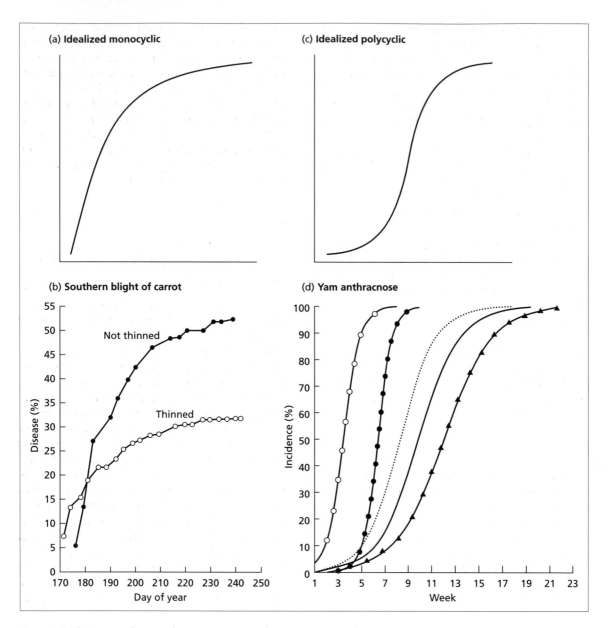

Fig. 5.2 Epidemic growth curves for monocyclic (a & b) and polycyclic (c & d) pathogens. Idealized curves (top) are compared with examples (below) from real epidemics. (b) Disease progress of southern blight of carrots, caused by the soil-borne fungus *Sclerotium rolfsii* in plots where the crop was either thinned or left at high density. (After Smith *et al.* 1988.) (d) Disease progress curves for foliar anthracnose of yams, caused by *Colletotrichum gloeosporioides*. Differences in the rate of disease development between five different sites in Barbados are mainly due to climatic factors, especially amounts of rainfall. (After Sweetmore *et al.* 1994.)

occurs. On susceptible hosts under favourable environmental conditions the multiplication of the pathogen is very rapid and consequently the rate of disease increase is exponential. In the example

shown, anthracnose disease of yams, caused by the foliar pathogen *Colletotrichum gloeosporioides*, differences in the slope of the curve are due to environmental differences between sites, and especially the amount of rainfall which influences the rate of spread and the infection process of this splash-dispersed fungus. This pattern of accelerating epidemic development is not commonly seen with monocyclic pathogens, although in perennial crops or where continuous cultivation of one crop is practised, disease may increase dramatically over a number of years (see Fig. 4.3).

The epidemiologist Vanderplank has compared these two patterns of disease increase to the different types of interest earned on money invested in savings accounts. Monocyclic pathogens increase in a manner analogous to simple interest, while polycyclic pathogens can be described more accurately by equations for compound interest. The epidemics can also be contrasted in terms of the relative birth rates and death rates of the pathogens involved. The main features of each type of epidemic are summarized in Table 5.1.

These concepts are useful when comparing the dynamics of different epidemics, but as usual there are exceptions to such generalizations. For instance, soil-borne pathogens like *Pythium* and many species of *Phytophthora* may reproduce rapidly on plant roots to release new inoculum in the form of zoospores capable of infecting further hosts.

Mathematical description of epidemics

With monocyclic pathogens, the disease growth curve (Fig. 5.2) shows a steep early increase in incidence, followed by a gradual decline. With polycyclic pathogens, the growth curve is sigmoid, similar to the classic growth curve describing the increase in biomass of an individual organism, or the multiplication of a bacterial culture. One can distinguish three phases: an initial lag, a phase of exponential increase, and a final decline.

What factors determine these shapes? With monocyclic diseases, the principal determinants of disease are the amount of inoculum present at the start of an epidemic, and the number of host plants available for infection. This can be described by an equation, derived by assuming that if the number of infected individuals is q, then the rate of increase of disease dq/dt, at time t, is proportional to the number of uninfected individuals in the population. Thus if the total population size is q_A, then the rate of increase is proportional to $(q_A - q)$, so that:

$$\frac{dq}{dt} = r(q_A - q) \tag{5.1}$$

where r is the **infection rate**.

This can be solved to give the following equation:

$$q = q_A - (q_A - q_0)e^{-rt} \tag{5.2}$$

where q_0 is the initial number of infected individuals, and e is the exponential constant (the base for natural logarithms).

With polycyclic pathogens, the early lag phase is primarily due to the small amount of inoculum present at the start of an epidemic. In the case of potato late blight, the fungus overwinters in seed tubers or discarded tubers left in piles known as cull heaps. Epidemics are initiated by sporangia formed on shoots developing from such tubers. At first there may be very few infected host plants in the population; a high proportion of these disease foci will be passing through the **latent period** between infection of the host and production of new infective propagules by the pathogen. Once the amount of inoculum has increased, however, the epidemic moves into the exponential phase (Fig. 5.2). The availability of inoculum is no longer limiting and there are still numerous disease-free hosts within the crop. Eventually, the rate of disease increase begins to fall and the

Table 5.1 Some features of monocyclic and polycyclic epidemics.

	Monocyclic	Polycyclic
Rate of increase	Simple interest model	Compound interest model
Reproduction	Long reproductive cycle (Low birth rate)	Short reproductive cycle (High birth rate)
Survival	Inoculum long-lived (Low death rate)	Inoculum short-lived (High death rate)
Examples	Soil-borne diseases	Airborne diseases

epidemic declines. There may be several reasons for this, including the finite number of hosts available for infection and the onset of unfavourable environmental conditions. In the case of potato late blight, the pathogen may literally exhaust the supply of new host plants.

In Fig. 5.2 sigmoid epidemic curves were obtained by plotting numbers of diseased plants against time. For this purpose the quantity of disease may be measured in any convenient way, including number of plants affected, area of leaf or other tissue colonized, or weight loss as compared with control plants. The model used to describe the disease curve for polycyclic pathogens assumes that the rate of increase of disease is proportional to the number of infected individuals (q), and also to the proportion of uninfected individuals ($1 - q/q_A$) so that:

$$\frac{dq}{dt} = rq\left(1 - \frac{q}{q_A}\right) \tag{5.3}$$

During the initial phase of infection the proportion of the population infected is small, so that the rate of increase of disease is proportional to q, and hence:

$$\frac{dq}{dt} \simeq rq \tag{5.4}$$

and growth is exponential, with the number of diseased individuals being given by:

$$q = q_0 e^{rt} \tag{5.5}$$

During this phase of growth:

$$\ln q = \ln q_0 + rt \tag{5.6}$$

where r is the infection rate, q_0 is the quantity of disease present in the crop at the start of the epidemic, and the relationship between $\ln q$ and time is linear (Fig. 5.3).

The vital factor in these formulae, in terms of its significance in different host–pathogen–environment combinations, is the infection rate r. One can obtain a value for r in any field situation by measuring the amounts of disease present (q_1, q_2) at times t_1 and t_2 during the logarithmic phase of disease increase (Fig. 5.3). Equation 5.6 may then be rewritten:

$$\ln q_1 = \ln q_0 + rt_1 \tag{5.7}$$

and

$$\ln q_2 = \ln q_0 + rt_2 \tag{5.8}$$

If eqn 5.8 is subtracted from eqn 5.7 then:

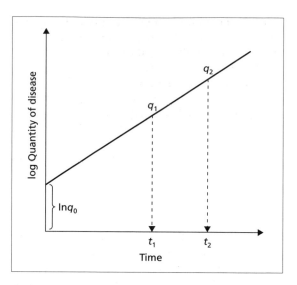

Fig. 5.3 Disease increase with time plotted on a logarithmic scale.

$$\ln q_2 - \ln q_1 = r(t_2 - t_1) \tag{5.9}$$

or

$$r = \frac{(\ln q_2 - \ln q_1)}{t_2 - t_1} \tag{5.10}$$

An important qualification must be made in applying these equations, as they only hold if there is no restriction of the development of the pathogen. In practice, of course, the population of healthy plants is declining as q increases. In this case the full solution of the growth equation can be used, which is the classic logistic curve:

$$q = \frac{q_A}{1 + \left(\dfrac{q_A - q_0}{q_0}\right)e^{-rt}} \tag{5.11}$$

This can be written in a form similar to eqn 5.6 in which $\ln q$ is replaced in the equation by $\ln(q/q_A - q)$ to give:

$$\ln\left(\frac{q}{q_A - q}\right) = \ln\left(\frac{q_0}{q_A - q_0}\right) + rt \tag{5.12}$$

Thus instead of using the amount of disease, q, to estimate the infection rate, we use $q/(q_A - q)$, the ratio of the quantity of diseased material to the quantity of healthy material (the previous equation assumes that the quantity of healthy material is almost the same as the total amount of material, i.e. the amount of diseased host is very small). This ratio becomes $q/(100 - q)$ if disease is expressed in percentage terms.

We can then estimate r using:

$$r = \left[\ln\left(\frac{q_2}{q_A - q_2} \right) - \ln\left(\frac{q_1}{q_A - q_1} \right) \right]\left(\frac{1}{t_2 - t_1} \right) \qquad (5.13)$$

or

$$r = \ln\left(\frac{q_2(1 - q_1)}{q_1(1 - q_2)} \right)\left(\frac{1}{t_2 - t_1} \right) \qquad (5.14)$$

Although the disease growth curve can provide useful generalizations about the progress of an epidemic, it is inevitably an over-simplification. We have assumed a uniform environment for multiplication and spread of the pathogen. In reality, environmental conditions fluctuate and the progress of an epidemic will be intermittent rather than continuous. Similarly, the spatial distribution of disease will be discontinuous and the rate of increase will differ depending upon the density of uninfected hosts and other factors. This temporal and spatial heterogeneity is reflected in more recent, non-linear models of epidemic development.

The influence of the environment on an epidemic is complex, with both direct effects on the latent period and indirect effects on the relative resistance of host plants. Temperature has a profound effect on the incubation period between infection and production of further inoculum. For instance, with black stem rust, *Puccinia graminis*, the disease cycle (from spore to spore) takes 15 days at 10°C, but only 5–6 days at 23°C. Obviously in this case the latent period is reduced at higher temperatures and the epidemic will progress more rapidly. Most fungal and bacterial pathogens require free water for infection, and the correlation between rainfall, surface wetness or relative humidity, and disease increase can be striking, and is used in many of the forecasting schemes already discussed in Chapter 4.

The progress of an epidemic may also depend on the date at which it begins. Practical experience shows that if a disease occurs before a certain crop growth stage it will have a far greater impact than later outbreaks. This may simply be because of the time available for the pathogen to multiply and spread during the remainder of the growing season.

Spatial spread of disease

As the pathogen population increases in size during an epidemic it also expands in space. Mathematical models have been developed to describe this process. The simplest models of disease spread assume that the amount of disease at a given distance x from the inoculum source can be described by an empirical function, for example:

$$\ln y = \ln a - b \ln x \qquad (5.15)$$

which is known as the inverse power law model, or the alternative negative exponential model:

$$\ln y = \ln c - kx \qquad (5.16)$$

Figure 5.4 shows a typical disease gradient obtained for a rust fungus dispersing from a single infected source plant in a field.

These empirical models do not take account of the biological mechanisms involved in disease spread. More complex mechanistic models have been proposed, based on diffusion-reaction equations, where, for large populations, inoculum can be considered to behave as a gas.

$$\ln y = \ln a - rx^2 \qquad (5.17)$$

Alternatively, stochastic models which incorporate probabilities of infection, dependent on the distance from an inoculum source, can be applied. The equa-

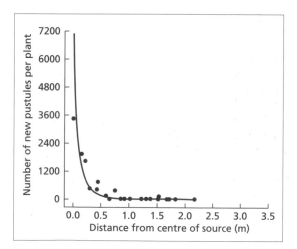

Fig. 5.4 Disease gradient for bean rust, *Uromyces phaseoli*, dispersing from a single infected source plant. (After Mundt & Leonard 1985.)

tions used in mechanistic and stochastic models are complex and beyond the scope of this book.

Epidemic modelling

It is essential that mathematical models of epidemics are valid and reliable, and that they are orientated towards a clearly defined goal. In addition, they should be simple in construction, logical and mathematically correct. Several types of model are used in epidemiology (Fig. 5.5), aimed at describing epidemic processes, predicting outcomes, or explaining the mechanisms underlying epidemic development. Descriptive models allow for the presentation of experimental data, which may be orientated towards the proof of a hypothesis. Predictive models are a subdivision of the empirical approach, in which the emphasis is placed on those variables which have the greatest value in foretelling the likely course of events. Mechanistic models involve an element of simplification in that all the mechanisms of the epidemic are represented by a single model or a series of linked sub-models.

Examples of descriptive models abound, although

they are not necessarily thought of in this context. A simple hypothesis described in mathematical terms is the epidemic threshold theorem. If i = the infectious period (i.e. the time during which a lesion is producing spores) and R_c = the basic infection rate (i.e. the number of new lesions produced per day), then an epidemic will commence when $iR_c > 1$.

Predictive models often involve the use of the apparent infection rate, r. Measurements of r can be used to determine the likely advantage of sanitation measures, which reduce the initial inoculum, and fungicide treatments, which reduce the infection rate (see Fig. 5.7). Furthermore, the impact of different resistance genes, either exposed alone or in combination, on the progress of an epidemic can be assessed by making measurements of r in various cropping schemes.

Mechanistic models attempt to simulate the whole process of epidemic development. For example, EPIMAY provides an account of epidemics of southern corn leaf blight, *Bipolaris maydis*, and EPIDEM is a program modelling epidemics of early blight of potato, *Alternaria solani*. In both these systems the infection cycle of the pathogen is described in qualita-

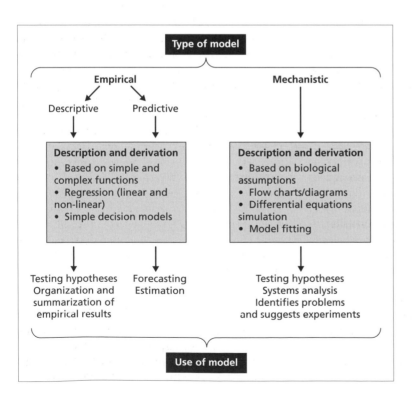

Fig. 5.5 Mathematical models used in epidemiology. (After Kranz & Royle 1978.)

tive and quantitative terms and the biological and environmental factors affecting each stage in the process are added into the computer program. If data on weather and crop development are then presented to the computer, it will provide a forecast. Despite the obvious attractions of these systems it must be noted that such simulators are only as good as the quality of the basic data on which they are founded and there is still a distinct shortage of such information for many important pathogens. These simulations have, however, the virtue of exposing gaps in our knowledge of the interactions between host and pathogen under field conditions, and help in the design of experiments to acquire the necessary data.

Population structure of epidemics

The mathematical methods and models discussed so far in this chapter are concerned primarily with the population dynamics of plant pathogens, and developing quantitative estimates of disease progress. There are also, of course, important qualitative questions about epidemics, concerning for instance the original source(s) of infection, and the specific properties of the pathogen involved in the disease outbreak. Does the multiplication and spread of disease within a field or across a region represent an increase in a single pathotype, or are several different pathotypes involved? In other words, what is the composition of the pathogen population during epidemic development? These questions focus on the genetic structure of the pathogen population, the uniformity or diversity of pathogen genotypes present, and the extent of gene flow between them.

Until recently, information on pathogen population structure was based on certain characteristics, or phenotypes, which could be easily determined. Two examples, already mentioned in Chapter 4, are the virulence of different pathogen isolates on a range of host cultivars differing in resistance, and sensitivity to chemicals such as fungicides. Sampling the pathogen population during a disease outbreak, at different sites and different stages of epidemic development, provided insights into the extent of variation present. These surveys helped to determine the influence of particular selection pressures, such as host resistance genes, or fungicide treatments, on the pathogen population (see Chapters 11 and 12).

The advent of techniques for detecting variation at the genomic level, based on differences in DNA sequences, means that it is now possible to analyse pathogen populations with a much higher degree of resolution. DNA fingerprinting (Fig. 5.6) can, for instance, detect different variants within a population, and sometimes even distinguish between different individuals. A range of different molecular markers are now available for analysing genetic diversity in populations, including isoenzymes, DNA probes, and amplification of random or defined DNA sequences via the polymerase chain reaction (PCR). Detailed discussion of these various marker techniques is beyond the scope of this book; instead, some examples of the application of biochemical and molecular markers to the analysis of particular epidemics should illustrate the potential value of this approach.

Global epidemiology of potato late blight

The genetic centre of origin of the late blight pathogen, *Phytophthora infestans*, is believed to be the central highlands of Mexico. In this region there is a high degree of genetic diversity in the pathogen population, due to the occurrence of both mating types of the fungus (A1 and A2), which allows sexual reproduction to occur. The first global migration of the pathogen during the 1840s involved only the A1 mating type. Additional evidence based on isozyme and DNA markers suggests that this pandemic, and until recently most subsequent late blight epidemics, have been caused by one dominant genotype of the pathogen, spreading outside Mexico as a single clonal lineage. Since the 1970s, however, the A2 mating type has been detected in many different countries, and the composition of the late blight population has undergone a dramatic change. The original dominant genotype has been displaced by new genotypes with different isozyme patterns and DNA fingerprints, and the genetic diversity of the late blight population has increased. This indicates that a second, global migration of the pathogen is taking place. The full implications of this change are not yet clear, but the increase in genetic variation may pose an additional threat to the potato crop.

The recent history of Dutch elm disease

Two pandemics of Dutch elm disease have occurred in the northern hemisphere this century. The first spread widely in Europe and North America but caused only limited mortality in the elm population.

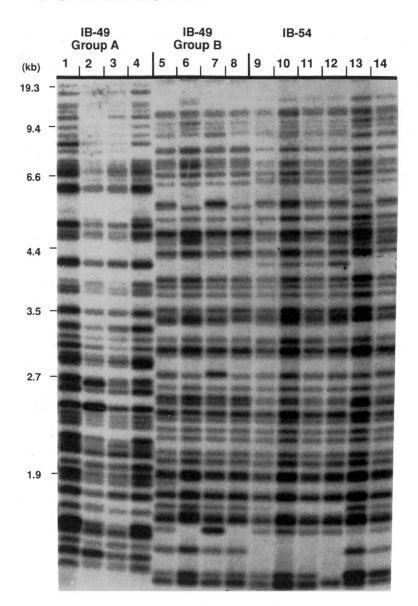

Fig. 5.6 DNA fingerprints of 14 isolates of the rice blast pathogen, *Magnaporthe grisea*, obtained using a probe which hybridizes with a repetitive DNA sequence. IB-49 and IB-54 are two different pathotypes of the fungus. The patterns obtained show that pathotype IB-49 includes two distinct groups of isolates, one of which is very similar to IB-54. Such fingerprinting can trace the history of particular variants and distinguish between different pathogen clones occurring in epidemics. (From Levy *et al.* 1991. © American Society of Plant Physiologists. Reprinted by permission of the publisher.)

A second, more serious outbreak started some time during the 1940s and has been spreading ever since. Detailed analysis of populations of the pathogen from different countries, using biological characters as well as DNA markers, has clarified the nature of these different outbreaks, and provided fascinating insights into the evolution of a plant disease epidemic. First it has been demonstrated that the two outbreaks have been caused by two related but distinct fungal species. The initial epidemic involved *Ophiostoma ulmi*, and the second a more aggressive species, *O.* *novo-ulmi*, which has been responsible for the high mortality rate among elms. Furthermore, the current pandemic is due to two distinct strains or races of *O. novo-ulmi*. One, the Eurasian (EAN) race, appears to have arisen in eastern Europe, and subsequently spread eastwards and westwards. The other, known as the North American (NAN) race, is believed to have evolved in the USA on susceptible elms and was introduced into western Europe in the 1960s. Since then it has spread eastwards so that in places the two races occur together. These highly aggressive strains have rapidly replaced the original, non-aggressive species.

The use of molecular markers has, in this case, helped to clarify the possible origins and history of a disease epidemic. It has also provided information on the genetic structure of the pathogen population at different stages of epidemic development. Where the disease is spreading into a new area, the so-called epidemic front, there is low genetic diversity, suggesting that the population is clonal. In areas where the disease has been established for longer, the population is more heterogeneous, linked to the development of variant types of the fungus which differ in their ability to undergo genetic exchange. There is therefore a mixture of genotypes present, between which gene flow is restricted.

Gene flow in epidemics

DNA markers have proved a valuable tool for estimating the degree of gene flow between different populations, and for determining the extent to which sexual recombination is occurring. With *Mycosphaerella graminicola (Septoria tritici)* on wheat, for example, comparison of pathogen populations between seasons has revealed a high level of genetic diversity, confirming that epidemics originate from sexual ascospore inoculum in which recombination has taken place. This contrasts with epidemics of several other foliar pathogens of cereals, such as many rusts and powdery mildew, where asexual spores are often the primary inoculum, and disease is due to multiplication of a single, clonal population of the pathogen.

Relevance of epidemiology to control

Rational disease control measures are aimed at altering one or more of the three prerequisites for an epidemic to occur, namely susceptible hosts, virulent inoculum and a favourable environment. For instance, one can plant resistant crop varieties, inoculum can be reduced by crop sanitation or seed sterilization, and the environment can be altered in such a way that infection no longer occurs. The use of fungicides or other chemicals may be regarded as exemplifying ways in which such unfavourable environments can be established. These applied aspects of epidemiology will be discussed further in Chapters 11–14.

The distinction between monocyclic and polycyclic diseases also has relevance to methods of control. With monocyclic disease the most effective approach is often to reduce the initial inoculum as there is a direct relationship between inoculum density and the amount of disease produced (see Fig. 13.3). Fungicidal seed treatments are effective in controlling several cereal smut diseases because they kill most of the seed-borne inoculum. However, when dealing with polycyclic diseases, merely reducing the initial inoculum is usually ineffective. This is because in these diseases even a tiny amount of inoculum can quickly multiply to damaging proportions. Control measures for polycyclic diseases are usually designed to prevent the pathogen from generating fresh inoculum. In other words, the latent period between infection and sporulation is increased. This is usually achieved by using fungicides or by growing resistant hosts in which the disease cycle is slowed down. Figure 5.7 shows the effects of various fungicide sprays on the progress of a potato blight epidemic. Although none of the treatments actually prevented the epidemic from occurring, three of them (Bordeaux mixture, zineb, maneb) extended the lag phase sufficiently to prevent significant reductions in crop yield. When the phase of exponential increase in disease is delayed in this way, bulking-up of the tubers can take place before foliar damage reaches a critical level (see Fig. 4.9).

Hence the real value of quantitative estimates of disease progress is that the effect of different control strategies on epidemic development can be evaluated. Such estimates are also useful in correlating amounts of disease at a particular stage in the season with differences in cultivar resistance, and with eventual effects on crop yield. Parameters such as the area under the disease progress curve (AUDPC) can, for instance, be used to compare epidemic development on different genotypes of the same crop. This is a simple descriptor of an epidemic which takes account of the time of onset of disease, and the shape of the curve, by calculating the area beneath the curve over a specified period. Epidemics which start earlier, and increase more rapidly, will obviously give higher AUDPC values than those which are delayed and develop more slowly. Comparisons with yield data can then be made and used to define critical thresholds for economic loss, and the need for control actions such as fungicide treatment.

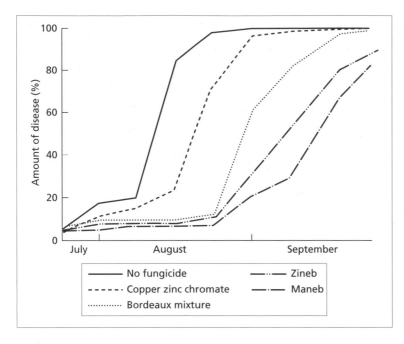

Fig. 5.7 Progress of a potato late blight epidemic and the effects of several foliar fungicides applied weekly from mid-July. (Data from Hooker 1956.)

Quarantine

The essential feature of these various strategies is that, while the pathogen is not completely eradicated, the progress of an epidemic is delayed. An alternative approach is to try to prevent the pathogen from coming into contact with the crop. In this respect the aim is to stop pathogens becoming **endemic**, i.e. permanently established in a particular geographical area.

A regular feature of the domestication and exploitation of plants has been the introduction of crops into new geographical areas. In some instances such crops have remained free of their most damaging pathogens and it is a matter of concern that this situation should continue. Consequently, strict quarantine measures are enforced to prevent the introduction of diseases. The cultivation of rubber in Malaysia provides a good example. This tree is a native of the Amazon Basin and was introduced into Malaysia, via the Royal Botanic Gardens at Kew, in the 1870s. Fortuitously, the destructive leaf blight disease caused by *Dothidella ulei* was not present in the introduced stock and quarantine measures have ensured that the area has remained free from the disease ever since. In South America, leaf blight has limited the commercial exploitation of the rubber tree.

Similar quarantine schemes operate in most countries, backed up by national and international legislation. The efficacy of such quarantine measures depends, however, on the existence of effective natural barriers, such as oceans and mountain ranges. Even then, exclusion may prove impossible as long-distance dispersal sometimes occurs (see p. 41). A good example of the failure to contain important pathogens is provided by the destructive coffee rust pathogen, *Hemileia vastatrix*, which has dramatically extended its range to include all the major coffee-growing regions of the world (Fig. 5.8). One significant landmark in this global spread was the first appearance of the disease in Brazil in 1970. International travel most likely plays a large part in such outbreaks. For example, yellow stripe rust of barley (*Puccinia striiformis* f.sp. *hordei*) was recorded for the first time in Colombia in 1975, and has subsequently spread to many South American countries, causing crop losses estimated at millions of dollars. The genetic strain of the pathogen responsible for this outbreak is almost identical to race 24, a type found in Europe, and one theory is that spores of the fungus were transported on the clothing of a plant breeder from South America who visited the UK to attend a conference.

One important safeguard preventing the world-

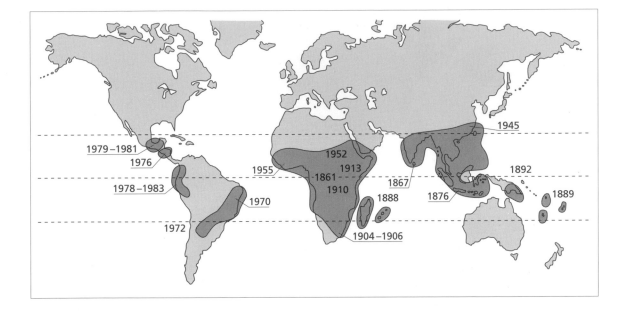

Fig. 5.8 Present distribution of the coffee rust fungus, *Hemileia vastatrix*, showing dates when the disease was first recorded. (After Schieber & Zentmyer 1984.)

wide distribution of pathogens is to ensure that all propagation material is free from infection. This has long been the aim of testing schemes which assess the extent to which seed is contaminated. The results of these tests are then used to decide whether the seed should be exported, or if it can be successfully treated with chemicals. Alternatively, export of seed from regions where particular diseases are endemic may be restricted. This is a more contentious issue, as illustrated by the recent debate in the USA about karnal bunt of wheat, caused by the smut fungus *Tilletia indica*. This disease, first described from India, is now widespread and occurs in Mexico, close to the wheat-growing areas of southern USA. In 1983 the US Department of Agriculture imposed a strict quarantine on wheat seed from Mexico, intended to prevent import of the pathogen. However, there have been a number of outbreaks in the USA since this embargo, probably due to airborne spread of smut spores. Since Mexico is an important centre for cereal breeding and the distribution of new wheat germplasm, there is now concern that this quarantine is limiting work on genetic improvement of the crop,

rather than providing an effective barrier to spread of the disease.

It was noted in Chapter 3 that viruses are commonly transmitted by vegetative propagation. For this reason propagation material such as budwood used for the multiplication of woody perennials like citrus is carefully screened to ensure that it is free from viruses, viroids and phytoplasmas. Similarly, the production of 'seed' potatoes is subject to close scrutiny by government officials. Contamination of seed stocks themselves is reduced by propagating them in geographic areas which are relatively free from the insect vectors transmitting potato viruses.

In recent years, advances in botanical techniques, such as tissue culture, have been exploited to improve crop hygiene. It is now possible to regenerate unlimited clones of whole plants from excised tissues or even from single cells. Cultures made from the shoot tip meristem are used to provide virus-free planting material which ensures high yields for at least the first few seasons. A heat treatment (see p. 236) may be included to improve the efficiency of eradication. An additional advantage of these techniques is the possibility of maintaining virus-free genetic stocks under sterile conditions for an indefinite period. Tissue culture is gradually supplanting the traditional methods of vegetative propagation. It is already used for many ornamentals, soft fruits such as strawberries, and several woody perennials.

Don't bring plants carrying pests and diseases into Britain.

To protect the health of Britain's crops and gardens, there are strict rules covering the import of plant material. Before travelling abroad, make sure that you know them.

Further information can be obtained from:

Plant Health Division
Ministry of Agriculture, Fisheries and Food,
Room 504, Ergon House, c/o Nobel House, Smith Square, London SW1P 3JR
Telephone: 071-238 6477 or 6479

Plant Health

© Crown Copyright April 1990 PB0239

Disease legislation

For quarantine to be effective there must be strict rules governing the movement of plants and their products between countries. There also needs to be a level of awareness of the threat posed by imported pests and diseases among not only those who deal with plant materials but the general public as well. Figure 5.9 shows a poster used in the UK to alert travellers to the rules concerning plant imports.

Legislative measures exist at both a national and international level. Agencies such as the Food and Agriculture Organization (FAO) of the United Nations aim to set international standards and to develop guidelines for quarantine in member countries by means of agreements such as the International Convention on Crop Protection. Individual countries also have their own legal framework to restrict the spread of disease. As an example, potato wart disease (Fig. 5.10) has been controlled in the UK by a series of legislative measures which date from 1908. Initially, this pathogen was made a scheduled pest, which meant that farmers had to notify outbreaks to the Ministry of Agriculture. Subsequent government orders prohibited the import of wart-affected potatoes and compelled farmers to grow only resistant cultivars on contaminated land. At present, wart disease occurs only rarely in commercial crops in the UK and outbreaks are mainly confined to plants grown in allotments and gardens.

Continued vigilance is required to prevent other potentially serious diseases from becoming established in new areas. When outbreaks do occur a stringent eradication policy is often adopted. A good example of this is citrus canker, caused by the bacterium *Xanthomonas campestris* pv. *citri*. Citrus canker can be managed by a combination of chemical sprays, host resistance, and cultural measures, but as the pathogen is spread on fruit, it is important for producers to be free of the disease to avoid restrictions in export markets. In the early 1900s, a coordinated campaign, with regulatory powers imposed by both state and federal governments, eradicated canker from citrus groves in Florida. The state then

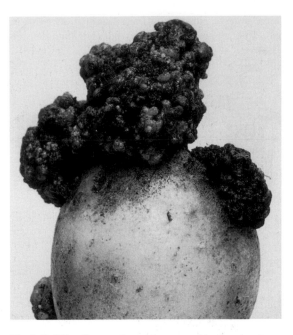

Fig. 5.10 Wart disease of potatoes, caused by *Synchytrium endobioticum*. Infection leads to abnormal cell division and enlargement, and wartlike outgrowths on tubers. (Photograph Crown Copyright.)

remained free of the disease for more than 50 years, largely due to port-of-entry inspections which intercepted infected material. In 1984 a new form of the pathogen was discovered on nursery stock in Florida, raising fears that the disease would become widely distributed. A strict quarantine was imposed on affected nurseries, with all citrus stock being burned. Where trees potentially exposed to the pathogen had been sent to new sites, all other trees within 40 m of such stock were also destroyed. By 1987 almost 20 million citrus trees had been removed, and the initial eradication policy had been revised to take account of the perceived risk of further outbreaks. There is continued monitoring of the situation, and to reduce risk, citrus imports are restricted to states on the Pacific coast such as Washington and Oregon, where the crop is not grown. Similar eradication campaigns for citrus canker have been carried out in Australasia, South Africa and parts of Brazil.

The unintentional introduction of a new pathogen or a more virulent strain of an endemic pathogen may have serious consequences for the indigenous wild species as well as cultivated crops. The decimation of

Fig. 5.9 (*Facing page*) Poster issued by the Plant Health Division of the Ministry of Agriculture, Fisheries and Food to publicize the risks of imported pests and pathogens. (Courtesy of MAFF. Crown Copyright 1990.)

the European elm population by an introduced strain of *Ophiostoma novo-ulmi* has highlighted this problem and raised fears that similar destructive pathogens, for example the oak wilt fungus *Cerato-cystis fagacearum*, may follow in its footsteps.

Further reading

Books

Campbell, C.L. & Madden, L.V. (1990) *Introduction to Plant Disease Epidemiology*. John Wiley & Sons, New York.

Ebbells, D.L. & King, J.E. (eds) (1979) *Plant Health*. Blackwell Scientific Publications, Oxford.

Gregory, P.H. (1973) *The Microbiology of the Atmosphere*, 2nd edn. John Wiley & Sons, New York.

Jeger, M.J. (ed.) (1989) *Spatial Components of Plant Disease Epidemics*. Prentice Hall, New Jersey.

Leonard, K.J. & Fry, W.E. (1986) *Plant Disease Epidemiology*. Vol. 1, *Population Dynamics and Management*. Macmillan, New York.

Leonard, K.J. & Fry, W.E. (1989) *Plant Disease Epidemiology*. Vol. 2, *Genetics, Resistance, and Management*. McGraw-Hill, New York.

Vanderplank, J.E. (1963) *Plant Diseases: Epidemics and Control*. Academic Press, New York.

Zadoks, J.C. & Schein, R.D. (1979) *Epidemiology and Plant Disease Management*. Oxford University Press, Oxford.

Reviews and papers

Brasier, C.M. (1990) China and the origins of Dutch Elm Disease: An appraisal. *Plant Pathology* 39, 5–16.

Dubin, H.J. & Stubbs, R.W. (1986) Epidemic spread of barley stripe rust in South America. *Plant Disease* 70, 141–144.

Fitt, B.D.L., McCartney, H.A. & Walklate, P.J. *et al.* (1989) The role of rain in the dispersal of pathogen inoculum. *Annual Review of Phytopathology* 27, 241–270.

Fry, W.E., Goodwin, S.B., Dyer, A.T. *et al.* (1993) Historical and recent migrations of *Phytophthora infestans*: Chronology, pathways and implications. *Plant Disease* 77, 653–661.

Gibson, G.J. (1996) Investigating mechanisms of spatiotemporal epidemic spread using stochastic models. *Phytopathology* 87, 139–146.

Gilligan, C.A. (1990) Comparison of disease progress curves. *New Phytologist* 115, 223–242.

Hill, S.A. & Torrance, L. (1989) Rhizomania disease of sugar beet in England. *Plant Pathology* 38, 114–122.

Levy, M., Romao, J., Marchetti, M.A. & Hamer, J.E. (1991) DNA fingerprinting with a dispersed repeated sequence resolves pathotype diversity in the rice blast fungus. *The Plant Cell* 3, 95–102.

McCartney, H.A. (1991) Airborne dissemination of plant fungal pathogens. *Journal of Applied Bacteriology Symposium Supplement* 70, 39S–48S.

McDermott, J.M. & McDonald, B.A. (1993) Gene flow in plant pathosystems. *Annual Review of Phytopathology* 31, 353–373.

Mundt, C.C. & Leonard, K.J. (1985) A modification of Gregory's model for describing plant disease gradients. *Phytopathology* 75, 930–935.

Schieber, E. & Zentmyer, G.A. (1984) Coffee rust in the Western Hemisphere. *Plant Disease* 68, 89–93.

Schoultes, C.L., Civerolo, E.L., Miller, J.W. *et al.* (1987) Citrus canker in Florida. *Plant Disease* 71, 388–395.

Smith, V.L., Campbell, C.L., Jenkins, S.F. & Benson, D.M. (1988) Effects of host density and number of disease foci on epidemics of southern blight of processing carrot. *Phytopathology* 78, 595–600.

Sweetmore, A., Simons, S.A. & Kenward, M. (1994) Comparison of disease progress curves for yam anthracnose (*Colletotrichum gloeosporioides*). *Plant Pathology* 43, 206–215.

Thresh, J.M. & Owusu, G.K. (1986) The control of cocoa swollen shoot disease in Ghana: an evaluation of eradication procedures. *Crop Protection* 5, 41–52.

Vloutoglou, I., Fitt, B.D. & Lucas, J.A. (1995) Periodicity and gradients in dispersal of *Alternaria linicola* in linseed crops. *European Journal of Plant Pathology* 101, 639–653.

Part 2
Host–Pathogen Interactions

'Far from being an insurmountable obstacle to the analysis of an organic system, a pathological disorder is often the key to understanding it.' [Konrad Lorenz, 1903–1989]

Given favourable environmental conditions, the progress of disease in a crop is determined by interactions between individual plants and the pathogen. Such interactions at the whole plant, cellular and molecular level are of fundamental interest, as a better understanding of the processes of disease development may eventually lead to more rational and effective control methods. This has already proved the case with infectious diseases of animals, where the discovery of the immune system facilitated major advances in disease prevention.

Unravelling the key processes in host–pathogen interaction is a difficult task, as two organisms are involved, each with its own separate structure and metabolism. Each partner may influence the activity of the other by a dynamic process of signal and response, so that the properties of the host–pathogen complex are distinct from those of either partner alone. Even in the case of virus infections, where the pathogen has a relatively simple molecular structure, and no independent metabolic activity, the biochemistry of disease is still not fully understood.

Several types of interactions are possible. Physical characteristics, such as the shape and texture of the host surface, the thickness of the cuticle, or stomatal morphology, may be important. The growth and reproduction of the pathogen may itself cause physical damage, for instance by blocking vascular elements, or tearing through epidermal cell layers. Biochemical interactions also occur (Fig. 1). In living cells, instructions encoded in DNA are copied to RNA, which then directs the synthesis of proteins. These may play a structural role in the cell, or alternatively act as enzymes influencing the rate or direction of metabolic pathways. Host–pathogen interactions can take place at any of these levels. For instance, preformed metabolites present prior to infection may act as inhibitors or toxins. Structural proteins or other polymers in host or pathogen cell walls or membranes

may interact as signals or receptors. Enzymes may be released or activated, altering metabolic pathways, or attacking substrates in cells. Finally, interactions may change the expression of genetic information in cells, by activating or repressing genes, or by introducing a new set of instructions. For instance, the crown gall pathogen *Agrobacterium tumefaciens* genetically engineers host cells to synthesize compounds serving as a source of nutrients for the bacterium (see p. 135). Similarly, viruses and related pathogens act as molecular pirates, subverting host cell synthesis to produce further copies of themselves.

Stages in host–pathogen interaction

The invasion of a plant by a pathogenic microorganism is in reality a continuous process, but certain key stages can be identified (Table 1). First the pathogen must locate the host; in many cases this is a random process depending upon chance contact between propagules and a susceptible plant. Many pathogenic fungi, for instance, produce airborne spores which are carried by wind currents (see p. 39). Similarly, bacterial cells may be splashed or blown between plants by wind-driven rain. In these cases, successful contact with a new host relies upon the huge numbers of propagules produced, which increases the probability that at least some will land on a suitable plant. Other pathogens, however, possess more sophisticated adaptations for locating the host, usually by responding to a particular signal. Such recognition cues may trigger germination of a dormant propagule, or attract hyphae or motile cells towards the host. Zoospores of root-infecting fungi such as *Pythium* or *Phytophthora* swim along concentration gradients of sugars or amino acids exuded by plant roots. These chemotactic responses are often non-specific, as both host and non-host roots are attractive, but some recognition cues may be more

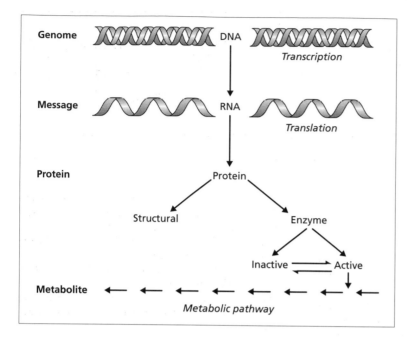

Metabolic pathway

Fig. 1 Scheme for molecular control of metabolism showing possible levels of interaction.

Table 1 Key stages in host–pathogen interaction.

Locating the host
Attachment/adhesion to host surface
Penetration and entry into host
Colonization of host tissues
Suppression or avoidance of host defence mechanisms
Reproduction of the pathogen
Dispersal of pathogen from host

finely tuned. Dormant sclerotia of *Sclerotium cepivorum* germinate in response to sulphur compounds released only by onion and a few related species (see p. 49), while the tiny seeds of the parasitic angiosperm *Striga* respond to a germination stimulant, strigol, present in root exudates from host plants such as cotton. This compound is active at remarkably low concentrations.

Pathogens dispersed by insects or other vectors take advantage of the host recognition mechanisms of the vector itself. Further examples of survival between hosts, and adaptations optimizing the chances of a pathogen contacting a living host, have already been discussed in Chapter 3.

Once contact has been made, a complex sequence of interactions between host and pathogen is initiated

(Table 1). The pathogen must first breach or bypass the outer defences of the host. Gaining access to suitable nutrient substrates in the host is usually a prerequisite for continued development of the pathogen and hence of the disease. The spores of biotrophic fungi, for instance, contain food resources sufficient only for germination and a limited amount of hyphal growth; unless the fungus is able to penetrate living host tissues during this brief phase of independence it will die. Such dependence on living cells reaches an extreme in the case of viruses. Multiplication can only occur if the virus gains access to the biosynthetic machinery of the host cell. By contrast, some necrotrophic parasites are able to flourish for an extended period without penetrating their hosts.

The encounter between a potential pathogen and a plant host has two possible outcomes. The pathogen may successfully invade the plant, multiply, and cause disease. Alternatively the invading microorganism may fail to establish itself in the host, and disease does not develop. In reality there are various gradations between these two extremes. The whole process may be viewed as a contest between host and pathogen, with the outcome depending upon the relative balance between microbial pathogenicity factors, and host defence mechanisms (Fig. 2).

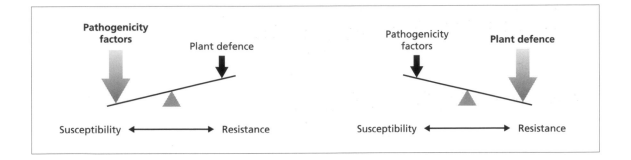

Fig. 2 Host–pathogen interactions as a balance between pathogenicity and defence.

What properties distinguish the pathogen from the majority of microorganisms which do not cause disease? Why do particular pathogens attack only a limited range of host plants? How do plants recognize and respond to potential pathogens? How does the successful pathogen evade or counter host defences? What physiological changes underlie symptom development? The following five chapters of the book will attempt to answer these questions.

Further reading

Books

Callow, J.A. (ed.) (1983) *Biochemical Plant Pathology.* John Wiley & Sons, Chichester.

Crute, I.R., Holub, E.B. & Burdon, J.J. (eds) (1997) *The Gene-for-Gene Relationship in Plant–Parasite Interactions.* CAB International, Wallingford.

Smith, C.J. (ed.) (1991) *Biochemistry and Molecular Biology of Plant–Pathogen Interactions.* Proceedings of the Phytochemical Society of Europe, Vol. 32. Oxford Science Publications, Oxford.

Verma, D.P.S. (ed.) (1992) *Molecular Signals in Plant–Microbe Communications.* CRC Press, Boca Raton, Fla.

Reviews and papers

Plant–Microbe Interactions (1996) *The Plant Cell* 8(10). [A special issue containing an introduction and 18 review articles on many aspects of plant–pathogen and plant–mutualist relationships.]

6 Entry and Colonization of the Host

'Whenever the little seeds of rust come to rest upon the same stalk, finding some open mouths of the exhaling vessels, there they enchase their minute radical fibers, and there they infiltrate in such a manner, that they graft into the tender and delicate arteries peculiar to the plant.' [G. Targioni-Tozzetti, 1712–1783]

To gain access to host nutrients and establish a parasitic relationship, the microorganism must first pass through the external protective layers of the host. Plant pathogens enter their hosts in a variety of ways. Some penetrate directly through the intact surface covering of the plant. Others pass through natural openings, or through regions where the external defences are especially thin. The most important such route is through the stomata, but other zones of weakness include glands, hydathodes, lenticels, nectaries and root tips. Many other pathogens enter the host via wounds resulting from physical or chemical damage, or from the activities of animal pests. Wounds can also be self-inflicted, for instance by the abscission of leaves or during the emergence of lateral roots.

A distinction may therefore be drawn between pathogens which enter directly through the protective barriers of the host, and those which bypass these defences (Fig. 6.1). Several major groups of pathogens, including the viruses, phytoplasmas and many fungi causing post-harvest diseases of fruit, are almost entirely dependent upon wounds to gain entry to the host.

The entry route is important in determining the nature of the initial host–pathogen interface formed. For instance, bacterial cells washed through stomata by rain can multiply initially in intercellular spaces, while pathogens penetrating directly through the host epidermis must cross cell walls and often grow within host cells. These differences will affect the types of nutrients available to the pathogen, and also the molecular events involved in recognition of the pathogen by the host.

The infection court

Let us assume that a pathogen has been successfully dispersed or has grown into contact with a potential host plant. Subsequent development on the surface of the host, penetration into the host and the very early stages of establishment within host tissues comprise the process of **infection**. This stage in the life cycle of the pathogen ends when the organism becomes dependent on the host for nutrients, at which point it begins to **colonize** tissues around the initial site of infection.

The initial site of contact between the pathogen and the surface of the host is described as the **infection court**. In any discussion of host penetration it is useful to distinguish at the outset between the aerial and subterranean surfaces of the plant. In one respect the problems confronting airborne and soil-borne pathogens are similar, in that both must breach the outer defensive layers of the host, but there are major differences between the two environments. Soil exerts a buffering effect against extremes of temperature, water availability, and other environmental fluctuations. A propagule landing on an aerial plant surface is exposed to wide daily fluctuations in temperature and hazards such as desiccation.

The leaf and root surfaces of plants are termed the **phylloplane** and the **rhizoplane**, respectively. The allied terms **phyllosphere** and **rhizosphere** describe the habitats adjacent to these surfaces. In recent years a great deal has been learned regarding the influence of physical, chemical and biological factors on pathogen behaviour in these two infection courts. Factors influencing the germination of fungal spores are of special significance and include humidity, dura-

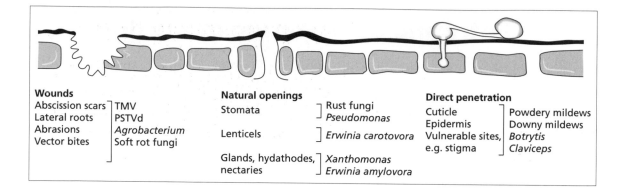

Wounds		Natural openings		Direct penetration	
Abscission scars	TMV	Stomata	⎤ Rust fungi	Cuticle	⎤ Powdery mildews
Lateral roots	PSTVd		⎦ *Pseudomonas*	Epidermis	Downy mildews
Abrasions	*Agrobacterium*	Lenticels	⎤ *Erwinia carotovora*	Vulnerable sites,	*Botrytis*
Vector bites	Soft rot fungi		⎦	e.g. stigma	⎦ *Claviceps*
		Glands, hydathodes,	⎤ *Xanthomonas*		
		nectaries	⎦ *Erwinia amylovora*		

Fig. 6.1 Some entry routes for plant pathogens. TMV, tobacco mosaic virus; PSTVd, potato spindle tuber viroid.

tion of leaf surface wetness, temperature, light, pH, nutrient availability and the quality and quantity of host exudates. Exudates from leaves and roots contain numerous chemicals such as sugars, amino acids, mineral salts, phenols and alkaloids; any of these may stimulate or inhibit germination and/or growth of pathogens. Root exudates are particularly significant in determining the behaviour of soil fungi which produce motile zoospores. These are chemotactically attracted to the elongating zone of host roots where they encyst prior to entry. The initial phases of development on the host surface would seem to represent an especially vulnerable stage in the life cycle of fungal pathogens, as witnessed by the efficacy of protectant fungicides in the control of many diseases.

The principal components of the aerial surfaces of herbaceous plants are summarized in Fig. 6.2. In practice, there are considerable physical and chemical differences between the outer layers of various plant species, and even between different parts of the same plant. Thus the cuticle may vary in chemical composition and thickness on leaves, flowers and fruits. Variations are also found in different regions of the same organ, for example between the upper (adaxial) and lower (abaxial) surfaces of leaves. Other factors influencing the structure and composition of external layers include the conditions under which the plant has grown, and the developmental stage that the plant has reached. Seedling tissues are particularly prone to infection by opportunist pathogens, such as the 'damping-off' fungus *Pythium*, whereas mature

plants are seldom attacked. A critical factor here is the relative ease with which the pathogen can penetrate the cuticle and epidermal layers in the young plant.

The outer cell walls of primary roots are usually impregnated with lipid materials, including suberin and cutin. These form a definite membrane comparable with the leaf cuticle, but the use of the term 'cuticle' to describe this root covering is not accepted by some scientists. There is also some doubt as to whether such protective layers are present in the physiologically active apical region of the root. The root-hair zone is especially vulnerable to pathogens, as it is in intimate contact with a large volume of soil. The necessity for efficient water and nutrient uptake by the root hairs means that mechanical barriers, which would perhaps deter pathogens, are absent. Root-cap cells secrete a mucilaginous gel which encloses the growing root and is a distinctive feature of the rhizosphere.

There are even greater differences between the surfaces of herbaceous tissues and the stems and roots of woody perennials. Periderm, commonly termed bark, is formed following secondary thickening and the accompanying increase in girth of the organs. It comprises three layers, with the outermost being composed of dead cork cells which have suberized walls. Suberin is a complex material containing mixtures of hydroxy acids and it is very resistant to microbial attack. This substance, together with lignin and cellulose in cork cell walls and resins in their lumina, ensures that bark is virtually impregnable to invasion by microorganisms. Similar protective layers also form over abscission wounds and damaged tissues which are exposed when large branches are broken off in storms.

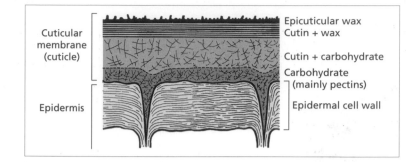

Fig. 6.2 The external layers bounding herbaceous plant organs. (After Jeffree *et al.* 1976.)

Adhesion

For airborne, wind- or splash-dispersed spores the first problem is effective adhesion to the host surface. Epicuticular wax is hydrophobic and repels water droplets and any microbial propagules they contain; firm attachment is essential to prevent the pathogen losing contact prior to infection. Molecules aiding microbial attachment are often described as **adhesins**. Spores of many pathogenic fungi produce an extracellular matrix which surrounds the spore and binds tightly to plant cuticles, as well as to inert hydrophobic surfaces such as Teflon, the coating of non-stick saucepans. This matrix may also protect spores from desiccation and provide a medium for immobilization of secreted enzymes. In the rice blast pathogen, *Magnaporthe grisea*, mucilage is released from the apex of conidia following hydration (Fig. 6.3a,b), serving as a form of biological 'glue'. Other pathogens which produce dry spores and which adhere to dry leaf surfaces, such as the rust and powdery mildew fungi, probably employ electrostatic mechanisms as well as adhesive materials to ensure attachment. Following contact with the host, conidia of the powdery mildew *Erysiphe graminis*, produce a small, short hypha (Fig. 6.4a). This primary germ tube helps to anchor the spore to the host surface, but may also be important in taking up water to aid pathogen growth.

Bacterial pathogens also synthesize extracellular molecules which promote adhesion. With animal-pathogenic species, small hairlike processes known as fimbriae are important in attaching cells to epithelial surfaces. Similar proteins are known to bind to plant cell walls, but their significance in plant infection is not clear. The crown gall pathogen, *Agrobacterium tumefaciens*, elaborates cellulose microfibrils which

help to secure the bacterium to host cells, as well as binding further bacteria.

Direct penetration

As shown in Fig. 6.2, direct penetration of herbaceous tissues requires entry through layers of wax, cutin, pectin and a network of cellulose fibrils impregnated with other wall polymers, before the pathogen makes contact with host protoplasm. This would seem to be a formidable obstacle. Nevertheless, many fungal pathogens are able to enter their hosts in this way. Biotrophic fungi, such as the rusts and the downy and powdery mildews, often gain access by growing down into the epidermis, but direct penetration is by no means restricted to this type of pathogen. Even necrotrophic fungi, such as *Botrytis*, can in suitable circumstances enter hosts directly by penetration through the cuticle.

Direct penetration of the host by fungi is frequently associated with the development of hyphal modifications known collectively as infection structures. Some examples are shown in Figs 6.3–6.5. Once the spore has germinated there follows a period of growth in which the germ tube extends over the leaf surface. This growth may appear to be random but some evidence points to the possibility that the germ tube is 'searching' for a favourable site for penetration (Fig. 6.4b). The length of germ tube developed varies but eventually extension growth ceases and the tip of the hypha swells to form an **appressorium**. This spherical or ovoid structure increases the area of contact and attachment between the fungus and the host surface (Fig. 6.3d). Penetration then takes place by the downward growth of a narrow hyphal thread or infection peg formed from the lower surface of the appressorium. There has been much debate as to the

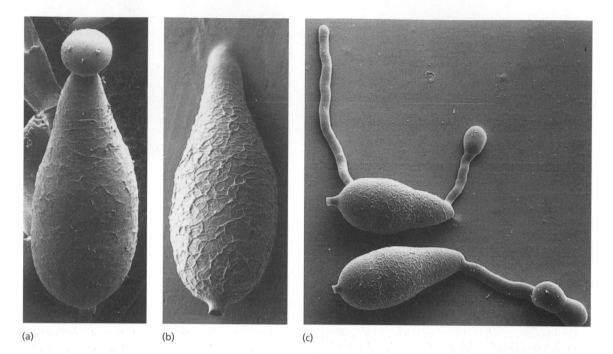

(a) (b) (c)

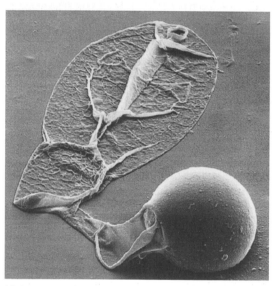

(d)

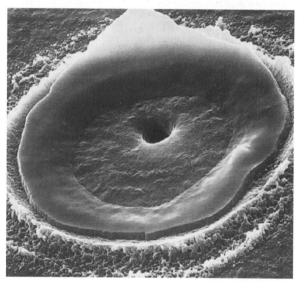

(e)

Fig. 6.3 Early infection stages of the rice blast fungus, *Magnaporthe grisea*. (a) Conidium with apical droplet of spore tip mucilage (×2900). (b) Conidium attached to substrate by adhesion of spore tip mucilage (×2900). (c) Germination of conidia and early stages of appressorium development, seen as swelling of germ tube apex (centre and lower right) (×1100). (d) Mature, globose, turgid appressorium attached to collapsed germ tube and conidium. A septum separates the appressorium from the germ tube (×3000). (e) Remnants of appressorium attached to a polyethylene surface. The upper part of the cell has been removed by sonication. What remains is the appressorial pore, with the dent made by mechanical force of the penetration peg clearly visible, and part of the smooth surrounding wall, composed of melanin. Note the halo of extracellular matrix material around the attachment site (×15 500). (a, d & e, From Braun & Howard 1994; b & c, from Howard 1994.)

(a)

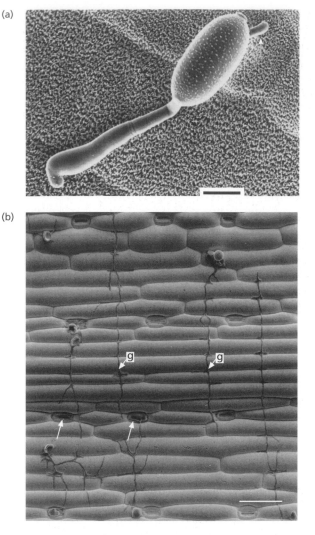

(b)

Fig. 6.4 Early development of fungal pathogens on the surface of barley leaves viewed by scanning electron microscopy. (a) Germinating spore of the powdery mildew *Erysiphe graminis* showing small, primary germ tube (arrow) and larger appressorial germ tube with hooklike tip. Scale bar = 10 µm. (Courtesy of Tim Carver.) (b) Germ tubes (g) of brown rust (*Puccinia hordei*) showing growth perpendicular to the orientation of epidermal cells, short branches formed at cell junctions, and appressoria (arrows) formed over stomata, which occur in rows. Scale bar = 100 µm. (From Read *et al.* 1992.)

and transmission electron microscope studies suggest that some pathogens degrade the cuticle and cell-wall polymers during penetration. Differential staining techniques have revealed localized dissolution of the cuticle and cell walls around infection pegs, implicating the action of hydrolytic enzymes in penetration by fungi.

The mechanisms employed by pathogenic fungi to penetrate plant surfaces have recently been analysed using elegant molecular and genetic techniques. Several fungi are known to produce cutinase, an enzyme able to degrade cutin; cutinase from the pea pathogen *Fusarium solani* f.sp. *pisi* was purified and used to raise antibodies specific for the enzyme. Such antibodies can be utilized either to detect production of the enzyme, for instance by immunolabelling in electron micrographs, or to inhibit enzyme activity. A combination of these approaches suggested that cutinase is a vital factor in breaching the host surface (Table 6.1). The gene coding for cutinase in *F. solani* has now been cloned and sequenced, and its regulation studied in detail. In germinating spores, cutinase synthesis is induced by breakdown products of cutin. Low levels of constitutive activity release cutin fragments which induce rapid expression of the cutinase gene. This system of regulation ensures that the enzyme is only synthesized in any quantity when pathogen spores contact a plant surface. Further evidence that cutinase is required for direct penetration has been obtained by introducing the *Fusarium* cutinase gene by transformation into another fungus, *Mycosphaerella*, which normally requires a wound to infect the host. Possession of the cutinase gene enabled the transformants to penetrate intact host surfaces.

More recently this apparently conclusive story has been questioned following experiments using a technique known as gene disruption. In this procedure the functional gene for cutinase was replaced by a defective copy unable to produce the enzyme. Fungal transformants containing the disrupted gene, and in which cutinase synthesis was abolished were still pathogenic to pea seedlings, and apparently able to penetrate the intact host surface. It is difficult to reconcile these conflicting results, as molecular genetic evidence suggests that there is only a single copy of the cutinase gene in the pathogen, and no detectable enzyme activity was present in the pathogenic transformants containing a defective gene.

The possibility that fungal enzymes such as cuti-

actual mechanics of penetration. Early workers showed that many fungi will successfully penetrate artificial materials such as gold leaf, suggesting that the process is entirely mechanical. However, scanning

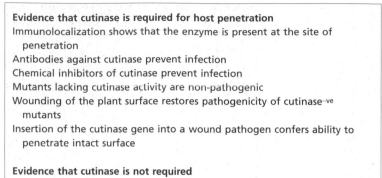

Table 6.1 The cutinase debate.

Evidence that cutinase is required for host penetration

Immunolocalization shows that the enzyme is present at the site of penetration

Antibodies against cutinase prevent infection

Chemical inhibitors of cutinase prevent infection

Mutants lacking cutinase activity are non-pathogenic

Wounding of the plant surface restores pathogenicity of cutinase-ve mutants

Insertion of the cutinase gene into a wound pathogen confers ability to penetrate intact surface

Evidence that cutinase is not required

Gene replacement, to disrupt a single cutinase gene, abolishes cutinase mRNA and enzyme activity, but does not alter pathogenicity

nase might be required for direct host entry does not preclude a role for mechanical forces in penetration. Confirmation of the importance of such forces has come, perhaps unexpectedly, from studies on the mode of action of certain fungicides. The compound tricyclazole (Table 11.2) effectively controls several fungi, including *Magnaporthe grisea*. This fungus penetrates rice plants directly from dark, pigmented appressoria; appressoria formed in the presence of the fungicide are non-pigmented, and no penetration takes place from them. The biochemical target of tricyclazole turns out to be melanin synthesis, so that production of the pigment is inhibited in treated appressoria. Normally melanin is deposited in the appressorial wall (Fig. 6.3e), making it rigid and impermeable to solutes. As the appressorium matures, hydrostatic pressure builds up inside until sufficient force is generated to push the infection peg down through the cuticle. Functional appressoria formed on inert plastic surfaces actually leave a microscopic dent at the point where the peg projects (Fig. 6.3e). In non-melanized appressoria the wall remains relatively thin, flexible and permeable, and the infection peg appears unable to breach the surface. Albino mutants of *M. grisea* which are unable to synthesize melanin are similarly incapable of achieving penetration from non-pigmented appressoria. Thus in this, and other similar pathogens with pigmented appressoria, mechanical force would appear to be the primary means for penetrating the host.

Root-infecting fungi also form infection structures which are generally more complex than those produced by fungi attacking aerial tissues. Some isolates of *Gaeumannomyces* produce appressoria in the form of short side branches from runner hyphae, beneath which narrow penetration hyphae enter the root cortex. *Rhizoctonia solani*, a versatile pathogen attacking a wide variety of hosts, forms both lobed appressoria and more complex aggregations of repeatedly branched hyphae, called infection cushions. The former tend to be produced on aerial tissues and the latter on roots and other subterranean organs. Multiple infection hyphae are produced from the lower surface of infection cushions, and enter the host. The eyespot fungus, *Pseudocercosporella* (*Tapesia*) *herpotrichoides*, infects the stem base of wheat by colonizing the coleoptile and then penetrating through successive leaf sheaths. Multicellular plates of mycelium, termed infection plaques, are produced on the surface of each leaf sheaf (Fig. 6.5), and these act as compound appressoria enabling the fungus to penetrate epidermal cells at numerous sites. Fungi attacking perennial hosts, such as trees, in which the surface is protected by a layer of bark or periderm, often infect from compound structures. When rhizomorphs of *Armillaria mellea* encounter a suitable host the concerted action of the numerous hyphae comprising these strands (see p. 32) is often sufficient to penetrate the intact surface layers.

It was noted above that the root hair zone is particularly vulnerable to invasion by pathogens. *Plasmodiophora brassicae* enters root hairs at an early stage in its life cycle (it subsequently returns to the soil, re-enters the root epidermis and proliferates in cortical cells causing the club root symptoms). The mode of entry of *Plasmodiophora* into root hairs

(a)

(b)

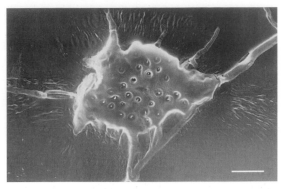

(c)

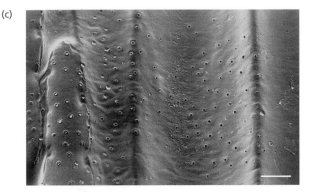

Fig. 6.5 Infection structures of the eyespot fungus, *Pseudocercosporella* (*Tapesia*) *herpotrichoides*, formed on wheat leaf sheaths. (a) Multicellular infection plaques. Scale bar = 100 μm. (b) View of underside of plaque, lifted from host surface, showing collar-like zones surrounding the infection pegs. Scale bar = 10 μm. (c) View of host surface exposed by removing infection plaque, showing multiple penetration sites. Scale bar = 10 μm. (a & c, From Daniels *et al.* 1991.)

appears to be unique. Zoospores of the pathogen encyst on the root hair wall. Entry begins when a bullet-shaped structure is suddenly forced from within the cyst through the wall into the root hair cell. The contents of the spore are then rapidly injected through the resulting puncture into the host cell (Fig. 6.6). This is a particularly dramatic example of mechanical penetration, as the actual infection process takes only about a second.

The bacterium *Rhizobium* also initiates infection through root hairs, one of the very few examples of direct penetration by a plant-infecting bacterium, although in this case the relationship is ultimately mutualistic, with the formation of nitrogen-fixing root nodules.

A further interesting example of a vulnerable site exploited by pathogens is the surface of the female organ, the stigma, which is adapted to trap pollen and permit penetration by pollen tubes to ensure fertilization. Several specialized pathogens, notably species of the ergot fungus *Claviceps*, produce airborne spores which germinate on the host stigma to form penetration hyphae which mimic pollen tubes and extend downwards to invade the ovary. The period of susceptibility to infection is brief, due to the short time during which the stigma is receptive to pollination. As was discussed in Chapter 3, some viruses, for example bean mosaic virus, may be transmitted to the ovules of healthy plants through infected pollen. This is a particularly interesting case as the virus takes advantage of a normal event in the life cycle of the plant to circumvent the structural defences of the host.

Penetration through natural openings

Entry through stomata

The surface layers of field-grown plants are rarely free from minor wounds, but even if they were, there are still a number of natural openings through which microbes can enter. The most important of these are stomata, via which many pathogens enter their hosts. The detailed morphology of these structures may determine whether or not infection can occur, as in citrus fruits where the conformation of the cuticle

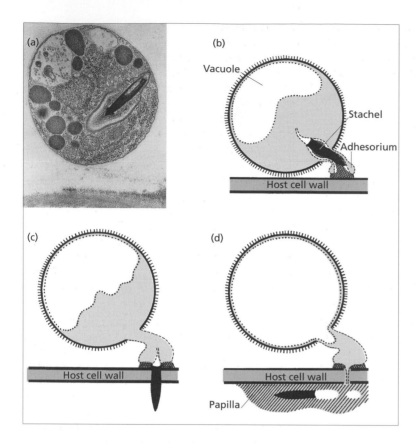

Fig. 6.6 (a) Electron micrograph section of a *Plasmodiophora brassicae* zoospore encysted on a root hair, showing bullet-like stachel (×26 500). (b–d) Diagrammatic summary of the penetration of a root hair: (b) vacuole enlarges and small adhesorium appears; (c) stachel punctures host wall; (d) penetration has occurred and the host protoplast has deposited a papilla at the penetration site. (From Williams *et al.* 1973.)

around the stoma either prevents or allows the passage of water droplets containing the bacterial pathogen *Pseudomonas syringae* pv. *citri*. Stomata are also the main site of entry of several important fungal pathogens. When a rust spore germinates on a cereal leaf the germ tube grows at right angles to the long axis of the leaf (Fig. 6.4b). This orientation of growth, which is an example of **thigmotropism**, is a contact response to the surface topography of cells, as similar tropisms occur on inert plastic replicas of leaves. Experiments with artificial surfaces etched or scratched with precise patterns suggest that the fungus recognizes a repetitive series of ridges spaced at intervals similar to the width of epidermal cells. Hyphal growth across the long axis of the cereal leaf ensures that the germ tube will sooner or later encounter a stoma, as these occur in longitudinal rows (Fig. 6.4b). Once a stoma is contacted the rust germ tube differentiates an appressorium and an infection hypha enters the substomatal cavity. With many rust fungi the signal for appressorium forma-

tion appears to be the shape of the stomatal guard cell, and in particular the stomatal lip. Bean rust, *Uromyces appendiculatus*, produces appressoria in response to small ridges about 0.5 µm in height, which corresponds closely to the dimensions of the stomatal lip of the host plant *Phaseolus*. On non-host leaves, extension growth of the germ tube continues indefinitely until the endogenous nutrient reserves are exhausted, and the germling dies. These precise morphogenetic responses to host surface features indicate that rust fungi possess a sophisticated contact-sensing system which aids location of natural openings for entry.

Several oomycete pathogens, including *Pseudoperonospora* on hops, *Plasmopara* on vines, and *Phytophthora* on potato, produce sporangia which germinate on host leaves by releasing motile zoospores. These zoospores are attracted to stomata where they encyst in a suitable position for their germ tubes to grow immediately between the guard cells. *Pseudoperonospora* zoospores are attracted to open stomata, but not to closed stomata. This attraction is based in part on recognition of the morphology of the open apertures and partly on a chemical stimulus connected with gaseous photosynthetic metabolites.

Lenticels, hydathodes and nectaries

Lenticels allow gas exchange to occur through bark on woody stems and secondarily thickened roots. These loosely packed openings in the periderm are also abundant on potato tubers, where they provide suitable sites for the entry of a number of microorganisms, especially the common scab pathogen *Streptomyces scabies*.

Glandular tissues which have especially thin surface barriers, such as hydathodes and nectaries, are also exploited by pathogens. Bacterial lesions on leaves often develop at the margin, at sites where water exudes through hydathodes. *Erwinia amylovora*, the bacterium responsible for the destructive fireblight disease of pears and apples, enters through nectaries at the base of flowers. In this case, the sugary secretions of nectar, when diluted by rain, provide a favourable medium for multiplication of the pathogen prior to penetration. Fireblight infections are also prevalent after severe thunderstorms, which suggests that the bacterium takes advantage of minor wounds caused by heavy rainstorms. Rainfall is an important predisposing factor in foliar infection by bacteria, as an external force sufficient to wash cells through natural openings into substomatal cavities and other internal tissues (see p. 51).

Penetration through wounds

For many pathogens, especially bacteria and viruses which are incapable of penetrating plants directly, wounds are the most frequent or only avenue of entry. Wounds are caused by human activities, as well as by natural agencies, including wind, hail, extremes of temperature and light, and by pests. The external barriers of the host may also be broken temporarily as a natural consequence of plant growth and development.

Many agricultural and horticultural practices involve accidental or even deliberate wounding. Grafting, pruning and picking spread pathogens through a crop or create wounds which can be exploited by opportunist fungi and bacteria.

Mechanical harvesters often increase the incidence of wounding of plant produce. Post-harvest rots of apples (caused by *Penicillium expansum*) and citrus fruits (caused by *P. digitatum* (Plate 8, facing p. 12) and *P. italicum*) are only important if the fruits are mechanically wounded during harvesting, packing or transport. Many important forest pathogens also enter through wounds. *Heterobasidion annosum*, which is a destructive pathogen of conifers (see Fig. 13.9), normally colonizes wounds caused by high winds, snow or other natural agencies. It has become a particularly damaging pathogen in plantations where it takes advantage of the stumps left after felling as sites for entry (see p. 240).

Leaf abscission provides opportunities for infection, as does any other point in the life cycle at which parts of the plant are detached. *Nectria galligena*, which causes apple canker, enters woody twigs through the vascular bundles that are exposed at leaf fall, and hence avoids the problem of penetrating intact bark. The vascular bundles in the leaf scar are, however, soon sealed off by the development of a cork layer and hence the pathogen must take immediate advantage of the infection sites created at leaf fall. Lateral roots emerge by breaking out through the cortex of the parent root. Lesions caused by root pathogens, such as soil-borne *Phytophthora* species, are often initiated at these sites. Soft rot pathogens of potatoes commonly enter tubers through the scar left during separation of the tuber from the parent plant (see Fig. 7.11).

As well as providing entry sites, wounds may release solutions rich in carbohydrates and amino acids, which stimulate germination of spores, or attract motile bacteria and fungal zoospores. The crown gall bacterium, *Agrobacterium*, is dependent on wounds to initiate tumours; exudates leaking from wounded cells have been shown to contain phenolic compounds such as acetosyringone, which serve as molecular signals activating virulence genes on the Ti plasmid (see Fig. 8.12). This specific recognition-response system ensures that virulence functions are expressed only in the presence of susceptible host cells.

Senescent tissues which remain attached to plants also serve as an entry route, and facilitate the invasion of adjoining healthy tissues by opportunist pathogens. The grey mould pathogen *Botrytis cinerea* colonizes vegetables such as courgettes (Plate 7, facing p. 12) or tomatoes by vegetative growth from the senescing remains of flowers, causing a disease known as blossom end rot.

Disease lesions may themselves allow the entry of other pathogens. In these instances the host is initially infected by a pathogen which may or may not itself cause serious damage. This pathogen, however, paves the way for more aggressive organisms. Potato late blight lesions in tubers may be exploited by soft rot bacteria such as *Erwinia carotovora* (see Fig. 8.2) which can destroy the tuber much more quickly than the blight fungus itself (see p. 125).

It was noted in Chapter 3 that many pests are also important vectors for plant pathogens. As well as dispersing the pathogen, their feeding activities cause wounds which serve as an entry route. The Dutch elm fungus, *Ophiostoma novo-ulmi*, is introduced directly into sapwood by its vector, the bark beetle. In this example the feeding tunnels not only breach the external protective layers of bark but also provide direct access to the vascular tissues in which the fungus can flourish. An even more elegant means of entry is provided by the aphid vectors of many viruses. The aphid stylet injects the virus into the sieve cells of the host with clinical efficiency, and subsequently the virus can spread freely via the phloem. Soil-borne viruses are often introduced via wounds caused by nematodes or fungal pathogens. The wide range of vectors exploited by viruses is paralleled by a similar variety of infection routes.

In some cases, wounds caused by animal pests can increase the incidence and severity of plant diseases, although the pest itself is not a vector. One of the best-known examples is the interaction between vascular wilt pathogens and nematodes. The fungus *Verticillium dahliae* causes a wilt disease in potatoes known as early dying, characterized by premature senescence of leaves and haulms. The pathogen survives in soil as microsclerotia, and there is a correlation between the number of pathogen propagules present in soil, and the incidence of early dying disease. If, however, the soil also contains significant numbers of nematodes capable of causing lesions on potato roots, the disease is much more severe. The most likely explanation for this synergistic effect is that feeding wounds caused by the nematodes provide enhanced access to the vascular tissues of host roots.

The host–pathogen interface

Once inside the plant, pathogens exhibit a wide variety of modes of growth within host tissues (Table 6.2). The site of contact between a pathogen and host cells is known as the host–pathogen interface. This zone is vital in understanding the nature of different host–pathogen interactions, as it is the site at which nutrient uptake by the parasite occurs, and also where molecular communication between the two partners takes place. It is likely, for instance, that recognition events determining active resistance or susceptibility to infection are initiated at this interface. Three main types of interface can be distinguished (Table 6.2):

1 intercellular, where the pathogen grows outside host cells;
2 partly intracellular, where limited penetration of cells by parasitic structures occurs;
3 intracellular, where growth and development takes place entirely within host cells.

These categories are not absolute as many pathogens which initially grow between host cells subsequently invade them once tissues become moribund.

Intercellular relationships are characteristic of bacteria and fungi (e.g. *Cladosporium fulvum*) that grow between cell walls and through intercellular spaces. Soluble nutrients such as sugars and amino acids are scavenged from the apoplast or released from cell walls through the action of secreted hydrolytic enzymes (see p. 125). Hence, there is no intimate contact with living host protoplasts. Often, host cells are killed in advance of invasion, through the action of enzymes or toxins. With such necrotrophic pathogens some kind of structurally defined interface is short-lived as host cells rapidly disintegrate. However, not all intercellular pathogens are so destructive; fungi such as the apple scab pathogen, *Venturia inaequalis*, grow for an extended period beneath the cuticle of infected leaves or fruits without causing apparent tissue damage (see Fig. 2.2).

Intracellular relationships typically involve a more permanent contact between the partners, and penetrated host cells may remain viable for an extended period of time. In these cases the host–pathogen interface is a living and dynamic zone, often involving the formation of modified membranes or specialized parasitic structures such as haustoria.

Structure and function of haustoria

Many biotrophic fungi form modified hyphae,

Table 6.2 Modes of growth of parasites within host tissues, and interfaces, with examples shown in Fig. 6.9.

Type	Pathogen	Host
Subcuticular	*Rhynchosporium*	Barley
	Venturia	Apple
Intercellular	*Cladosporium fulvum*	Tomato
	Sclerotinia	Bean
	Monilinia	Pear
	Most bacteria	Various
Vascular	*Fusarium*	Various
	Verticillium	Various
	Ophiostoma	Elm
	Some bacteria, phytoplasmas	
Haustorial		
Epiphytic with haustoria	Powdery mildews	Various
Intercellular with haustoria	Rust fungi	Various
	Peronospora	Cruciferae
Intracellular vesicle, with intercellular hyphae and haustoria	*Bremia*	Lettuce
	Phytophthora	Potato
Intracellular		
Vesicle and intracellular hyphae	*Colletotrichum*	Bean
	Pyrenophora	Wheat
Wholly intracellular	*Plasmodiophora*	Cruciferae
	Polymyxa	Cereals, beet
	Viruses	Various

known as haustoria, which enter host cells. Haustoria typically develop from intercellular hyphae as narrow branches which penetrate through the plant cell wall and then expand inside the cell (Fig. 6.7b,c). They are diverse in morphology, ranging from small, club-shaped extensions to much larger, lobed or branched structures (see Fig. 3.2). Other fungi, such as hemibiotrophic species of *Colletotrichum*, form intracellular vesicles (Fig. 6.7a.) and hyphae within initially penetrated cells; these structures have some similarities with haustoria, as the host–pathogen interface is a fungal cell in intimate contact with the host protoplast.

Although haustoria and equivalent structures are formed within plant cells, the host plasma membrane is not penetrated and remains intact as an invagination surrounding the fungal cell (Fig. 6.7b). The inter-face between host and pathogen is therefore a complex zone comprising the fungal plasma membrane, the fungal cell wall, and an extrahaustorial membrane, or EHM (Fig. 6.8). In addition, there is often an amorphous matrix, probably secreted by the host, between the EHM and the fungal cell wall. Typically, where the haustorial neck breaches the host cell wall, a collar of callose-like material is deposited (Fig. 6.7c). In an incompatible host–pathogen combination this may extend to form a sheath completely encasing the haustorium. Finally, in the majority of haustoria, a discrete, electron-dense ring is visible in the fungal cell wall in the neck region (Figs 6.7b & 6.8). This is not observed in haustoria formed by oomycete pathogens such as the downy mildews and *Albugo* (Fig. 6.7c).

The structure and location of haustoria, which

(a)

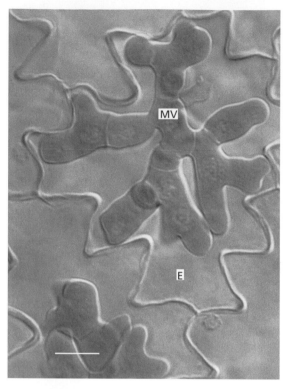

(b)

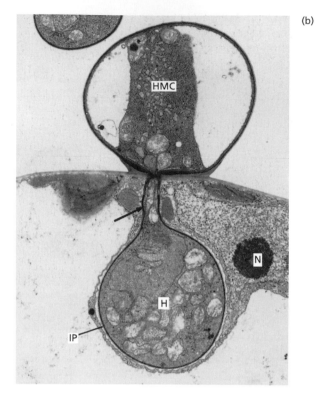

(c)

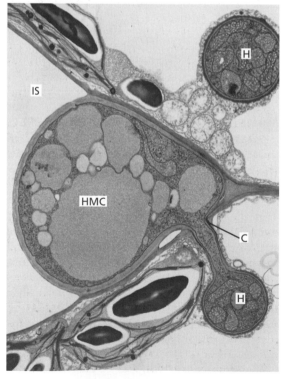

Fig. 6.7 Intracellular structures formed by biotrophic fungi. (a) Multilobed vesicle (MV) of *Colletotrichum destructivum* inside an epidermal cell (E) of the host plant, alfalfa. Scale bar = 10 μm. (From Latunde-Dada *et al*. 1997.) (b) Haustorium of flax rust, *Melampsora*. Note invaginated plasma membrane (IP) and host cell nucleus (N) adjacent to haustorium. A dark neck-ring (arrow) is also visible (×8000). (c) White blister rust, *Albugo,* in mesophyll tissues of cabbage, showing haustorial mother cell (HMC) in intercellular space (IS) and spherical haustorium (H) adjacent to chloroplast with dark starch grains. A collar (C) surrounds the penetration site. A second haustorium is visible in the adjacent host cell (×9600). (From Coffey 1975, 1976.)

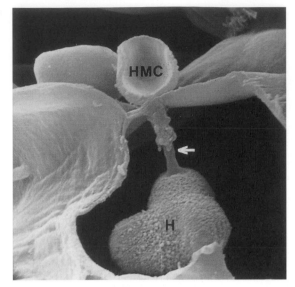

(a)

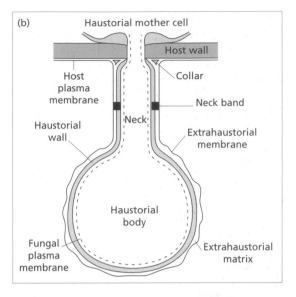

(b)

Fig. 6.8 The structure of haustoria. (a) Scanning electron micrograph of coffee leaf tissue infected by rust, *Hemileia vastatrix*. The tissue has been frozen and fractured to reveal a haustorium (H) within a mesophyll cell. Note a slight swelling (arrow) in the haustorial neck at the position of the neck band, and the haustorial mother cell (HMC) external to the penetrated host cell (×5000). (Courtesy of Rosemarie Honneger.) (b) Diagrammatic interpretation of haustorial structure, showing the main interfacial components.

provide an enlarged surface area of the parasite directly adjacent to nutrient sources such as chloroplasts (Fig. 6.7c), suggests that they play a role in nutrient uptake. Obtaining direct physiological evidence to confirm this idea has proved difficult. To date, most of the work on haustorial function has been conducted with powdery mildew fungi, as these epiphytic parasites form haustoria only in epidermal cells, and are therefore a convenient experimental system for analysis. Most of the fungal biomass can be stripped off the leaf and separated from the host tissues. If plants infected by powdery mildew are fed radiolabelled carbon as $^{14}CO_2$ a proportion of the carbon fixed in photosynthesis travels to the epiphytic hyphae and spores of the fungus. No significant uptake of radiolabelled solute occurs until after formation of the first haustoria. It has also proved possible to isolate intact powdery mildew haustoria from epidermal cells; such structures comprise the haustorial neck and body with the EHM still attached. Experiments with labelled sugars and amino acids have shown that solutes cross the EHM, and that the epidermal cell cytoplasm plays an essential role in transporting assimilates into haustoria. The main compound initially moving from the host to

the fungus is sucrose. Thus, with powdery mildew fungi, the pathway of carbon flow is from the source (chloroplasts in mesophyll cells) to the sink (epidermal cells which lack chloroplasts), and then into a secondary sink, the fungus, via haustoria in epidermal cells. The plant sugars are eventually converted into fungal metabolites such as mannitol and glycogen.

What is the actual mechanism by which solutes are removed from host cells? Electron micrographs of stained or freeze-fractured haustoria suggest that the invaginated region of the host membrane, the EHM, is altered in structure and composition by comparison with the rest of the host plasma membrane. In particular, the EHM lacks intramembrane particles, and ATPase, an enzyme involved in the active transport of solutes. ATPase activity can be detected in the host membrane where it lines the plant cell wall, and also in the fungal plasma membrane inside the haustorium, but not in the EHM. It appears therefore that both the host cell protoplast and the fungal protoplast are actively importing solutes, while the membrane enclosing the haustorium has diminished control of solute transport, and leaks nutrients into the extrahaustorial matrix, from where they are scavenged by the fungus. One further feature of this model is the electron-dense band of impermeable material (Figs 6.7b & 6.8) where the EHM contacts the haustorial neck. This is presumed to prevent solutes diffusing along the haustorial cell wall in the neck region. The extrahaustorial matrix and the haustorial wall are therefore a sealed compartment, and any solutes leaking across the EHM can only enter the fungus via active transport through the haustorial plasma membrane. In biotrophs lacking a haustorial neck band, such as the downy mildews, the plasma membrane of the penetrated cell still appears to comprise two functional domains, so there may be parallel processes of nutrient acquisition.

These experiments confirm the importance of haustoria in nutrient uptake by at least one group of biotrophic fungi. Calculations suggest, however, that with many biotrophs a significant proportion of nutrients can be also acquired from the host apoplast via intercellular hyphae. A detailed three-dimensional analysis of colonies of brown rust (*Puccinia recondita*) on barley estimated that the total length of intercellular hyphae in a colony is approximately 1 m. Haustoria occur at a frequency of one every 70 μm, giving a total number of more than 10 000 per colony. However, haustoria accounted for less than 20% of the total colony surface area, and the major area of contact between host and pathogen was therefore between intercellular hyphae and host cell walls. There may of course be other possible functions for haustoria, such as a regulatory role in manipulating host metabolism, or in the maintenance of compatibility between the two partners (see Chapter 10).

Further insights into haustorial function are now being gained by studies using monoclonal antibodies raised against isolated haustorial complexes, or the infection vesicles of hemibiotrophs such as *Colletotrichum* (Fig. 6.7a). Such antibodies have detected specific antigens which appear to be located only at the host–pathogen interface. This confirms that novel proteins or glycoproteins are produced at specific stages in the development of biotrophic fungi within their hosts. Identification of such stage-specific antigens, and the genes encoding them, should ultimately provide insights into the nature and regulation of biotrophic parasitism by fungi.

Some pathogens appear to grow preferentially at sites in the plant where nutrient transfer is occurring. A good example of this type of relationship is the

ergot fungus, *Claviceps purpurea*, which colonizes ovary tissues and diverts nutrients passing from transfer cells to the developing embryo. The pathogen competes with the embryo and ultimately replaces it with a fungal structure, the sclerotium. It has also been noted that many biotrophic fungi, such as rusts, invade host tissues adjacent to vascular elements, where loading or unloading of sugars into or from phloem cells is occurring. The term **transfer-intercept** infection has been used to describe such behaviour.

Intracellular pathogens

Intracellular relationships are typical of some mutualistic associations, for example the root nodule bacterium, *Rhizobium*, and some types of mycorrhizal fungi. In this context it is interesting to recall the theory that the chloroplasts and mitochondria of eukaryotic cells may have arisen from endosymbiotic microorganisms. A few pathogenic fungi also live inside host cells. The club root pathogen *Plasmodiophora* exists in the form of a naked cell, or **plasmodium**, and the interface consists simply of the plasmodial cell membrane surrounded by a second membrane which is presumed to originate from the host. An even more intimate contact is found in the parasitic chytrids such as *Olpidium*, where the fungal cell is not surrounded by a host membrane and is therefore in direct contact with the host cytoplasm. A similar type of relationship has been found in cells infected by phytoplasmas.

The ultimate examples of intracellular pathogens are the viruses and viroids. Virus particles occur within the cytoplasm, plastids and nuclei. Hybridization techniques used to locate and visualize specific nucleic acid sequences have recently shown that viroids accumulate in the nucleolus. Due to their unique properties viruses and viroids are not comparable to cellular pathogens regarding the nature of the host–pathogen interface. Successful replication of viruses requires removal of the coat protein, so that the interface during multiplication is between a nucleic acid molecule and the synthetic machinery of the host cell.

Development following infection

Following entry there are wide variations in the extent and pattern of colonization of host tissues (Fig.

6.9). Further development is related both to the nature of the parasitic relationship between the two partners, and to the relative success of host resistance mechanisms in limiting pathogen invasion (see p. 140). Broadly speaking, two main patterns of colonization occur, **localized** or **systemic**. In localized infection the pathogen multiplies or grows within a particular tissue or organ to give discrete lesions. In a systemic infection the pathogen spreads widely throughout the plant, and in extreme cases occurs in every part of the root and shoot system. Complete systemic colonization, in which literally every cell is infected, probably never occurs, as even viruses which spread efficiently from cell to cell are usually absent from meristems and from gametophyte tissue (see below).

Within these two broad categories there are numerous subtle variations in pathogen behaviour, and in the ultimate extent of damage to the host. Many pathogens exhibit **tissue specificity**; in other words they grow preferentially in certain host tissues. Vascular pathogens, for instance, including both bacterial and fungal examples, grow within the xylem, while phytoplasmas are usually confined to the phloem. The reasons for such behaviour are poorly understood, although the mode of nutrition of the pathogen is clearly important. The less specialized

necrotrophic pathogens tend to spread indiscriminately through plant organs, while biotrophs, in keeping with their more benign form of parasitism, grow selectively within certain well-defined host tissues. A further special case is where a pathogen induces major changes in the organization and morphology of host tissues, for instance tumours, in which it subsequently lives; the classic example of this mode of colonization is provided by *Agrobacterium* (see Fig. 8.11).

One might assume that there is a correlation between the extent of host colonization by a pathogen, and the eventual severity of disease, but this is by no means always the case. For instance, a localized pathogen may disrupt an essential physiological function, such as water transport, or produce a diffusible toxin which can act at a distance from the lesion itself. In an extreme example of this type of behaviour, the pathogen causing choke disease of grasses, *Epichloe typhina*, is restricted to a short section of leaf sheath tissue but its growth in this strategic position prevents the emergence of the flowering axis and hence its effect on the life cycle of the host is dramatic. Conversely, it is quite common to encounter systemic virus or viroid infections in which the host is asymptomatic. Such cryptic infections pose a particular problem when attempting to eradicate a pathogen from a crop.

The pattern of colonization can be determined by the infection route. Downy mildew fungi, such as *Peronospora* and *Plasmopara* species, typically infect leaf tissues, where growth is restricted by large veins, resulting in angular, localized lesions. Infection of the stem apex of young seedlings, however, leads to a more systemic mode of growth below the dividing meristem, causing severe stunting of the host. Some modern pea cultivars are more susceptible to this type of infection by *Peronospora pisi*, simply because they lack large stipules — leaf-like structures which normally enclose and protect the stem apex.

Only a few fungi, notably the smuts, are truly systemic. *Ustilago nuda*, causing loose smut of barley, occurs as a dormant mycelium in the embryo of infected grain. During germination of the seed the pathogen also resumes activity and grows intercellularly within the young seedling. As the plant matures the pathogen keeps pace just behind the apical meristem, and eventually invades the developing flower head to form a mass of black teliospores which replace the grain. The older

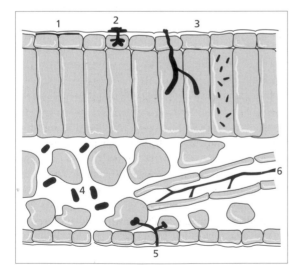

Fig. 6.9 Some patterns of pathogenic invasion of plant tissues. 1, Subcuticular; 2, epiphytic with haustoria; 3, intracellular; 4, intercellular; 5, intercellular with haustoria; and 6, vascular.

mycelium in the stem may break down as the host matures, but it often persists in the nodes. One interesting feature of smut diseases is that visible symptoms are not obviously manifest until the pathogen sporulates. Infected plants may, however, be slightly taller than normal.

Amongst those fungi which are specific to particular host tissues, the vascular wilt pathogens have a particularly interesting mode of spread within the host. Fungi such as *Fusarium oxysporum* and *Verticillium albo-atrum* enter in the apical region of the root. In this region the endodermis is not fully differentiated and the fungi are able to grow through it and reach the developing protoxylem. Further colonization of the living host is restricted to the xylem (see Fig. 7.9). Hyphae grow through the vessels and tracheids and pass from cell to cell via pit pairs. In addition, long-distance movement is accomplished by the production of microconidia, which are carried in the transpiration stream. This mode of spread is much more rapid than would be possible by mycelial growth; the Panama disease wilt pathogen, *Fusarium oxysporum* f.sp. *cubense*, can migrate from the bottom to the top of an 8 m-tall banana plant in less than two weeks. Because xylem tissues ramify throughout the plant, wilt pathogens can migrate into every part of their host. Thus although they are tissue specific these pathogens can become virtually systemic.

Colonization by bacteria

The morphology of bacterial cells limits their capacity for widespread growth through compact tissues. Thus, spread within the host is often accomplished by maceration of tissues or by the exploitation of natural channels.

For example, the bacterium *Erwinia carotovora* causes a common storage soft rot of potato tubers (see Fig. 8.2). The pathogen is unable to pass through intact periderm and therefore gains entry via lenticels or wounds, including those caused by other agents (see p. 99). Once inside the tuber, the bacterium spreads rapidly through the parenchyma giving rise to a soft, slimy, putrid lesion, and under favourable conditions it can quickly destroy the tuber. At all stages of colonization the bacterium occupies intercellular spaces, and its spread is facilitated by the production of pectolytic enzymes which degrade middle lamellae and thus macerate the host tissues. However, under aerobic storage conditions at low relative humidities, the rate of invasion is slower and the bacterium may be localized by a black oxidation zone. This is partly an anatomical barrier, involving the deposition of suberin in tuber cell walls, and partly a chemical barrier, consisting of oxidized polyphenols. Similar histological defence reactions are important in restricting the spread of many other relatively unspecialized pathogens within storage tissues.

Not all plant-pathogenic bacteria macerate tissues. *Agrobacterium tumefaciens* grows biotrophically within crown gall tumours without separating or killing cells, while the bean halo blight pathogen, *Pseudomonas syringae* pv. *phaseolicola*, proliferates in intercellular spaces, causing a water-soaked lesion in which host cells initially remain alive. Chlorotic symptoms in this case are associated with the production of a toxin (see p. 133), but toxin-deficient strains of the bacterium are equally capable of multiplying in bean tissues. Fireblight, *Erwinia amylovora*, invades flowers, leaves and stem tissues, spreading through vascular elements causing a necrotic die-back. The bacterium produces only low amounts of cell-wall-degrading enzymes, and host cell death appears to be due instead to a toxin.

A number of bacteria can cause serious vascular wilt diseases, in which the symptoms parallel those caused by fungal pathogens (see Fig. 7.9). Vascular wilt bacteria are classified in the genera *Clavibacter*, *Erwinia*, *Pseudomonas*, *Ralstonia* and *Xanthomonas*. They are important in both tropical and warm-temperate regions, and include *Erwinia tracheiphila*, which is responsible for bacterial wilt of wild and cultivated species of the Cucurbitaceae, and *Ralstonia solanacearum*, which attacks a number of plants including banana (Moko disease), tobacco, tomato and potato. These bacteria enter their hosts in diverse ways, such as through wounds created by vectors or cultivation practices, through damaged tissues caused by the emergence of lateral roots or via hydathodes and stomata. Their unicellular morphology then makes them ideally suited for transport within xylem vessels. In some instances, the bacteria then spread rapidly into adjoining parenchyma tissue. In this respect several of these bacteria differ from the typical vascular wilt fungi, which do not grow out from the xylem until after the host has died.

Colonization by viruses

The spread of viruses within their hosts is unique in that they can only multiply within cells and are small enough to behave as subcellular particles. As such they are able to move directly from cell to cell through plasmodesmata; virus particles have been observed within plasmodesmata by electron microscopy. This short-distance cell-to-cell spread is fairly slow, with the virus taking four or five hours to move from one cell to the next. There is now evidence that the movement of viruses through plasmodesmata is mediated by 'movement proteins' encoded by the virus itself, for instance the P30 protein of tobacco mosaic virus (TMV) (see p. 37). Antibodies specific for the movement protein show that it becomes localized to plasmodesmata; this has been recently confirmed by experiments in which a green fluorescent marker protein, derived originally from jellyfish, was linked to the virus protein, and the subcellular distribution of fluorescence studied by microscopy. Highly localized bright-green sites were seen associated with host cell walls, coinciding with pit fields where plasmodesmata perforate the wall. The exact mechanism involved is not yet known, although these proteins may modify plasmodesmata to alter the exclusion size and hence allow free passage of the virus.

Much faster long-distance spread takes place via the phloem; here the rate of movement has been estimated as high as several centimetres per hour. Phloem transport plays an important part in the development of systemic virus infections, although how the transported form of the virus leaves the sieve tubes is a mystery. Literally every cell in the plant may become infected, although the small numbers of infected seeds and pollen grains in most virus diseases suggests that movement into gametophyte tissue of the developing embryo is restricted. Often, meristematic tissues are also virus free; this fact has been put to good use in the production of virus-free plants by meristem culture.

Further reading

Books

Cole, G.T & Hoch, H.C. (eds) (1991) *The Fungal Spore and Disease Initiation in Plants and Animals*. Plenum Press, New York.

Mendgen, K. & Lesemann, D.E. (eds) (1991) *Electron Microscopy of Plant Pathogens*. Springer-Verlag, Berlin.

Morris, C.E., Nicot, P.C. & Nguyen-The, C. (eds) (1996) *Aerial Plant Surface Microbiology*. Plenum Press, New York.

Nicole, M. & Gianninazzi-Pearson, V. (eds) (1996) *Histology, Ultrastructure and Molecular Cytology of Plant–Microorganism Interactions*. Kluwer Academic Publishers, Dordrecht.

Reviews and papers

Carver, T.L.W. & Thomas, B.J. (1990) Normal germling development by *Erysiphe graminis* on cereal leaves freed of epicuticular wax. *Plant Pathology* **39**, 367–375.

Carver, T.L.W., Thomas, B.J. & Ingerson-Morris, S.M. (1994) The surface of *Erysiphe graminis* and the production of extracellular material at the fungus–host interface during germling and colony development. *Canadian Journal of Botany* **73**, 272–287.

Chasan, R. (1992) Cutinase — Not a weapon in fungal combat? *The Plant Cell* **4**, 617–618.

Deom, C.M., Lapidot, M. & Beachy, R.N. (1992) Plant virus movement proteins. *Cell* **69**, 221–224.

Dickman, M.B., Podila, G.K. & Kolattukudy, P.E. (1989) Insertion of cutinase gene into a wound pathogen enables it to infect intact host. *Nature* **342**, 446–448.

Gay, J.L. (1984) Mechanisms of biotrophy in fungal pathogens. In: *Plant Diseases. Infection, Damage and Loss* (eds R.K.S. Wood & G.J. Jellis), pp. 49–59. Blackwell Scientific Publications, Oxford.

Green, J.R., Pain, N.A., Cannell, M.E. *et al.* (1994) Analysis of differentiation and development of the specialized infection structures formed by biotrophic fungal plant pathogens using monoclonal antibodies. *Canadian Journal of Botany* **73**, S408–S417.

Hamer, J.E., Howard, R.J., Chumley, F.G. & Valent, B. (1988) A mechanism for surface attachment in spores of a plant pathogenic fungus. *Science* **239**, 288–290.

Howard, R.J. & Valent, B. (1996) Breaking and entering: Cell biology of host penetration by the fungal rice blast pathogen *Magnaporthe grisea*. *Annual Review of Microbiology* **50**, 491–512.

Howard, R.J., Ferrari, M.A., Roach, D.H. & Money, N.P. (1991) Penetration of hard substrates by a fungus employing enormous turgor pressures. *Proceedings of the National Academy of Sciences of the USA* **88**, 11281–11284.

Huang, J.-S. (1986) Ultrastructure of bacterial penetration in plants. *Annual Review of Phytopathology* **24**, 141–157.

Kolattukudy, P.E., Podila, G.K. & Mohan, R. (1989) Molecular basis of the early events in plant–fungus interaction. *Genome* **31**, 342–349.

Lee, Y.-H. & Dean, R.A. (1993) cAMP regulates infection structure formation in the plant pathogenic fungus *Magnaporthe grisea*. *The Plant Cell* **5**, 693–700.

Mendgen, K. & Deising, H. (1993) Infection structures of fungal plant pathogens—a cytological and physiological evaluation. *New Phytologist* **124**, 193–213.

Mims, C.W. & Richardson, E.A. (1988) Ultrastructure of appressorium development by basidiospore germlings of the rust fungus *Gymnosporangium juniperi-virginianae*. *Protoplasma* **148**, 111–119.

Money, N.P. & Howard, R.J. (1996) Confirmation of a link between fungal pigmentation, turgor pressure and pathogenicity using a new method of turgor measurement. *Fungal Genetics and Biology* **20**, 217–227.

Oparka, K.J., Boevink, P. & Santa Cruz, S. (1996) Studying the movement of plant viruses using green fluorescent protein. *Trends in Plant Sciences* **1**, 412–418.

Padgett, H.S., Epel, B.L., Kahn, T.W., Heinlein, M., Watanabe, Y. & Beachy, R.N. (1996) Distribution of tobamovirus movement protein in infected cells and implications for cell-to-cell spread of infection. *The Plant Journal* **10**, 1079–1088.

Read, N.D., Kellock, L.J., Knight, H. & Trewavas, A.J.

(1992) Contact sensing during infection by fungal pathogens. In: *Society for Experimental Biology Seminar Series 48: Perspectives in Plant Cell Recognition* (eds J.A. Callow & J.R. Green), pp. 137–172. Cambridge University Press, Cambridge.

Roberts, A.M., Mackie, A.J., Hathaway, V., Callow, J.A. & Green, J.R. (1993) Molecular differentiation in the extrahaustorial membrane of pea powdery mildew haustoria at early and late stages of development. *Physiological and Molecular Plant Pathology* **43**, 147–160.

Saeed, I.A.M., MacGuidwin, A.E. & Rouse, D.I. (1997) Synergism of *Pratylenchus penetrans* and *Verticillium dahliae* manifested by reduced gas exchange in potato. *Phytopathology* **87**, 435–439.

Spencer-Phillips, P.T.N. (1997) Function of fungal haustoria in epiphytic and endophytic infections. *Advances in Botanical Research* **24**, 309–333.

Stahl, D.J. & Schäfer, W. (1992) Cutinase is not required for fungal pathogenicity on pea. *The Plant Cell* **4**, 621–629.

Storey, G.W. & Evans, K. (1987) Interactions between *Globodera pallida* juveniles, *Verticillium dahliae* and three potato cultivars, with descriptions of associated histopathologies. *Plant Pathology* **36**, 192–200.

7   The Physiology of Plant Disease

'It is of the first importance to understand that disease is a condition of abnormal physiology, and that the boundary lines between health and ill health are vague and difficult to define.' [H. Marshall Ward, 1854–1906]

The invasion of the host by a foreign organism leads, sooner or later, to changes in host physiology. If the organism is a pathogen, these changes will eventually prove deleterious to the host. Alternatively, where the pathogen fails to establish itself, these changes may be important in preventing the pathogen from gaining a foothold. In practice it is often difficult to distinguish between post-infectional changes which are linked with resistance processes, and those which are related to pathogenesis, i.e. disease development. Plants infected by quite different types of pathogens often exhibit very similar physiological symptoms. These similarities at first sight suggest that plants employ a common pathway of response to infection. However, many of the physiological effects also result if plants are subject to other forms of stress, such as mechanical or chemical injury. Some of the gross changes in the physiology of diseased plants represent a non-specific response to cellular damage inflicted by physical, chemical or microbial agents. An analysis of post-infectional changes in host physiology may, nevertheless, clarify the ways in which pathogens cause disease, and help to elucidate the mechanisms by which plants resist pathogenic attack.

Post-infectional changes in host physiology

Respiration

As parasitism involves a nutritional relationship much of the work on the physiology of diseased plants has been concerned with energy metabolism. One of the most prominent changes which occurs following infection is a substantial increase in respiration rate. This is equally true for diseases involving fungi, bacteria and viruses, although most of the available information has been obtained from plants infected by biotrophic fungi. Figure 7.1 shows the rate of oxygen uptake in barley leaves infected by two different fungi: one, *Pyrenophora teres,* is a nectrotroph; the other, *Erysiphe graminis*, is a biotroph. In both instances oxygen consumption is increased, but with the nectrotroph the respiration rate peaks earlier and then declines as the host tissues are progressively destroyed. With the biotroph the rate increases steadily as the pathogen sporulates and remains high until the host tissues senesce. This pattern inevitably raises the question as to whether the increase is simply due to the additional respiration of the pathogen itself, rather than to a genuine host response. This, like many other apparently simple questions concerning host–pathogen physiology, is not easy to answer. Several lines of evidence suggest that while the pathogen makes some contribution to the increase, the greater part of it cannot be explained on this basis.

Powdery mildew fungi are only in intimate contact with their host where the haustoria enter epidermal cells. It is possible to peel off the epiphytic mycelium of the pathogen and measure the respiration rate of the host leaf with only an insignificant portion of the fungus (namely, the haustoria) remaining. Experiments like this have shown that the increase in respiration is maintained even after removal of the pathogen. There are alternative ways of approaching this problem; for instance, one can measure the respiratory rate of uninfected tissues adjacent to lesions or, in the case of necrotrophic pathogens, one can examine the effects of toxic factors produced in culture on the respiratory metabolism of host cells.

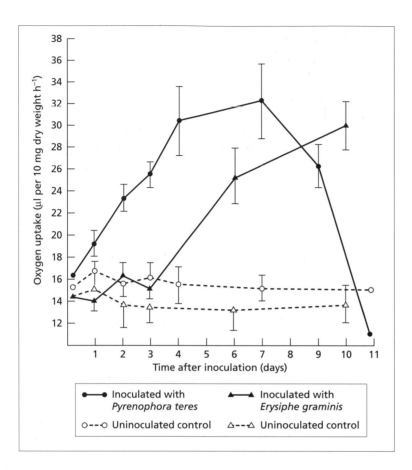

Fig. 7.1 Time course of respiration in two susceptible barley cultivars, Wing and Sultan, inoculated respectively with the net blotch pathogen *Pyrenophora teres* (a necrotroph) and the powdery mildew pathogen *Erysiphe graminis* f.sp. *hordei* (a biotroph). Vertical bars represent standard errors. (After Smedegaard-Petersen 1984.)

Both approaches have demonstrated enhanced rates of host respiration.

Perhaps the most convincing argument in support of the idea that increased post-infectional respiration is due to a stimulation of host metabolism comes, however, from studies on virus diseases. Viruses, being non-cellular, possess no respiratory apparatus of their own, and yet a similar stimulation of respiration rate is found in a variety of viral infections. For example, the development of necrotic local lesions in *Nicotiana glutinosa* inoculated with tobacco mosaic virus (TMV) is accompanied by a pronounced increase in respiration. In systemically infected hosts, an increase may accompany symptom development, but the change is less marked.

All in all, the rise in respiration rates following infection would seem mainly to represent a response by host tissues. This response bears similarities to the transitory increase in respiration observed in plants subjected to mechanical injury.

The mechanism of respiratory increase

While measurements of the gross respiratory rate (in terms of either oxygen uptake or carbon dioxide evolution) indicate that the physiology of the host is altered by infection, this information is, in isolation, of limited value. It does not tell us anything about the mechanism of the increase, or how the pathogen stimulates the host. Unfortunately, most of our present knowledge of disease physiology has not progressed much beyond this sort of 'tip of the iceberg' observation. There are, however, a number of theories which seek to explain the enhanced respiration rate in diseased plants (Fig. 7.2).

In healthy cells, respiration is regulated by a number of factors, the most important of which is the availability of adenosine diphosphate (ADP). Agents such as 2,4-dinitrophenol (DNP) stimulate the respiration rate by 'uncoupling' electron transfer from oxidative phosphorylation. In essence, this means

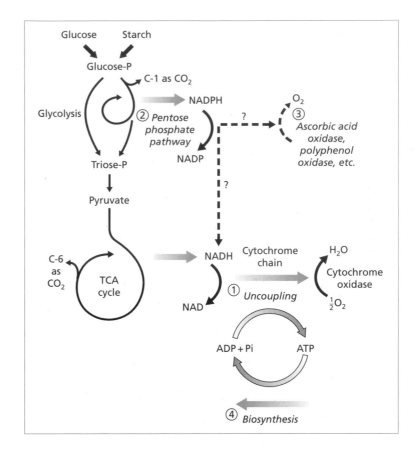

Fig. 7.2 Theories concerning stimulation of respiration in infected plants. ADP, adenosine diphosphate; ATP, adenosine triphosphate; NAD, nicotinamide adenine dinucleotide; NADH, reduced NAD; NADP, nicotinamide adenine dinucleotide diphosphate; NADPH, reduced NADP; Pi, inorganic phosphate; TCA, tricarboxylic acid. See text for further details of ①–④.

that while electron flow continues, the regeneration of adenosine triphosphate (ATP) from ADP and inorganic phosphate no longer takes place. Due to the continued consumption of ATP in cellular metabolism, the pool of ADP is replenished and the usual feedback mechanism based on the availability of ADP no longer operates.

It has been suggested that pathogens may uncouple host respiration (① in Fig. 7.2). This hypothesis is based on evidence that respiration in diseased tissues is no longer stimulated by treatment with DPN. In addition, the level of ATP is often lower in infected tissues. Presumably some compound, for instance a toxin, produced by the pathogen acts as an uncoupler in host cells. While there is some limited experimental support for this hypothesis, other explanations have now gained ground.

The major metabolic pathway for the degradation of glucose to pyruvate is glycolysis, otherwise known as the Embden–Meyerhof pathway. An alternative route for the production of pyruvate, the pentose

phosphate pathway, is generally considered to be of less importance although it does provide intermediates for the biosynthesis of many vital cellular materials, including nucleic acids. Assessment of the relative contribution of each pathway relies largely on data from labelling studies in which radioactive carbon is incorporated into either the C-6 or the C-1 position of the glucose molecule. Subsequent measurement of the ratio of labelled carbon dioxide released from each source during respiration indicates which pathway is predominant; activation of the pentose phosphate pathway leads to an increased contribution from the C-1 position, and hence lowers the C-6 : C-1 ratio. In infected plants, the C-6 : C-1 ratio is typically lower than the values obtained from healthy tissues, suggesting that there is increased participation of the pentose route (② in Fig. 7.2). Assays of several pentose phosphate pathway enzymes support this idea, their activity being higher in diseased tissues. It should be borne in mind, however, that most of the available data have been obtained from host–pathogen systems involving fungi. The pentose phosphate pathway operates at a higher level in fungi than in green plants, and the increased contribution in infected tissues may simply reflect this feature of fungal metabolism.

Alterations in the pathway of glucose degradation would not necessarily result in a rise in gross respiration rate, although the pentose phosphate pathway is marginally less efficient in generating ATP. It now seems likely that the real significance of the switch to the pentose pathway is linked to its role in the biosynthesis of various compounds. As well as providing pentoses for the biosynthesis of nucleic acids, pentose phosphate intermediates are involved in the production of numerous aromatic compounds, notably phenols and their derivatives. Many of these compounds are associated with host defence reactions, a topic which will be discussed further in Chapter 9.

In addition to changes in respiratory pathways, alternative terminal oxidation systems may operate in diseased tissues (③ in Fig. 7.2). Apart from the usual cytochrome system terminating with cytochrome oxidase, systems involving phenol oxidases and ascorbic acid oxidase have been detected in plants. Both of these enzymes appear to be activated in diseased tissues. The precise mechanisms and significance of such oxidation systems are not clear, but phenol oxidases play a part in the production of phenolic compounds, an observation consistent with the general pattern of post-infectional metabolism. It also seems likely that the reduced coenzyme NADPH generated by the pentose pathway is oxidized by one of these enzymes, rather than by the cytochrome system. As far as is known, these alternative oxidases are unable to participate in the formation of ATP during oxygen uptake.

While evidence exists in support of each of the above theories regarding respiratory changes in the infected host, the most satisfactory explanation for the increased rate of metabolism in infected plants is perhaps the most obvious. With biotrophic pathogens the increased respiration is associated with enhanced synthetic, rather than degradative, metabolism; in other words, there is a general increase in the biosynthetic activities of the host (④ in Fig. 7.2). This increase in turn requires more rapid utilization of ATP and thereby removes the restraints imposed by the availability of ADP. It is significant that in many diseases caused by fungi the major increase in

respiration coincides with the onset of sporulation by the pathogen. This is precisely the time when the fungus will be exerting the maximum drain on host nutrients due to the considerable energy requirement for the production of spores or other propagules. The biosynthesis of defence compounds by the host will also consume energy in the form of ATP.

In conclusion, an increase in host respiration, which in turn implies a general stimulation of host metabolism, is one of the most prominent physiological consequences of infection by pathogens. The basis for this stimulation presumably resides in the increased activity of host enzymes, either through activation or the depression of host genes. The molecular basis of these changes, and the pathogen-produced factors which may induce them, will be considered in Chapters 8 and 9.

Photosynthesis

Photosynthesis is the most distinctive physiological activity of green plants. The capture of solar energy by chlorophyll and its subsequent utilization to fix carbon dioxide into organic compounds is the basis of life on this planet. However, in spite of its fundamental importance, comparatively little is known about the effects of pathogens on photosynthesis.

Any pathogen which attacks green aerial tissues is likely to affect crop yield. In many cases the harmful effects of a pathogen can be directly attributed to the destruction of photosynthetic tissues. A serious outbreak of potato blight can completely defoliate an entire field, while *Botrytis fabae* can cause necrotic patches which occupy over 50% of the leaf area of broad beans. It is therefore obvious that one major result of pathogen invasion is a reduction in the photosynthetic capacity of a plant through the destruction of green tissue, and hence reduced interception of solar radiation.

However, this simple conclusion ignores the possibility that there are effects in adjacent uninfected tissues, or changes in the photosynthetic process itself. For instance, with the example above, *Botrytis* on beans, measurement of the relative growth rate of diseased plants has shown that it is similar to the growth rate of healthy plants, even when 30–40% of the leaf area is removed. The implication here is that the photosynthetic efficiency of the remaining leaf tissues is enhanced to compensate for the loss in area. However, there is eventually an effect on yield, due to a reduction in the number of pods formed per plant.

Chlorosis is one of the most common symptoms of plant disease, and is indicative of a reduction in the chlorophyll content of green tissues. A reduced chlorophyll content could be due to the breakdown of chlorophyll, inhibition of chlorophyll synthesis, or a reduction in the number of chloroplasts. In chlorosis associated with some virus infections, higher levels of the enzyme chlorophyllase have been detected, suggesting that chlorophyll is being degraded by the enzymatic reaction:

$$\text{Chlorophyll} \xrightarrow{\text{Chlorophyllase}} \text{Chlorophyllide} + \text{Phytol}$$

Symptoms of virus infection in leaves often include characteristic mosaics of green and yellow areas in which chloroplasts are reduced in number, or show ultrastructural abnormalities such as swelling and fewer lamellae, which are the sites of photochemical reactions. The chlorotic areas are often rich in starch, suggesting that virus infection affects both photosynthetic capacity and carbon partitioning within diseased leaves.

In leaves infected by biotrophic fungi there is a progressive loss in overall photosynthetic activity, although this is usually only noticeable in the later stages of infection, when premature senescence of the leaf may set in. In diseases caused by rust fungi, the reduction in photosynthesis cannot be accounted for by loss of photosynthetic area, but may be correlated with a decreased chlorophyll content. Electron microscopic studies have shown that chloroplast ultrastructure is altered in the later stages of infection. Plastid membranes break down and the overall changes are similar to those seen in senescent cells. Figure 7.3 shows the net photosynthetic rate of oak leaves infected by the powdery mildew pathogen *Microsphaera alphitoides*. There is a slight initial stimulation of photosynthesis in inoculated leaves, but this is followed by a gradual decline. The rate of $^{14}CO_2$ uptake by leaf discs from sugarbeet infected by powdery mildew is also reduced compared with healthy tissues (Fig. 7.4). Chloroplasts isolated from mildewed beet leaves show a reduced capacity to form ATP by non-cyclic photophosphorylation. More recent work on powdery mildew-infected barley leaves has shown that there is also a progressive reduction in the activity of several key enzymes of the Calvin cycle. This down-regulation may be linked

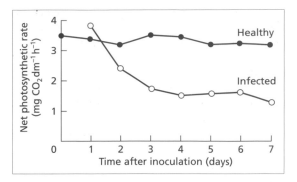

Fig. 7.3 Changes in photosynthesis of oak leaves following infection by the powdery mildew fungus *Microsphaera alphitoides*. (Data from Hewitt & Ayres 1975.)

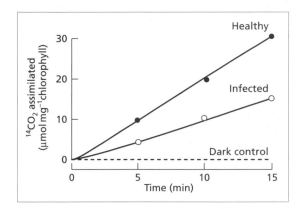

Fig. 7.4 Effect of *Erysiphe polygoni* on the rate of photosynthetic $^{14}CO_2$ assimilation by sugarbeet leaf discs. (Data from Magyarosy *et al.* 1976.)

to changes in the concentration of soluble carbohydrates in infected tissues (see below), which in turn affects the rate of photosynthetic CO_2 fixation. In *Arabidopsis* leaves infected by the biotrophic pathogen *Albugo candida*, reductions in the rate of photosynthesis are paralleled by decreases in the amounts of Rubisco (ribulose biphosphate carboxylase) protein present in host tissues.

Although the general pattern of photosynthesis in diseased plants seems to involve a reduction in activity, there are exceptions. In uninfected bean leaves on plants infected by *Uromyces phaseoli*, fixation of $^{14}CO_2$ is actually stimulated compared with rust-free controls. A similar phenomenon has been reported in uninfected leaves of pea plants inoculated with powdery mildew. These reports are of particular interest as they show that post-infectional changes in photosynthesis may occur at a distance from the infection site.

The most important exception to the usual sequence of chlorosis and reduced photosynthetic activity is also seen in diseases caused by the rust and powdery mildew fungi. This is the so-called 'green island' effect, where tissues in the vicinity of fungal pustules are green even though the surrounding areas of the leaf are chlorotic (Fig. 7.5). There has been much discussion of the significance of these green islands because the selective retention of chlorophyll around infection sites suggests that the pathogen exerts some degree of control over host physiology. The similarities between this delay in senescence and the effects of hormonal factors, such as cytokinins, has prompted the view that the pathogen secretes hormonally active compounds. In diseases caused by powdery mildew fungi, evidence suggests that chlorophyll is initially degraded but subsequently resynthesized in areas of the leaf adjacent to disease lesions. Green islands are probably associated with the redirection of host nutrients which occurs in diseases caused by biotrophic parasites. Necrotrophic pathogens are generally less subtle in their effects and rapidly break down host organelles, including chloroplasts.

Translocation of nutrients and water

The damage caused by biotrophic pathogens is due, to some extent at least, to their ability to redirect host nutrients for their own use. The idea that fungal colonies act as 'metabolic sinks' in their hosts is supported by radioisotope tracer experiments in which labelled carbon accumulates preferentially in disease lesions (Fig. 7.6). In this way, photosynthate originally destined for developing host tissues, such as new shoots or roots, is instead utilized by the pathogen. The reduced root growth and grain yield of cereals infected by rusts and powdery mildews is due to this disturbance in the nutrient balance of the plant. As is shown in Fig. 7.6, a large part of the imbalance seems to be caused by the retention of sugars and amino acids in the diseased older leaves. This has been confirmed by feeding a 'pulse' of $^{14}CO_2$ to rust-infected leaves and then comparing the amount of carbon remaining with that retained by healthy leaves

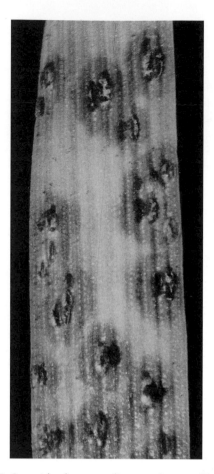

Fig. 7.5 Green islands surrounding pustules caused by brown rust, *Puccinia recondita*, on wheat leaf.

(Fig. 7.7). Efflux of fixed carbon from the rusted leaf is substantially reduced. Tracer experiments have also shown that carbon originally present in host sugars, such as sucrose, can be subsequently detected in typical fungal metabolites like the sugar alcohols, mannitol and sorbitol. The conversion of host sugars to fungal metabolites could in fact serve to maintain a concentration gradient and ensure a continued flow from host to fungus.

The enzyme invertase is believed to play a central role in carbohydrate metabolism through effects on the relative levels of hexose sugars, such as glucose, and the principal translocated sugar, sucrose. Plant tissues infected by biotrophic fungi show substantial increases in invertase activity (Fig. 7.8). Increased hydrolysis of sucrose may have effects

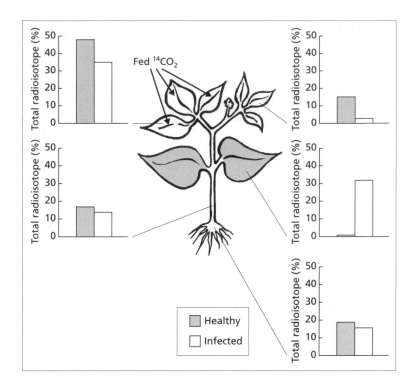

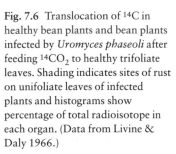

Fig. 7.6 Translocation of ^{14}C in healthy bean plants and bean plants infected by *Uromyces phaseoli* after feeding $^{14}CO_2$ to healthy trifoliate leaves. Shading indicates sites of rust on unifoliate leaves of infected plants and histograms show percentage of total radioisotope in each organ. (Data from Livine & Daly 1966.)

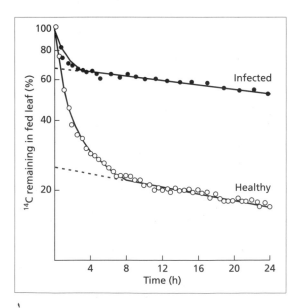

Fig. 7.7 Kinetics of efflux of ^{14}C from healthy and rusted first leaves of barley, following a pulse of labelled CO_2. (After Owera *et al.* 1983.)

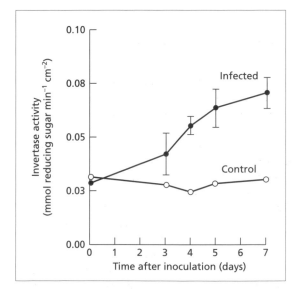

Fig. 7.8 Invertase activity in barley leaves infected by powdery mildew, *Erysiphe graminis*. (After Scholes *et al.* 1994.)

both on photosynthesis and carbon partitioning in diseased leaves. The accumulation of glucose and fructose is one possible explanation for the down-regulation of the photosynthetic Calvin cycle noted above, due to end-product inhibition. It may also affect the availability of soluble sugars for the fungus. One suggestion is that the pathogen can compete more effectively with the plant for hexose sugars, rather than sucrose, as the uptake and translocation of carbon in phloem tissues involves a sucrose transport system. Changes in invertase activity might therefore have major effects on the pattern of carbohydrate translocation in diseased plants.

The multiplication of viruses in plant cells is entirely at the expense of the host. Sequestration of host metabolites to make more virus particles may explain some of the deleterious effects of virus infection. In severe infections the virus particles themselves come to represent about 10% of the dry weight of leaf tissues, and multiplication on this scale must severely tax the synthetic capacities of the host. Virus replication requires the synthesis of two components, nucleic acid and coat protein, and one major side effect of infection may therefore be a deficiency of inorganic nutrients, in particular phosphorus and nitrogen, available to the host. There is evidence in virus-infected plants for a shift in photosynthetic products from sugars to amino acids, perhaps reflecting the biosynthetic demand for coat protein.

The nutrient stress imposed by a redirection of host nutrients to satisfy the energy and biosynthetic needs of the pathogen is very different from that caused by pathogens which actually colonize the transporting tissues of the plant. Vascular wilt pathogens impair the flow of water and mineral salts through the xylem. Translocation of sugars through phloem tissues is also disrupted by some pathogens; a number of virus infections cause necrosis of phloem elements, resulting in nutrient imbalances in the host. For instance, potato plants infected by leaf roll virus have higher than normal carbohydrate levels in their leaves, while that of the tubers is reduced. Two possible explanations for this symptom are either the inhibition of translocation of sugars, or the breakdown of the phloem tissue. High concentrations of phytoplasmas often build up in the phloem elements of diseased plants but in this case the symptoms, such as wilting and stunting, are thought to be due to

production of a diffusible toxin rather than to occlusion of the sieve tubes.

The interdependence of physiological processes such as ion uptake and the translocation of water and nutrients should be emphasized. In take-all disease of cereals, invasion of the root cortex by the pathogen does not significantly affect ion uptake or translocation. Instead the crucial stage appears to be the subsequent colonization of phloem tissues by the fungus. This reduces the translocation of nutrients to the apical meristems, with the result that the distal portions of the root cease to function and ion uptake is impaired.

The wilt syndrome

The most pronounced effect of vascular wilt pathogens is on the water economy of the host. In tomatoes infected by *Fusarium oxysporum* f.sp. *lycopersici* the resistance to water flow through the xylem is substantially increased compared to the resistance of uninfected stems. This effect can be partially explained on the basis of physical obstruction of the vessels by hyphae, but the vascular wilt syndrome is complex and involves host responses to infection as well as direct effects of the pathogen and its products (Fig. 7.9). Blockages caused by the growth of the pathogen are compounded by its secretion of polysaccharides and pectolytic enzymes; in turn the host responds by producing gums and mucilages and by forming tyloses in the vessels. The end result is that water flow may be reduced to less than 5% of that in healthy plants. A further factor interfering with xylem function may be gas bubbles, or embolisms, breaking the water column; this has been observed in sapwood colonized by the Dutch elm pathogen, *Ophiostoma novo-ulmi*. As well as causing severe water stress, infection by wilt pathogens also reduces the passage of essential mineral ions to the leaves. The overall consequences of vascular blockage are, however, difficult to assess, as these pathogens also secrete toxins which have physiological effects throughout the plant.

Many plant pathogenic bacteria can enter the vascular system through wounds, and spread and multiply in xylem vessels. Proliferation of bacterial cells may occlude vessels, while secretion of extracellular slime or other high-molecular-weight materials further reduces flow by plugging pit membranes.

(a)

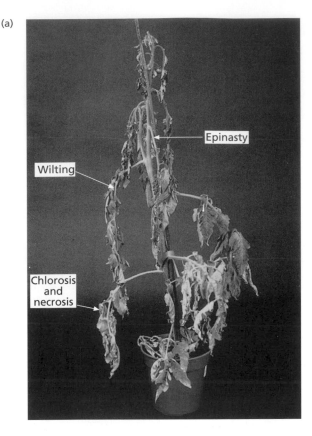

(b)

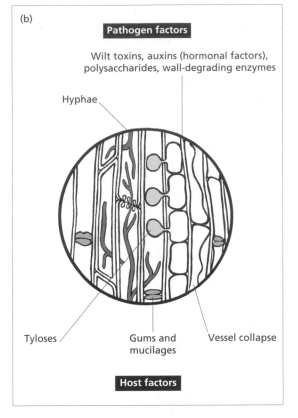

Fig. 7.9 (a) Wilted tomato plant infected by *Verticillium albo-atrum*. (b) Diagrammatic longitudinal section of vascular tissue from an infected stem.

Transpiration

Wilting is one of the most common disease symptoms in plants, but the physiological basis of the symptom is not the same in all cases. The water economy can be disrupted through reduced water uptake, reduced flow rate or increased water loss through transpiration. Root rot pathogens destroy root tissues, and therefore reduce the surface area available for uptake and disrupt the transport of water through the root system. The vascular wilt fungi reduce flow and at the same time reduce the transpiration rate. This second effect can be explained on the basis of water stress in the leaves, coupled with stomatal closure. Many other pathogens increase the transpiration rate (Fig. 7.10). This effect is predictable inasmuch as any pathogen which damages the surface layers of the leaf

will increase cuticular transpiration and this may, in fact, be the major source of the increased rate of water loss. Damage of this sort is often restricted to the reproductive phase of the pathogen's life cycle; the rust fungi form pustules which tear through the host epidermis prior to release of the spores (Fig. 7.5 & Plate 4, facing p. 12). At this point wilting often occurs for the first time. The physical damage inflicted on the host during sporulation is a good example of the indirect and harmful effects that biotrophic fungi have on the plant.

It is often difficult, however, to assess the basic reasons for increased transpiration in diseased plants. The data shown in Fig. 7.10 are from a relatively early stage of barley infection by *Rhynchosporium* when the host cuticle is still intact. A complicating factor is the stomatal behaviour of diseased leaves. In this example, a higher proportion of stomata remain open in the dark in infected plants, which no doubt contributes to the increased level of water loss. Abnormal opening of stomata also occurs in potato

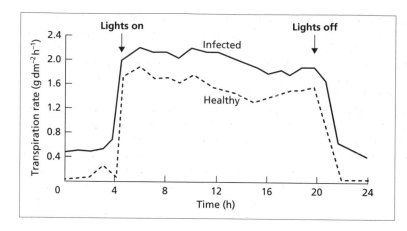

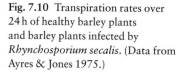

Fig. 7.10 Transpiration rates over 24 h of healthy barley plants and barley plants infected by *Rhynchosporium secalis*. (Data from Ayres & Jones 1975.)

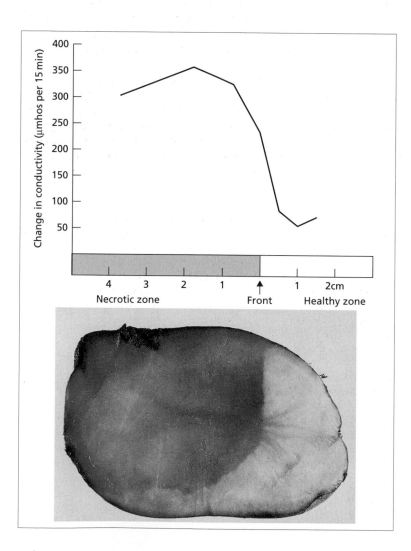

Fig. 7.11 Effect of *Phytophthora erythroseptica* on electrolyte leakage from potato tuber tissues. Arrow indicates advancing edge of necrotic lesion seen in the photograph below of a tuber showing typical pink rot symptoms.

leaves infected by *Phytophthora infestans* (Plate 1, facing p. 12). Here the effect is confined to a zone surrounding the necrotic lesions. Stomata within this zone open abnormally wide and remain open in the dark; they therefore do not present an obstacle to the developing fungal sporangiophores, which characteristically emerge through the stomata at night. The physiological consequences of this alteration in stomatal behaviour are of particular interest. Fixation of $^{14}CO_2$ is enhanced in the same zone, and a causal relationship between increased photosynthesis and higher rates of gas exchange through the open stomata has been suggested.

The few studies which have been made on the water relations of virus-infected plants indicate that the transpiration rate is reduced, and the total water content of the host is lower. This is especially true in severely diseased plants.

Cell water relations

In addition to effects on the water economy of the whole plant, pathogens may also influence the water relations of individual cells. It is generally accepted that cell water relations are controlled by the plasma membrane, although in plant cells the rigid wall is also important in maintaining turgor. Due to its semi-permeable properties the plasma membrane is of central importance in regulating the passage of ions and organic molecules into and out of the cell. This membrane is a complex and dynamic structure which maintains a suitable intracellular environment for metabolism.

It has been known for some time that one of the most common effects of pathogens on plant cells is to increase their permeability. The membrane apparently loses its semipermeable properties, and mineral ions and other electrolytes leak out into the external medium. This effect is pronounced in soft rot diseases caused by necrotrophic pathogens, in which host cell necrosis is a major feature (Fig. 7.11). This is predictable as a loss of cellular compartmentalization is likely to take place in moribund cells. The actual cause of membrane damage is, however, less obvious; possible culprits include toxins and hydrolytic enzymes such as pectinases and phospholipase (see Chapter 8).

It has been shown that biotrophic pathogens also increase the permeability of host tissues. Cells penetrated by haustoria can still be plasmolysed, so the integrity of the host membrane is maintained. Nevertheless there is a definite increase in leakage of electrolytes from diseased tissues, and this change in the semipermeable properties of the plasma membrane is

Table 7.1 Some pathogens which cause abnormal growth in the host plant.

Pathogen	Host	Disease
Fungi		
Plasmodiophora brassicae	Crucifers	Club root
Synchytrium endobioticum	Potato	Wart disease
Gibberella fujikuroi	Rice	Bakanae (foolish seedling)
Crinipellis perniciosa	Cocoa	Witch's broom
Taphrina deformans	Peach	Leaf curl
Ustilago maydis	Maize	Smut
Bacteria/phytoplasmas		
Agrobacterium tumefaciens	Various	Crown gall
Pseudomonas syringae pv. savastanoi	Olive	Knot disease
Phytoplasma	Peanut	Witch's broom
Viruses		
Cocoa swollen shoot virus (CSSV)	Cocoa	Swollen shoot
Potato leaf roll virus (PLRV)	Potato	Leaf roll
Beet necrotic yellow vein virus (BNYVV)	Sugarbeet	Rhizomania

(a)

(b)

Fig. 7.12 Growth abnormalities induced by plant pathogens. (a) Cocoa plants infected with cocoa swollen shoot virus. Nodal swellings (A) are associated with apical die-back. Leaf symptoms include vein clearing (C), banding (B) and fern pattern (D). (Courtesy of J.T. Legg.) (b) Gall symptoms caused by the corn smut pathogen *Ustilago maydis* on a maize cob. (Courtesy of ICI Plant Protection.)

related to the uptake of nutrients by the pathogen. Recent models of haustorial function suggest that the extrahaustorial membrane loses control of nutrient transport so that sugars pass freely to the fungus (see p. 103). Leakage of electrolytes has also been recorded in virus-infected tissues, especially roots, but this does not seem to be a common symptom in these diseases.

The importance of endomembrane systems in the regulation of cell metabolism has only recently been fully appreciated. It is not surprising therefore that the effects of pathogens on cell membranes is a topic of major interest to plant pathologists. In addition to the plasma membrane itself, the cell also contains membrane-bound organelles, such as plastids, mitochondria, peroxisomes and lysosomes, and alterations in one or several of these types of organelle may be involved in many aspects of disease physiology.

Growth regulation

All plant pathogens affect the growth and development of their hosts to a greater or lesser extent. The diversion of nutrients or the destruction of host tissues will inevitably lead to reduced performance, and in some cases may severely stunt the plant. These effects on plant growth are, however, essentially indirect and therefore different from the specific growth abnormalities induced by a variety of pathogens (Table 7.1).

Symptoms such as galls and tumours, excessive branching, leaf epinasty, abnormal induction of adventitious roots and premature leaf abscission are all associated with changes in the control of plant growth and differentiation. Such deranged growth is characteristic of many diseases involving microbial pathogens (Fig. 7.12).

Although plant morphogenesis is influenced by environmental conditions, control of the basic processes of cell division and differentiation is mediated by hormonal compounds such as indoleacetic acid, gibberellins, cytokinins, abscisic acid and ethylene. Changes in the concentration or distribution of these hormones have widespread effects on the physiology of plants. Because alterations in growth regulation can often be attributed to the production of hormonally active compounds by the pathogen, discussion of diseases involving growth abnormalities will be deferred until the next chapter.

Conclusion

It should be apparent from this brief review that many of the measurable changes in the physiology of infected plants are common to a variety of diseases. In the search for a unifying concept in host–pathogen interaction it is tempting to interpret these similarities as evidence for a common pathway or sequence of biochemical events following infection. However, many of these gross alterations, for example increased respiration and permeability changes, are also characteristic of plants damaged by non-microbial agents. In view of this, the explanation for pathogenesis and host resistance may be more to do with changes unique to diseased plants, rather than to these general responses to stress conditions. This explanation must ultimately be sought in terms of molecular interactions occurring in the host–pathogen complex.

Further reading

Books

Ayres, P.G. (ed.) (1992) *Pests and Pathogens: Plant Responses to Foliar Attack*. Bios Scientific Publishers, Oxford.

Goodman, R.N., Király, Z. & Wood, K.R. (eds) (1986) *The Biochemistry and Physiology of Plant Disease*. University of Missouri Press, Columbia, Mo.

Mace, M.E., Bell, A.A. & Beckman, C.H. (eds) (1981) *Fungal Wilt Diseases of Plants*. Academic Press, New York.

Misaghi, I.J. (1982) *Physiology and Biochemistry of Plant–Pathogen Interactions*. Plenum Press, New York.

Reviews and papers

Ayres, P.G., Press, M.C. & Spencer-Phillips, P.T.N. (1996) Effects of pathogens and parasitic plants on source-sink relationships. In: *Photoassimilate Distribution in Plants and Crops* (eds E. Zamski & A.A. Schaffer), pp. 479–499. Marcel Dekker, New York.

Farrar, J.F. (1984) Effects of pathogens on plant transport systems. In: *Plant Diseases: Infection Damage and Loss* (eds R.K.S. Wood & G.J. Jellis), pp. 87–104. Blackwell Scientific Publications, Oxford.

Farrar, J.F. (1985) Carbohydrate metabolism in biotrophic plant pathogens. *Microbiological Sciences* 2, 314–317.

Farrar, J.F. (1992) Beyond photosynthesis: the translocation and respiration of diseased leaves. In: *Pests and Pathogens: Plant Responses to Foliar Attack* (ed. P.G. Ayres), pp. 107–127. Bios Scientific Publishers, Oxford.

Kneale, J. & Farrar, J.F. (1985) The localization and frequency of haustoria in colonies of brown rust on barley leaves. *New Phytologist* **101**, 495–505.

McGrath, M.T. & Packer, S.P. (1990) Alteration of physiological processes in wheat flag leaves covered by stem rust and leaf rust. *Phytopathology* **80**, 677–686.

Scholes, J.D. & Rolfe, S.A. (1996) Photosynthesis in localized regions of oat leaves infected with crown rust (*Puccinia coronata*)—Quantitative imaging of chlorophyll fluorescence. *Planta* **199**, 573–582.

Scholes, J.D., Lee, P.J., Horton, P. & Lewis, D.H. (1994) Invertase—understanding changes in the photosynthetic and carbohydrate-metabolism of barley leaves infected with powdery mildew. *New Phytologist* **126**, 213–222.

Smedegaard-Petersen, V. (1984) The role of respiration and energy generation in diseased and disease-resistant plants. In: *Plant Diseases: Infection Damage and Loss* (eds R.K.S. Wood & G.J. Jellis), pp. 73–85. Blackwell Scientific Publications, Oxford.

Técsi, L.I., Maule, A.J., Smith, A.M. & Leegood, R.C. (1994) Complex, localized changes in CO_2 assimilation and starch content associated with the susceptible interaction between cucumber mosaic virus and a cucurbit host. *The Plant Journal* **5**, 837–847.

8   Microbial Pathogenicity

'We have little exact knowledge of the chemico-physiological processes in the life of the parasitic fungi because the symbiotic relation puts great complications and difficulties in the way of their precise investigation.' [Anton De Bary, 1831–1888]

The majority of microorganisms, if inoculated into a plant, fail to grow or to cause disease. Pathogenicity is the exception rather than the rule. The successful pathogen must possess special properties enabling growth and multiplication in the host to the extent that disease develops.

Until quite recently, identifying the factors determining microbial pathogenicity relied upon a combination of microscopy and painstaking biochemical analyses of microbial cultures or extracts from infected plants. Observation of pathogens on or in plants, especially by electron microscopy, provided clues as to how host defences are breached, or tissues colonized and damaged. Evidence obtained in this way is useful but usually inconclusive; for instance, changes in the ultrastructure of cell walls or membranes might suggest the action of enzymes or toxins, but the factor(s) responsible remain unidentified. Alternatively, the pathogen may be grown in culture, and the filtrate analysed for the presence of substances which reproduce disease symptoms when introduced into the host. Obviously this approach is limited to culturable microorganisms. Quite apart from the difficulty of purifying molecules of interest from complex mixtures, there is the additional problem of proving that a factor produced *in vitro* is also produced in the host plant. There are many examples of biologically active compounds produced by pathogens in culture which have never been detected in infected plants, and which probably play little or no part in pathogenesis. Even greater problems are encountered when trying to identify key factors in the host–pathogen complex itself. The crucial events may be localized to only a few cells, and the molecules of interest may be unstable or present in low concentrations. Improvements in the sensitivity and accuracy of analytical methods, for instance by using antibodies to detect molecules *in situ*, have extended the scope of such studies, but many problems remain.

A molecular genetic approach

The development of molecular techniques for isolating DNA from cells, cutting it into smaller pieces with restriction enzymes, and cloning these pieces in a suitable vector (usually a bacterial plasmid or a bacteriophage) has provided a new way of identifying the factors responsible for microbial pathogenicity. The basic idea in this approach is straightforward (Fig. 8.1). Starting with a wild-type pathogenic strain of the microorganism one generates mutants which have lost the ability to infect the host. These non-pathogenic mutants are identified by inoculation tests onto a normally susceptible plant. At the same time a **genomic library** of DNA from the pathogenic wild type is prepared; this consists of pieces of the pathogen genome isolated and cloned in a phage or bacterium to produce multiple copies of each piece. Each piece or **clone** is now transferred individually back into the non-pathogenic mutant, and the transformed strains are tested on plants. Provided the library contains clones representing the whole genome of the pathogen, sooner or later a DNA sequence complementing the mutation, and thereby restoring pathogenicity, will be introduced. This clone should contain a gene, or genes, encoding a product essential for pathogenicity. Once the specific piece of DNA is found, the gene can be sequenced and the gene product identified.

The virtue of this approach is that specific functions can be inactivated or restored, thereby provid-

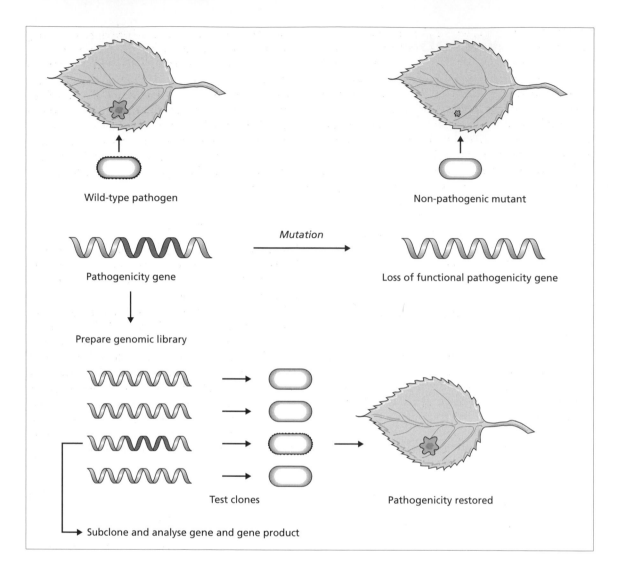

Fig. 8.1 Experimental procedure for identifying genes encoding pathogenicity functions.

ing conclusive evidence that a particular factor is important for infection to take place. There are, however, some potential drawbacks. The first is that mutations affecting certain functions not directly responsible for pathogenicity, for instance those influencing growth or nutrition of a microorganism, are also likely to interfere with the infection process. Careful controls are required to eliminate this possibility. Secondly, random or 'shotgun' cloning of DNA from the pathogen can be a hit-and-miss process; not

all of the genome may be represented in the library, and testing each of the random clones for ability to restore pathogenicity is a laborious process. One solution to this problem is to generate mutations in such a way that the site of the inactivated gene can be readily identified. The most common strategy is to use **transposons**, DNA sequences that insert at a random site in the genome; if such a sequence inserts into a gene essential for pathogenicity, then the gene is disrupted, and pathogenicity is reduced or lost. Transposons usually carry a recognizable genetic marker, such as resistance to an antibiotic, and can therefore be easily detected. The DNA fragment containing the inactivated pathogenicity gene can be cloned by selecting for the resistance marker encoded by the transposon. The gene of interest is 'tagged' by the transposon which disrupts it. This and related methods of insertional mutagenesis are now the preferred approach for identifying genes determining pathogenicity or host specificity (see p. 174).

Molecular genetic techniques have already proved to be powerful tools for analysing pathogenicity, especially in bacteria, which have relatively small, haploid genomes. With viruses and viroids, which have even smaller genomes, it is possible to manipulate part or all of the nucleic acid, and to define gene functions by inserting mutations at strategic points in the genome. This approach was used, for instance, to identify the role of the tobacco mosaic virus (TMV) P30 protein in virus movement between cells (see p. 37).

Enzymes and microbial pathogenicity

Free-living bacteria and fungi produce a wide variety of enzymes which are secreted into the external environment, and play an important role in the utilization of nutrient substrates. Many plant-pathogenic species also produce extracellular enzymes (Table 8.1). Whether at least some of these are determinants of pathogenicity has been debated for many years; with the application of molecular techniques, such as

those described above, this debate is now entering a new phase. The vital role played by enzymes in the penetration of intact host surfaces has already been described (see p. 95). During colonization, enzymes may also modify host cell walls and aid the spread of the pathogen through tissues. Enzymes that kill host cells may be important in suppressing defence responses which depend upon active metabolism. Alternatively, some microbial enzymes can detoxify inhibitors produced by the host plant (see p. 143).

Cell-wall-degrading enzymes

One group of plant pathogens in which enzymes have been extensively studied are the soft rot bacteria and fungi which typically attack storage tissues and cause spreading, necrotic lesions.

More than 100 years ago, Anton de Bary demonstrated that extracts of such rotted tissue can macerate firm, healthy tissues. Further work in the early part of this century by William Brown at Imperial College showed that the maceration of host tissues is due to the action of cell-separating enzymes produced by the pathogen. These enzymes degrade the pectic substances in the middle lamella between cells, thereby facilitating the colonization of host tissues. The ability to produce pectic enzymes is widely dis-

Table 8.1 Some extracellular enzymes produced by plant pathogens.

Enzyme	Substrate attacked	Comments
Cutinase	Cutin	Important in host penetration?
Pectic enzymes		
Pectin esterase	Pectin	Cell-separating activities in soft
Polygalacturonase	Pectate	rot diseases
Pectate lyase	Pectate	
Hemicellulases		
Xylanase	Xylan	Predominant polymers in
Arabanase	Araban	monocot cell walls
Cellulase	Cellulose	May be detected in later stages of infection
Ligninase	Lignin	Produced by timber decay fungi
Phospholipase	Phospholipids	Breaks down membranes
Protease	Protein	Significance uncertain
β-Glucosidase	Glucosides	Degrades plant inhibitors

tributed amongst fungi and bacteria and is characteristic of necrotrophic pathogens such as *Botrytis*, *Sclerotinia* and *Erwinia* (Fig. 8.2).

Brown recognized that the soft rot syndrome involved two processes: host cells within the lesion were separated through the action of enzymes, and then the cells died. Brown proposed two alternative theories to explain these processes.

1 Host cells are separated and killed by the same substance; i.e. the macerating factor and the lethal factor are identical.

2 Host cells are separated by a macerating factor, and subsequently killed by a different lethal factor, for instance a toxin of some kind.

In his experiments Brown was unable to separate macerating activity from lethal activity, a result which favoured the first hypothesis.

Resolving this question has proved difficult, because soft rot pathogens produce not one but a whole series of wall-modifying enzymes, each with a different mode of action. For instance, pectin methyl esterase removes the methyl groups of pectin to yield pectic acid, which is then more susceptible to chain-splitting enzymes such as pectate lyase and polygalacturonase, which cleave the polymer into fragments. The complete digestion of pectic polymers is thus a multienzyme process. Synthesis of pectic enzymes may be induced or repressed by catabolites in a highly specific manner, so that some are only produced in the presence of a suitable carbon source, such as pectin or other constituents of the cell wall. As well as pectolytic activities, there are enzymes which attack other wall polymers, such as cellulases, hemicellulases and ligninases (Table 8.1). The latter are particularly important in diseases involving wood decay, such as heart rots of trees. In combination these enzymes can degrade all the polymers present in higher plant cell walls.

To return to the question of the lethal factor, it is now known that certain pectic enzymes, when purified, can in isolation kill host cells as well as macerate tissues. An example is shown in Fig. 8.3, where the activity of a single enzyme from *Erwinia*, isolated by electrophoresis, is shown to correspond with lethal activity towards potato cells. The way in which the enzyme actually kills host cells is, however, still debated. Evidence favours the idea that cell walls in plant tissues treated with pectic enzymes lose their ability to support the plasma membrane, particularly under osmotic stress. Direct effects of reaction products of the enzyme on the cells do not appear to be responsible for cell death, although such products may be important as signal molecules triggering host defence responses (see p. 171).

The relative importance of different pectic enzymes for the virulence of soft rot pathogens has been intensively studied in the bacterium *Erwinia* which, as a close relative of *Escherichia coli*, is a convenient model for genetic manipulation. Genes encoding individual enzymes have been isolated from *Erwinia* and cloned in *E. coli,* thereby allowing analysis of their structure and regulation. Mutants deficient in particular enzymes can also be produced. Several interesting discoveries have emerged from this work.

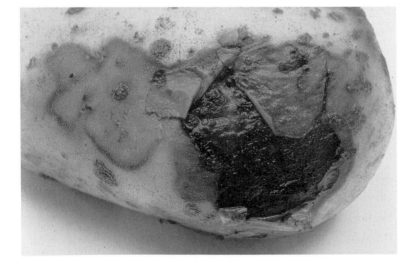

Fig. 8.2 Soft rot lesion in a potato tuber caused by the bacterium *Erwinia carotovora*.

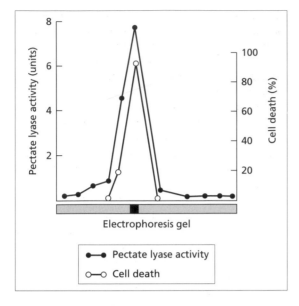

Fig. 8.3 Separation of a pectolytic enzyme (endopectate lyase) from *Erwinia* by acrylamide gel electrophoresis. The enzyme has migrated in the gel as a single band. Note that high enzyme activity coincides with greatest lethal activity towards potato cells. (Data from Basham & Bateman 1975.)

Firstly, some pectic enzymes are coded for by several genes; in *E. chrysanthemi,* for instance, there are at least five pectate lyase genes (designated *pelA* to *E*), each producing a different pectate lyase isoenzyme. The genes are grouped into two clusters (Fig. 8.4). Secondly, not all of the genes are essential for tissue

maceration and host colonization. Mutations in *pelE* have the greatest effect on pathogenicity (Fig. 8.4). Thirdly, non-pathogenic mutants have been identified which still produce the full repertoire of pectic enzymes, but fail to secrete them to the external environment. These Out⁻ bacteria have mutations in genes encoding proteins which are essential components of a secretion pathway exporting enzymes and other proteins across the outer membrane. Thus pathogenicity requires not only the synthesis of certain enzymes, but also effective export out of the cell. A similar requirement for enzyme export has been found in other plant-pathogenic bacteria such as *Xanthomonas campestris.*

While our understanding of soft rot pathogenesis has advanced considerably since de Bary's original observations, the degree of complexity revealed by these recent studies suggests that simple molecular models, based on single factors, cannot explain the whole disease process. Recent work on *Erwinia carotovora* (Fig. 8.2), for instance, shows that mutants deficient in production of another type of wall-degrading enzyme, cellulase, are reduced in virulence. Perhaps surprisingly, saprophytic fungi such as *Aspergillus* can produce levels of pectic enzymes equivalent to those produced by *Erwinia*, and if large numbers of spores are inoculated into wounds, can induce lesions in plant tissues. Under normal circumstances, however, such fungi are not plant pathogens. Clones of *E. coli* expressing *pel* genes introduced from *Erwinia* are able to macerate potato tissue, but only if abnormally high numbers of the bacterium are applied. The dividing line between parasitic and

Fig. 8.4 Diagrammatic version of the organization of pectate lyase (*pel*) genes in *Erwinia chrysanthemi,* showing two gene clusters producing different pectate lyase isoenzymes (PL), and the effect of mutations disabling each gene on the virulence of the pathogen.

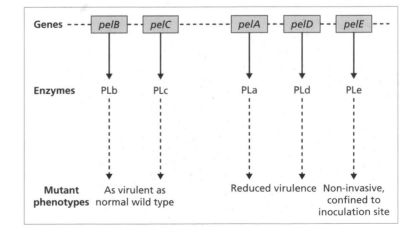

saprophytic microorganisms thus appears to be less clear than previously supposed, certainly in terms of biochemical capabilities. The main difference may turn out to be in the regulation of particular genes and processes, rather than in the presence or absence of such genes.

Cell-wall-degrading enzymes have been implicated in several other types of plant disease, such as vascular wilts, but for many pathogens their overall significance is less clear. Typically, biotrophic fungi do not disrupt host tissues, but nevertheless these fungi penetrate cell walls, for instance when haustoria are formed. The crucial difference between such biotrophs and tissue-destructive necrotrophs is not necessarily in the enzymes each produce, but in the amount produced and the way the enzyme is targeted. Biotrophs deploy their enzymes at localized sites, often as cell-bound activity, and thus avoid indiscriminate damage to tissues. The extent of wall degradation is precisely controlled. One especially interesting group in this context are the hemibiotrophs (see p. 22), which initially grow in the host without killing cells, but later cause necrotic lesions. Following infection, growth of the bean anthracnose fungus, *Colletotrichum lindemuthianum*, is virtually asymptomatic for several days, but a spreading, water-soaked lesion then develops. It is likely that some metabolic switch takes place at a particular stage in host colonization, with the induction and secretion of lytic enzymes which cause necrosis.

Enzymes degrading host cell walls are usually produced by pathogens in a specific sequence: first pectic enzymes, followed by hemicellulases, then cellulases. This is consistent with a stepwise digestion of the cell wall, pectic matrix polymers being removed first to expose the microfibrillar skeleton (Fig. 8.5). The relative contribution of each activity will depend upon the chemistry of the wall. Monocotyledons, for instance, have cell walls rich in hemicellulose, hence pathogens colonizing these hosts, such as the stem-base-infecting fungi *Pseudocercosporella (Tapesia)* spp., *Fusarium culmorum* and *Rhizoctonia cerealis* produce significant amounts of xylanase and other hemicellulases.

Ability to degrade lignin is rare, and restricted mainly to specialized parasites of woody hosts, such as the white rot fungi. The mode of action of ligninase is unusual, involving the oxidative production of highly reactive free radicals which break the diverse chemical bonds in the lignin molecule, a process

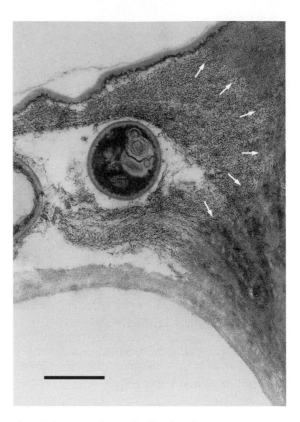

Fig. 8.5 Intramural growth of hypha of *Pseudocercosporella* showing degradation of wall adjacent to hypha and unmasking of wall polymers (arrows). Scale bar = 2 µm. (Courtesy of Alison Daniels.)

described as 'enzymic combustion'. The lack of ligninase activity in most bacteria and fungi explains why lignified cell walls are a highly effective barrier to microbial penetration (see p. 146).

Enzymes degrading host inhibitors

A further group of microbial enzymes which may play an important role in pathogenicity are those with activity towards plant antibiotics, such as preformed inhibitors, and phytoalexins produced in response to infection. For instance, β-glucosidases can cleave toxic glucosides present in plants, thereby inactivating such compounds and allowing the pathogen to establish itself in the host. Detailed discussion of these interactions is postponed until after the plant defence factors involved have been described (see p. 141).

Infection-specific proteins

An alternative approach to isolating mutants altered in pathogenicity is to try to identify genes which are expressed only during infection and colonization of the host plant. The logic here is simple. Microorganisms have tightly regulated metabolic processes and it is unlikely that they will be synthesizing products that are not required in their particular habitat. Hence genes switched on during growth in the plant may well encode products essential for pathogenesis. Finding such genes requires comparison of messenger RNA extracted from the pathogen, from the uninfected host plant, and from infected plant tissues. Differential screening of genomic libraries can then be carried out to detect sequences which are abundantly expressed only in the host–pathogen complex.

This approach has now successfully identified several fungal genes which are induced during growth in the host. These include a number of *ipi* (*in* *p*lanta *i*nduced) genes from the potato late blight pathogen *Phytophthora infestans*, and *MPG1* from the rice blast fungus *Magnaporthe grisea*. *MPG1* is of particular interest as it has a complex expression pattern, being switched on during the early stages of host penetration, and then expressed again later during symptom development. The gene encodes a small, secreted, hydrophobic polypeptide. Replacement of the *MPG1* gene with a defective copy unable to produce this hydrophobin protein reduces the pathogenicity of such mutants, and also affects spore production by the fungus. The effect on pathogenicity appears to be due to a malfunction in appressorium formation. It seems likely that the protein interacts with hydrophobic surfaces, such as the plant cuticle, and serves as a sensor inducing appressorial development. Hydrophobins are also produced by other fungi, including some plant pathogens, but these proteins are quite diverse, and probably fulfil several different roles in disease development. For instance the toxic protein cerato-ulmin, produced by the Dutch elm pathogen *Ophiostoma* (see below), is also a hydrophobin.

Toxins

The idea that pathogenesis might be due to the production of poisons by the pathogen is by no means new, as the potent bacterial toxins involved in human diseases such as diphtheria and tetanus were discovered more than 100 years ago. The importance of toxins in plant pathology has taken longer to establish, but there are now many examples of bacterial and fungal products known to play at least some role in symptom development in plants. In theory the term **toxin** can refer to any pathogen product which is harmful to the host, including the hydrolytic enzymes described in the previous section. In practice it is usually restricted to low-molecular-weight compounds which do not attack the structural integrity of plant tissues but instead affect metabolism in some other, more subtle fashion. A useful working definition is 'a metabolite of pathogen origin which is involved in plant disease'. Two important properties of toxins are: (i) that they are active in very low concentrations; and (ii) that they are mobile within the plant and may therefore act at a distance from the site of infection.

Experimental proof that a toxin is involved in a particular disease is tricky for a number of reasons. Many pathogens growing *in vitro* secrete substances which, when introduced into the host plant, reproduce some or all of the symptoms associated with infection by that pathogen. This in itself is not conclusive evidence as there is no guarantee that the same compound is produced *in vivo*. To overcome this problem a set of rules, somewhat akin to Koch's postulates, have been proposed to confirm the involvement of a toxin in a disease syndrome. In essence these rules insist that the suspected toxin must be isolated from the diseased host, purified, and reintroduced with subsequent development of the same symptoms. Needless to say these rules are of limited use in practice because a toxin may only be present in tiny amounts or may be so unstable that it breaks down during extraction. Purification of a suspected toxin from culture, studies on its structure and mode of action, and detection of the same chemical in the diseased host, are alternative steps towards confirmation that a toxin is involved. The ultimate proof may depend on identifying the genes responsible for toxin production and then disrupting them to assess effects on the ability of the pathogen to cause disease.

Several schemes have been proposed for the classification of toxins produced by plant pathogens, based on biological activity rather than the chemical structure of the compounds. Two main groups are usually distinguished (Table 8.2):

Table 8.2 Some toxins involved in plant disease.

Toxin	Pathogen	Host
Host-specific (selective)		
Victorin (HV-toxin)	*Helminthosporium (Cochliobolus) victoriae*	Oats
T-toxin	*H. maydis (C. heterostrophus)*	Maize
HC-toxin	*H. (C.) carbonum*	Maize
HS-toxin	*H. (C.) sacchari*	Sugarcane
PC-toxin	*Periconia circinata*	Sorghum
Ptr-toxin	*Pyrenophora tritici-repentis*	Wheat
AK-toxin	*Alternaria kikuchiana*	Pear
Host non-specific (non-selective)		
Fusicoccin	*Fusicoccum amygdali*	Almond
Cercosporin	*Cercospora* spp.	Various
Tentoxin	*Alternaria* spp.	Various
Tabtoxin	*Pseudomonas syringae* pv. *tabaci*	Tobacco
	P.s. pv. *coronafaciens*	Cereals
Phaseolotoxin	*P.s.* pv. *phaseolicola*	Bean
Syringomycin	*P.s.* pv. *syringae*	Various
Coronatine	*P.s.* pv. *glycinea* and others	Soybean

• host-specific (selective) toxins, which only affect plants susceptible to the pathogen producing them;
• host non-specific (non-selective) toxins, which also affect other plants as well.
Pathogen compounds which act directly as plant growth regulators are often treated separately, although it should be noted that certain toxins may possess hormone-like properties.

Host-specific toxins

In 1946 oat crops in the USA were affected by a seedling blight caused by *Cochliobolus (Helminthosporium) victoriae*. The fungus was especially damaging on an oat cultivar known as Victoria. The pathogen itself was typically localized to the basal part of infected plants, but symptoms extended into the leaves, which often collapsed. Suspicions that a mobile toxin might be involved were confirmed when fungal culture filtrates were shown to cause the same symptoms, and an active compound was isolated and given the name victorin (HV-toxin). The most interesting property of this toxin, subsequently characterized as a peptide linked to a tricyclic amine (Fig. 8.7), was that oat cultivars resistant to the fungus were unaffected by it, while susceptible cultivars such as Victoria were sensitive to extremely low concentra-

tions (Fig. 8.6). Sensitivity to the toxin therefore showed the same specificity as the pathogen itself. Furthermore, there was a direct correlation between toxin production and pathogenicity; genetic analysis showed that a single gene in the fungus determines toxin production, and in crosses between toxin-producing and non-producing strains, progeny lacking the toxin gene were only weakly pathogenic. All this evidence points to victorin as the major determinant of the disease (Table 8.3).

Victorin has widespread physiological effects in the host. It stimulates respiration, the increase being proportional to the concentration of the toxin applied to the plant. Toxin-treated tissues leak electrolytes, and electron micrographs of treated cells reveal apparent damage to the plasma membrane. But the primary lesion caused by victorin remained unknown, until recently. Then, a protein which binds victorin was isolated from susceptible oat tissues, and work was started to determine whether this is the toxin receptor, and how it differs from a very similar protein found in resistant oat cultivars. The victorin-binding protein turned out to be a subunit of an enzyme complex (glycine decarboxylase) located in mitochondria. The toxin is a potent inhibitor of this enzyme in susceptible oats, but how inhibition of the enzyme leads to the lethal effects of the toxin is not yet clear.

Fig. 8.6 Effects of victorin on resistant (R) and susceptible (S) oat seedlings. Toxin was added to the nutrient solution where indicated, 3 days before the photograph was taken. (From Scheffer & Yoder 1972.)

Table 8.3 Evidence for involvement of a host-specific toxin in disease.

> Purified toxin reproduces all disease symptoms in host
> Toxin affects susceptible but not resistant host genotypes
> Tox+ strains of the pathogen are pathogenic
> Tox- strains are non-pathogenic
> In crosses between Tox+ and Tox- strains there is co-segregation among progeny for toxin production and pathogenicity

One further intriguing aspect of the victorin story is that susceptibility to the toxin is correlated with resistance to an unrelated pathogen, the rust fungus *Puccinia coronata*. The genes responsible for toxin sensitivity and rust resistance are therefore closely linked or even identical. The importance of this observation is that by locating the gene encoding the toxin receptor protein it should also be possible to find a gene conferring specific resistance to a pathogenic fungus, and thereby analyse its function in host–pathogen recognition (see Chapter 10).

Several other *Helminthosporium (Cochliobolus)* species have been shown to produce host-specific toxins, including *H. maydis,* the causal agent of Southern corn leaf blight, and *H. sacchari,* responsible for eyespot disease of sugarcane (Table 8.2). These toxins differ in chemical structure and apparent mode of action. Hmt-toxin, for instance, consists of several linear polyketols (Fig. 8.7), which interfere with mitochondrial function, while HS-toxin is a sesquiterpene linked to galactose units, for which a putative plasma membrane receptor has been proposed. Such diversity in chemical structure raises interesting questions about the evolution of specificity in these host–pathogen interactions.

Recently, the biochemical basis of resistance to a host-specific toxin has been defined. The fungus *Helminthosporium (Cochliobolus) carbonum* causes leaf spot and ear mould of maize; specific pathogenicity depends upon the production of a cyclic tetrapeptide compound called HC-toxin (Fig. 8.7). This compound has a distinctive mode of action: rather than killing host cells it inhibits an enzyme which modifies histones, the proteins associated with DNA in chromatin. The main effect might therefore be on gene expression. Resistant maize plants possess a dominant gene specifying an enzyme which inactivates the toxin by degrading it. Susceptible maize genotypes lack this enzyme, and are therefore unable to inactivate the toxin. This work is significant as it confirms the central role played by a host-specific toxin in disease, and also assigns a specific biochemical function to a plant resistance gene (see Chapter 10).

The discovery of host-specific toxins raised hopes that the biochemical basis of pathogenicity, and conversely host resistance, might turn out to be simple, determined by a single compound of major effect. Overall these hopes have not been fully realized, as relatively few of the toxic compounds investigated to date have proved to be host-specific. Instead, most toxins appear to be non-specific in effect, and to play a less prominent role in the disease process.

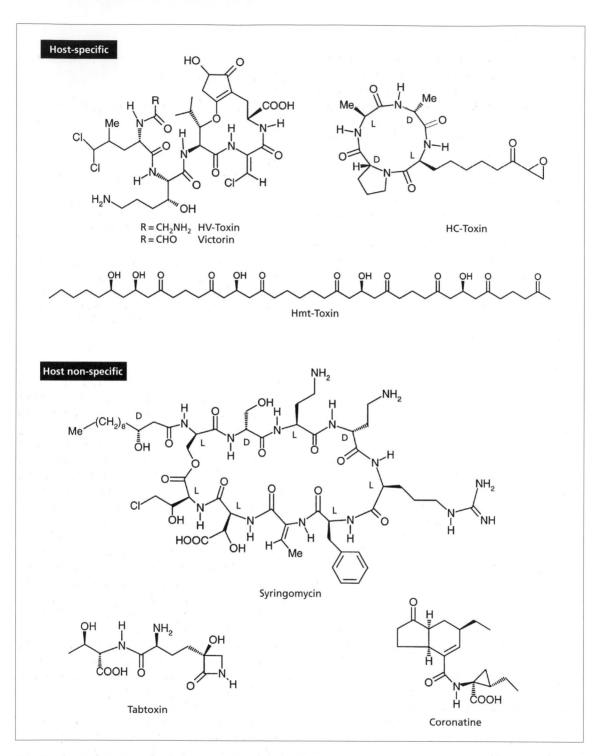

Fig. 8.7 Chemical structure of some toxins involved in plant disease.

Non-specific toxins

A wide variety of metabolites from plant-pathogenic fungi and bacteria have been implicated in disease, occurring in the infected host where they account for at least some of the disease symptoms. These toxins, which do not seem to be the sole determinants of pathogenicity, can instead be regarded as factors contributing to the virulence of a pathogen.

Non-specific toxins have effects on plant species other than the natural host. Tentoxin, produced by the fungus *Alternaria*, is a broad-spectrum inhibitor affecting chloroplast function; this compound induces chlorosis in a wide range of green plants, including pteridophytes, mosses and green algae, as well as seed plants. Cercosporin, produced by several plant-pathogenic species of *Cercospora*, appears to have a non-specific mode of action, involving generation of toxic oxygen radicals in the presence of light. Damage may occur to several cell components, including proteins, nucleic acids and membrane lipids.

The precise role of such non-specific toxins in a disease is often controversial. For instance, the correlation of toxin production with virulence may be questionable. This aspect has been investigated in detail in different pathovars of the bacterium *Pseudomonas syringae*, which are known to produce several potentially toxic compounds (Table 8.2). Wildfire disease of tobacco, caused by *P.s.* pv. *tabaci*, is characterized by lesions on leaves which are surrounded by a conspicuous chlorotic halo. Formation of this chlorotic zone has been attributed to a dipeptide, tabtoxin (Fig. 8.7), which inhibits the enzyme glutamine synthetase, leading to the accumulation of toxic concentrations of ammonia. Bacterial mutants not producing tabtoxin (Tox⁻) do not induce chlorotic lesions, but are still able to grow in host leaves, albeit less well than wild-type toxigenic strains (Fig. 8.8). Conversely, if genes enabling synthesis of tabtoxin are introduced into saprophytic strains of *P. syringae* found growing harmlessly on the leaf surface, the transformants are unable to cause disease symptoms. Together these observations indicate that colonization of the host plant depends upon factors other than toxin production. Rather similar results have been reported for other *Pseudomonas* toxins, such as syringomycin and coronatine (Fig. 8.7), where Tox⁻ mutants induced by transposon mutagenesis generally form smaller lesions, and multiply to a lesser extent in host tissues, than toxin-producing

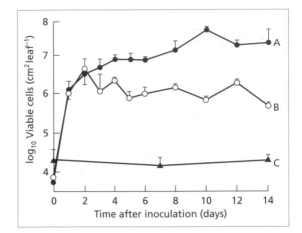

Fig. 8.8 Multiplication of the wildfire disease bacterium, *Pseudomonas syringae* pv. *tabaci* in tobacco leaves. A is a toxin-producing wild-type strain, while B is a non-toxin-producing mutant. A saprophytic bacterium, *P. putida* (C), is included for comparison. (After Turner 1984.)

strains. Not all such experiments have, however, provided convincing evidence that toxigenicity is linked to virulence, and mutations in biochemical pathways leading to toxin production may produce varying effects on final disease reaction type, ranging from complete abolition of symptoms, to little discernible change in disease severity. There is the additional difficulty that a transposon may insert in the bacterial genome at a point where important functions other than toxin production are also affected.

The general conclusion from these studies is that at least some non-specific toxins play an important role in disease development. In *Pseudomonas syringae*, genes for toxin production seem to be conserved amongst groups of strains, suggesting that they are important for survival of the bacteria in the plant environment. It is also becoming apparent that certain plant metabolites may activate expression of some toxin genes, thus acting as signal molecules. Such a regulatory system might provide a degree of specificity in the plant–pathogen interaction.

What advantage might production of a toxin confer on a potential pathogen? One possibility is that disruption of host cells may free nutrients for uptake by the pathogen. Another is that killing cells might prevent an active defence response by the host (see Chapter 10). Either way it should be obvious that this strategy is not an option for biotrophic pathogens, which require living host cells to survive.

High-molecular-weight compounds and pathogenesis

Not all phytotoxins produced by pathogens are small molecules. Several compounds implicated in vascular wilt diseases have been shown to be high-molecular-weight polysaccharides or glycoproteins. *Clavibacter insidiosum*, the causal agent of bacterial wilt in alfalfa, elaborates a large glycopeptide in culture which induces wilt symptoms in bioassays, and has also been detected in diseased plants. These substances may impair water flow simply by virtue of their size. Experiments have shown that synthetic polymers of molecular mass less than 50 kDa can be transported through the xylem and act in the leaves; if the molecules have a mass greater than 50 kDa the polymers remain in the xylem and contribute to physical plugging of the vessels. Strictly speaking, such disruption of the water economy of the host, while damaging to the plant, is not equivalent to a direct toxic effect on metabolism.

The role played by toxins in vascular wilt diseases is a particularly controversial research topic. A diverse array of toxic substances has been isolated from the culture filtrates of wilt pathogens, ranging from simple organic acids to highly complex macro-molecules. After starring briefly in the pathological literature, most have subsequently been relegated to a subsidiary role in disease causation. However, there are exceptions; a toxic peptide from *Verticillium dahliae* seems to show some host-specificity, as tomato and potato cultivars resistant to the fungus are less sensitive to the toxin. A toxin may also contribute to the devastating effects of the Dutch elm fungus, *Ophiostoma novo-ulmi*, on the host tree. The fungus produces a small hydrophobin protein, cerato-ulmin, which causes morphological and physiological symptoms similar to those seen in the disease. By using an antibody specific for the protein, it has proved possible to detect cerato-ulmin in elm wood in the early stages of disease development. The amount of toxin produced by different isolates of the fungus correlates, in most cases, with their degree of pathogenicity, so it seems likely that cerato-ulmin plays a significant role in the disease.

Ralstonia solanacearum causes a lethal wilt disease of crops such as potato and tomato. When grown in culture the bacterium produces an extracellular polysaccharide (EPS) which forms an amorphous slime around each cell. When the bacterium is multiplying in the host vascular system, this EPS may con-

Fig. 8.9 Relationship between extracellular polysaccharide (EPS) production and pathogenicity of *Ralstonia solanacearum* to tomatoes. (a) Time course of wilting caused by wild-type EPS⁺ strain, and mutants lacking EPS (EPS⁻) or the pectolytic enzyme β-1,4 endoglucanase (GLU⁻). (b) Extent of stem colonization and bacterial multiplication in tomato plants inoculated by soil infestation with wild-type and mutant strains, 5 days after soil inoculation. R, sample taken from root and first centimetre of tissue at the stem base. (After Saile *et al.* 1997.)

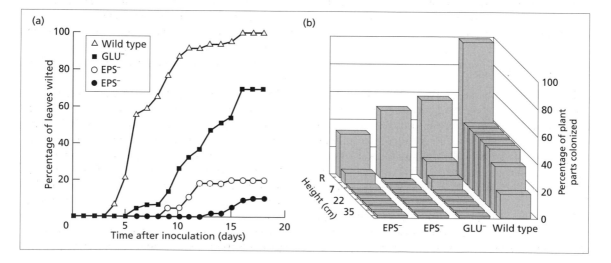

tribute to wilting by occluding xylem vessels and thereby reducing water flow. In culture, bacterial colonies forming EPS have a smooth, fluid appearance, but spontaneous mutants with a 'rough' colony morphology regularly occur. These mutants, which lack EPS, are usually less virulent than the 'smooth' strains, and cause milder symptoms in the host. The relationship between EPS production and virulence has, however, proved difficult to establish, as avirulent EPS⁻ mutants are often altered in other properties, and there are also some instances of EPS⁻ mutants which retain virulence. Recent analyses using bacterial strains with more precisely defined mutations reducing EPS production have shown that while the effects on virulence may vary between mutants, all cause less severe wilting than EPS⁺ strains (Fig. 8.9a). Infection tests in which the mutants were inoculated into tomato plants through stem wounds initially suggested that these strains retain the ability to multiply normally in the host, but when a more natural soil infestation test was used, it was shown that colonization of the host was in fact impaired (Fig. 8.9b). Hence EPS production not only contributes to symptom development but also influences the ability of the pathogen to invade and move within host tissues. The current consensus is therefore that extracellular polysaccharide is a significant virulence factor in this, as well as certain other, bacterial pathogens. However, mutants of *R. solanacearum* defective in other properties, such as pectolytic enzyme production (Fig. 8.9), or export of extracellular proteins, are also compromised in their ability to cause wilt symptoms and are less pathogenic than normal strains, even if polysaccharide production is unaffected. The overall picture concerning pathogenesis in this, and several other wilt diseases, remains complex.

Hormones and pathogenesis

Table 7.1 listed some examples of diseases in which invasion by the pathogen leads to abnormal growth of the plant. Such symptoms are consistent with some alteration in the hormonal status of the host. Four main classes of compound are known to be involved: (i) auxins, such as indole-3-acetic acid (IAA); (ii) gibberellins; (iii) cytokinins; (iv) ethylene (a volatile hormone).

Evaluation of the role played by individual growth regulators in pathogenesis is complex because their physiological effects in many cases overlap, and

normal plant growth and development is regulated by the integrated action of several types of hormone rather than by changes in the concentration of a single compound. There is also the difficulty of determining whether any change in hormone levels is due to the accumulation or loss of hormones produced by the plant itself, or to the production of similar compounds by the pathogen.

A typical feature of several of these diseases is an increase in the concentration of one or more hormones in affected tissues. Many plant pathogens can synthesize hormonally active compounds in culture; for instance *Ustilago maydis*, *Taphrina deformans* and several other fungal as well as bacterial pathogens produce IAA or related auxins. In only a few cases, however, has a direct link been established between hormone production and pathogenicity.

Hormones and plant tumours

Pseudomonas syringae pv. *savastanoi* infects woody hosts such as olives, oleander and privet, causing tumour-like overgrowths in which cell division, cell enlargement and abnormal differentiation of vascular elements takes place. Hormones produced by the bacterium are the main determinants of these symptoms; the biosynthetic pathway for conversion of the amino acid tryptophan into IAA has been defined (Fig. 8.10), and genes coding for the two key enzymes involved have now been identified in the pathogen. Interestingly, in some bacterial strains these genes are located on a plasmid, rather than the bacterial chromosome. Loss of this plasmid, so that the bacterium is unable to synthesize IAA, prevents gall formation, although some limited invasion of host tissues can still take place. IAA-producing strains cause tissue proliferation and hence create a niche in which the pathogen can multiply to high population levels.

The classic example of a plant tumour is crown gall disease caused by *Agrobacterium tumefaciens*. The closely related *A. rhizogenes* causes hairy root disease, characterized by the formation of abnormal, root-like processes on aerial parts of the plant (Fig. 8.11). Both pathogens have a very wide host range, although under natural conditions crown gall is mainly a problem in woody plants such as fruit trees or roses.

Crown gall has long attracted interest as, following initial infection via a wound, secondary galls are

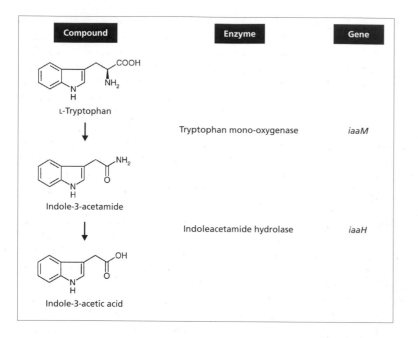

Fig. 8.10 The pathway for auxin synthesis from tryptophan in olive knot disease, *Pseudomonas syringae* pv. *savastanoi*, showing the enzymes involved and the genes encoding the enzymes. (After Shaw 1987.)

formed which do not contain the pathogen. These tumours are therefore autonomous and bear some resemblance to animal cancers. The identity of the factor which transforms healthy cells into rapidly dividing tumour cells was for many years unknown, and described only as the 'tumour-inducing principle'. More recently, intensive study of both the pathogen and transformed tissues has identified many of the mechanisms underlying crown gall pathogenesis, to the extent that more is now known about its molecular basis than, arguably, any other plant disease.

Virulent *A. tumefaciens* and *A. rhizogenes* strains contain large plasmids, described as either tumour-inducing (Ti) or root-inducing (Ri) plasmids, respectively. During infection, part of the plasmid is transferred to the plant cell, where it integrates into the host genome. This process is shown diagrammatically in Fig. 8.12. The transferred (or T) region of plasmid DNA carries several genes which, when expressed in the host cell, cause the characteristic tumour or hairy root symptoms. The *onc* (as in oncogenic—cancer-causing) genes specify the synthesis of auxins and cytokinins, leading to uncontrolled cell division, while other genes code for the production of unusual amino acids known as opines, which cannot be utilized by the plant, but are instead catabolized by the pathogen. *Agrobacterium* therefore acts as a 'genetic engineer', programming host cells to divide rapidly, and then to produce amino acids which serve as a carbon and nitrogen source for the pathogen.

Recent work has identified other important components of the transformation process. Prior to transfer, the T-DNA must first be excised from the circular plasmid. This is regulated by another set of plasmid genes in the *vir* (virulence) region. Activation of the *vir* genes takes place in response to certain wound chemicals, such as the phenolic compound acetosyringone (Fig. 8.12b), leaking from damaged plant cells. Thus the pathogen initiates mobilization of the T-DNA in response to a specific plant signal.

Because of its distinctive properties the Ti plasmid has been widely used as a vector for the deliberate introduction of foreign genes into plant cells. Clearly, in any genetic engineering programme designed to improve plants, it would be undesirable to introduce the *onc* genes responsible for hormone synthesis and disordered growth. These are therefore deleted prior to use, to give a 'disarmed' plasmid which, while retaining the ability to stably integrate into the plant genome, produces none of the pathogenic effects of a natural infection with the bacterium.

Other growth disorders

Spectacular symptoms sometimes result if pathogens invade host meristematic tissues. The fungus *Crinipellis perniciosa* infects cocoa trees; invasion of young shoots activates multiple meristems, leading to repeated branching and formation of a characteristic 'witch's broom'. The role of specific hormones in this process is not yet clear.

(a)

(b)

Fig. 8.11 Crown gall, *Agrobacterium tumefaciens*, and hairy root disease, *A. rhizogenes*, on tomato, showing (a) tumour at stem base, and (b) hairy roots formed on stem. (Courtesy of M.R. Davey.)

Cytokinins participate in the control of cell division, and have therefore been implicated in diseases such as witch's broom, as well as tumours, including the unsightly galls induced on maize plants by the smut fungus *Ustilago maydis* (see Fig. 7.12). Strains of the bacterium *Erwinia herbicola* which cause galls on ornamentals such as *Gypsophila* (*E. herbicola* pv. *gypsophilae*) produce significant amounts of cytokinin in culture. Production of the hormone is due to a gene for cytokinin biosynthesis located on a plasmid in a cluster together with genes for auxin production. Bacterial strains with mutations in this cytokinin gene, which do not produce the hormone, induce much smaller galls than the pathogenic wild-type strain. Hence in this host–pathogen system there appears to be a correlation between cytokinin production and pathogenicity. Peach leaves affected by the leaf curl fungus, *Taphrina deformans*, exhibit higher cytokinin activity than healthy leaves. Chromatographic analysis has shown that three cytokinins found in healthy leaves are more active in diseased leaves; an additional cytokinin not present in healthy leaves has also been detected, and this compound may be secreted by the pathogen.

Gibberellins are to some extent synonymous with plant disease as this class of physiologically active compounds was first identified in culture filtrates of the fungus *Gibberella fujikuroi*, the causal agent of 'foolish seedling' disease of rice. Infected plants outgrow normal plants and become weak and spindly, often collapsing. Secretion of gibberellins by the pathogen in the host is believed to cause this. Conversely, the stunting caused by some viruses can be reversed by applying exogenous gibberellins, suggesting that the reduced growth of these plants might be due to a decreased gibberellin content.

The role of ethylene in disease is even harder to assess, as this volatile hormone is produced by plants as a natural response to injury and many other forms of stress. Increased amounts of ethylene are evolved from plant tissues infected by fungi, bacteria and viruses. Ethylene is active at very low concentrations, and has a wide variety of physiological effects, including activation of plant defence responses, and wound healing (see p. 159). It may also act as a synergist, for example by increasing the severity of symptoms induced by toxins. Many plant pathogens are themselves capable of producing ethylene, but the origin of the gas in infected tissues is difficult to determine with certainty.

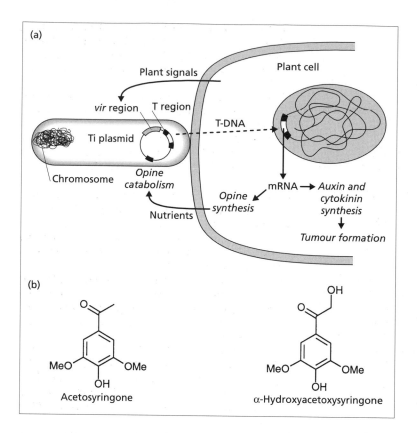

(a)

Plant signals

Plant cell

vir region   T region

Ti plasmid

Chromosome

Opine catabolism

T-DNA

Nutrients

Opine synthesis

mRNA → *Auxin and cytokinin synthesis*

Tumour formation

(b)

Acetosyringone

α-Hydroxyacetoxysyringone

Fig. 8.12 (a) Molecular processes underlying tumour induction in crown gall disease. The bacterium attaches to the plant cell wall, where plant signal molecules such as acetosyringone and related compounds (b) activate the *vir* region of the Ti plasmid. Transfer of the T-DNA and integration into the plant nuclear genome then takes place. Expression of the T-DNA leads to the production of opines and plant hormones. (After Melchers & Hooykaas 1987; by permission of Oxford University Press.)

Finally, mention should be made of growth inhibitors, such as abscisic acid (ABA). It has been suggested that ABA might play a part in the stunting seen in many plant diseases, but conclusive evidence is lacking. ABA levels rise in wilted plants, including those infected by vascular wilt pathogens, but this appears to be a host response, rather than a cause of the symptom itself.

Microbial pathogenicity in summary

Plant pathogens possess a variety of features which enable them to colonize and exploit host plants. Some of these are specific structures enabling entry to the host, such as appressoria and infection hyphae (Chapter 6), while others are extracellular secretions, enzymes, toxins or hormones, many of which contribute to pathogenesis. Export systems play an important role by transporting biologically active molecules across the host–pathogen interface. Disruption of genes encoding such putative pathogenicity factors has, however, given inconclusive results.

Such genetic 'knock-outs' often suggest that loss of a particular product, for instance a single enzyme, does not necessarily reduce the capacity of the pathogen to cause disease. The integrated action of several pathogenicity factors may therefore be important. The differences between pathogens and non-pathogens might, in fact, relate more to differences in the regulation of molecular processes required for infection, rather than to the presence or absence of a specific factor. There is increasing evidence that many virulence functions are only switched on in the presence of the host, often in response to a specific signal, which therefore serves as a host 'signature'. Host induction of microbial pathogenicity factors ensures that the biosynthetic investment required to produce them is expended only in situations where they will aid the growth and survival of the pathogen. From a practical viewpoint, identification of these signal compounds and pathways regulating pathogenicity should provide new opportunities for intervening in the disease process.

Further reading

Book

Durbin, R.D. (ed.) (1981) *Toxins in Plant Disease.* Academic Press, New York.

Reviews and papers

Alfano, J.R. & Collmer, A. (1996) Bacterial pathogens in plants: life up against the wall. *The Plant Cell* 8, 1683–1698.

Brosch, G., Ransom, R., Lechner, T., Walton, J.D. & Loidl, P. (1995) Inhibition of maize histone deacetylases by HC toxin, the host-selective toxin of *Cochliobolus carbonum*. *The Plant Cell* 7, 1941–1950.

Chasan, R. (1995) Victorin's secret. *The Plant Cell* 7, 385–387.

Collmer, A. (1987) Pectic enzymes and bacterial invasion of plants. In: *Plant–Microbe Interactions. Molecular and Genetic Perspectives,* Vol. 2 (eds T. Kosuge & E.W. Nester), pp. 253–284. Macmillan, New York.

Dean, R.A. & Timberlake, W.E. (1989) Production of cell wall-degrading enzymes by *Aspergillus nidulans*: a model system for fungal pathogenesis of plants. *The Plant Cell* 1, 265–273.

Expert, D., Enard, C. & Masclaux, C. (1996) The role of iron in plant host–pathogen interactions. *Trends in Microbiology* 4, 232–237.

Fenselau, S., Balbo, I. & Bonas, U. (1992) Determinants of pathogenicity in *Xanthomonas campestris* pv. *vesicatoria* are related to proteins involved in secretion in bacterial pathogens of animals. *Molecular Plant–Microbe Interactions* 5, 390–396.

Gaudin, V., Vrain, T. & Jouanin, L. (1994) Bacterial genes modifying hormonal balances in plants. *Plant Physiology and Biochemistry* 32, 11–29.

Hamer, J.E. & Holden, D.W. (1997) Linking approaches in the study of fungal pathogenesis: a commentary. *Fungal Genetics and Biology* 21, 11–16.

He, S.Y., Lindeberg, M., Chatterjee, A.K. & Collmer, A. (1991) Cloned *Erwinia chrysanthemi out* genes enable *Escherichia coli* to selectively secrete a diverse family of heterologous proteins to its milieu. *Proceedings of the National Academy of Sciences of the USA* 88, 1079–1083.

Hensel, M. & Holden, D.W. (1996) Molecular genetic approaches for the study of virulence in both pathogenic bacteria and fungi. *Microbiology* 142, 1049–1058.

Kang, Y., Huang, J., Mao, G., He, L.-Y. & Schell, M.A. (1994) Dramatically reduced virulence of mutants of *Pseudomonas solanacearum* defective in export of extracellular proteins across the outer membrane. *Molecular Plant–Microbe Interactions* 7, 370–377.

Lichter, A., Barash, I., Valinsky, L. & Manulis, S. (1995) The genes involved in cytokinin biosynthesis in *Erwinia herbicola* pv. *gypsophilae*—characterization and role in gall formation. *Journal of Bacteriology* 177, 4457–4465.

Morris, R.O. (1986) Genes specifying auxin and cytokinin biosynthesis in phytopathogens. *Annual Review of Phytopathology* 37, 509–538.

Navarre, D.A. & Wolpert, T.J. (1995) Inhibition of the glycine decarboxylase multienzyme complex by the host-selective toxin victorin. *The Plant Cell* 7, 463–471.

Oliver, R. & Osbourn, A. (1995) Molecular dissection of fungal phytopathogenicity. *Microbiology* 141, 1–9.

Orchard, J., Collin, H.A., Hardwick, K. & Isaac, S. (1994) Changes in morphology and measurement of cytokinin levels during the development of witches' brooms on cocoa. *Plant Pathology* 43, 65–72.

Osbourn, A., Bowyer, P., Lunness, P., Clarke, B. & Daniels, M. (1995) Fungal pathogens of oat roots and tomato leaves employ closely related enzymes to detoxify different host plant saponins. *Molecular Plant–Microbe Interactions* 8, 971–978.

Pieterse, C.M.J., Riach, M., Bleker, T., Van den Berg Venthuis, G.C.M. & Govers, F. (1993) Isolation of putative pathogenicity genes of the potato late blight fungus *Phytophthora infestans* by differential hybridization of a genomic library. *Physiological and Molecular Plant Pathology* 43, 69–79.

Quinn, F.D., Newman, G.W. & Kind, C.H. (1997) In search of virulence factors of human bacterial disease. *Trends in Microbiology* 5, 20–26.

Rahme, L.G., Stevens, E.J., Wolfort, S.F., Shao, J., Tompkins, R.G. & Ausubel, F.M. (1995) Common virulence factors for bacterial pathogenicity in plants and animals. *Science* 268, 1899–1902.

Saile, E., McGarvey, J.A., Schell, M.A. & Denny, T.P. (1997) Role of extracellular polysaccharide and endoglucanase in root invasion and colonization of tomato plants by *Ralstonia solanacearum*. *Phytopathology* 87, 1264–1271.

Talbot, N.J., Kershaw, M.J., Wakley, G.E., de Vries, O.M.H., Wessels, J.G.H. & Hamer, J.E. (1996) *MPG1* encodes a fungal hydrophobin involved in surface interactions during infection-related development of *Magnaporthe grisea*. *The Plant Cell* 8, 985–999.

Walker, D.S., Reeves, P.J. & Salmond, G.P.C. (1994) The major secreted cellulase, CelV, of *Erwinia carotovora* ssp. *carotovora* is an important soft rot virulence factor. *Molecular Plant–Microbe Interactions* 7, 425–431.

Walton, J.D. (1996) Host-selective toxins: agents of compatibility. *The Plant Cell* 8, 1723–1733.

Wolpert, T.J. & Macko, V. (1989) Specific binding of victorin to a 100-kDa protein from oats. *Proceedings of the National Academy of Sciences of the USA* 86, 4092–4096.

9   Plant Defence

'Plants have evolved with pathogens and insect pests for millions of years. It is therefore not surprising that a particular plant is resistant to most of them.'
[N.T. Keen, 1992]

Plants are continually exposed to insects, nematodes and other potentially damaging pests, as well as to a wide variety of parasitic microorganisms. Yet the majority of plants remain healthy most of the time. This observation suggests that plants must possess highly effective mechanisms for preventing parasitism and predation, or at least limiting their effects.

The topic of plant defence has assumed extra significance with the recent development of techniques for genetically engineering crops. If we can identify genes encoding defence functions, then it should be possible to clone such genes and transfer them to new host species, thereby extending the options available to the plant breeder (see p. 228).

Types of plant defence

Like animals, plants possess several lines of defence against potential invaders. Structural and chemical barriers effectively exclude the majority of organisms; should these be breached then a sensitive surveillance system can detect foreign cells and trigger a rapid response to injury or microbial attack. Plant defence systems can therefore be classified as either **passive** (**constitutive**) or **active** (**inducible**), depending upon whether they are pre-existing features of the plant, or are switched on after challenge (Fig. 9.1). These categories can be further subdivided into structural and chemical mechanisms, although such divisions are not mutually exclusive; for instance, a chemical inhibitor might be a component of a structural barrier. The picture of plant defence emerging from

recent research is of a finely tuned and integrated system in which different components act together in a coordinated and complementary fashion to contain infection.

Passive defence mechanisms

Passive anatomical features, such as the cuticle and bark, represent highly effective obstacles to penetration by most microorganisms (see p. 93). Plant surface structures may also play a role in defence against insects; for instance the sharp hairs (trichomes) found on some wild potato relatives impale and kill aphids. Every plant cell is itself surrounded by a substantial obstacle, the cell wall. Where this is impregnated with chemicals such as suberin and lignin it forms an even more effective barrier against pathogens.

Preformed inhibitors

Plants synthesize a vast array of secondary metabolites, many of which are toxic to potential pests and pathogens. Compounds such as phenols, alkaloids, glycosides, saponins, tannins and resins all possess antibiotic properties and may therefore contribute to resistance. The remarkable chemical diversity of the plant kingdom has been partly linked to this role in defence, with each plant family evolving a different set of metabolites capable of repelling herbivorous animals or inhibiting pathogens. The well-known resistance of some plant materials, especially wood, to microbial decay is due in part to the presence of inhibitors such as tropolones and stilbenes (Fig. 9.2), which act as natural preservatives.

A direct correlation between the presence of a toxic compound in plant tissues and resistance to a pathogen has been established in relatively few cases. One of the first examples concerned smudge disease

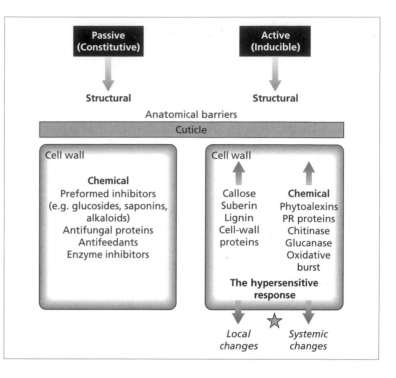

Fig. 9.1 Classification of types of plant defence, based on existing anatomical or biochemical features, or active changes induced after challenge by pathogens.

of onions, caused by *Colletotrichum circinans*, where the resistance of bulbs to infection was linked to two phenolic compounds, catechol and protocatechuic acid (Fig. 9.2), found mainly in the pigmented outer scales. Another preformed phenolic implicated in plant defence is chlorogenic acid, found in potato tubers resistant to common scab, *Streptomyces scabies*.

Inhibitors in plant tissues often occur as non-toxic precursors, which are converted to active forms following cell damage or exposure to enzymes. Apple and pear leaves contain glycosides such as arbutin, which are hydrolysed by the enzyme β-glucosidase to form glucose, plus toxic quinones inhibitory to the fireblight bacterium *Erwinia amylovora*. Other plants, including some legumes, contain cyanogenic glycosides which release the familiar poison hydrogen cyanide as a breakdown product. Brassica crops such as cabbage are rich in sulphur-containing mustard oils, which degrade to form volatile isothiocyanates inhibitory to several fungal pathogens. The cellular disruption caused by pathogens as they grow through host tissues may therefore initiate chemical reactions leading to a highly toxic local environment. These changes are induced by infection, but as they involve

conversion of preformed compounds they are distinct from defence responses involving active host metabolism.

Evidence that preformed inhibitors may play a key role in host–pathogen interaction has been obtained from detailed studies on the host specificity of pathogens. The take-all fungus, *Gaeumannomyces graminis*, occurs as several pathogenic varieties which differ in their ability to attack different cereal hosts. *G. graminis* var. *tritici* (Ggt) is pathogenic to wheat, but not oats, whereas var. *avenae* (Gga) can attack oats as well as wheat. Young oat roots, the normal site of infection by the take-all fungus, contain several chemically similar fluorescent saponin compounds known as avenacins (Fig. 9.2), which are toxic to many fungi. Such saponins act to disrupt cells by complexing with sterols in membranes. Enzymatic removal of the sugar residues from avenacin reduces its toxicity. Gga produces significant amounts of the enzyme, avenacinase, which carries out this reaction, whereas Ggt produces none. The ability of the oat-attacking form of the fungus to colonize this host can therefore be linked to its ability to detoxify a pre-formed inhibitor.

Confirmation that production of avenacinase is a

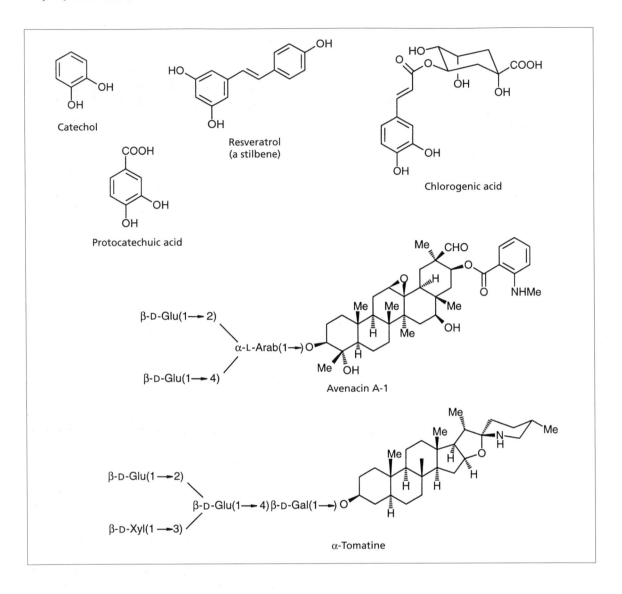

Fig. 9.2 Some preformed antimicrobial compounds from plant tissues.

key determinant of the host range of the take-all fungus has recently been obtained by isolating the gene encoding the enzyme from Gga, and then disrupting it to make a non-functional version which was reintroduced into the fungus. Pathogen strains in which the original functional avenacinase gene had been replaced by a disrupted version no longer produced the enzyme, and had simultaneously lost the ability to infect oat roots (Fig. 9.3). However,

these mutant strains were still able to attack wheat roots, showing that their pathogenicity was unimpaired.

Interestingly, enzymes with structural, biochemical and genetic sequence similarities to avenacinase have now been detected in some other groups of pathogenic fungi. For instance *Septoria lycopersici*, a pathogen infecting tomato leaves, produces a related enzyme, tomatinase, active against the alkaloidal saponin tomatine (Fig. 9.2) present in tomato leaf tissues. Thus, unrelated fungal pathogens appear to possess similar enzymatic mechanisms for dealing with the toxins they encounter when colonizing

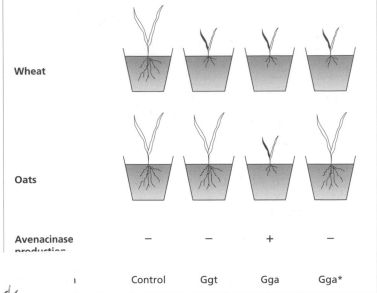

Fig. 9.3 Relationship between production of the saponin-detoxifying enzyme avenacinase and host range in the take-all fungus *Gauemannomyces graminis*. Variety *tritici* (Ggt) does not produce the enzyme and is non-pathogenic to oats, which contain avenacin. Variety *avenae* (Gga) produces the enzyme and is able to attack oats. Gga strains in which the gene for avenacinase production has been disrupted (Gga*) lose the ability to

Wheat

Oats

Avenacinase production

−	−	+	−
Control	Ggt	Gga	Gga*

[handwritten annotation:]

Plant defence proteins
→ e.g antibiotic effect in seeds
    presence of potent antifungal
        proteins
such proteins enhance resistance
of cell walls to enzymic attack +
digest to partially degraded oligomers
    − many of which are active inducers
        of plant defence responses

seeds prevents growth of a fungal pathogen in the zone surrounding the seed. This antibiotic effect helps to protect the emerging seedling during the vulnerable early stages of development. Analysis of such exudates has shown that the biological activity is due to the presence of potent antifungal proteins which are released from the seed as it imbibes water.

In fact, plants contain several types of proteins which are believed to serve a defensive function (Table 9.1). They include the antifungal peptides found in seeds, toxic cell-wall proteins known as thionins found in cereal leaves, hydrolytic enzymes such as glucanases and lysozyme, and various classes of enzyme inhibitors. The latter are often abundant in storage tissues and show activity against digestive enzymes such as proteases occurring in the digestive tract of insects. For instance, cowpea seeds contain a trypsin-inhibitor (CpTI) that may contribute to the resistance of such seeds to predation by insects. Other proteins may specifically inhibit microbial enzymes produced during tissue colonization. In the previous chapter we discussed the role played by pectolytic enzymes in the digestion of plant cell walls; enzymes such as polygalacturonase are central to this process. Recent analysis has identified proteins present in the cell walls of many plants, for instance beans, which inhibit the action of polygalacturonases produced by fungi, thus interfering with the infection process. Such proteins not only enhance the resistance of plant cell walls to enzymic attack, but also influence the nature of any end-products released. Rather than being digested completely to monomers, pectic substrates are partially degraded to oligomers, many of which are active inducers of plant defence responses (see p. 171). These inhibitor proteins therefore contribute to an early-warning system whereby plant cells can detect the presence of an invading fungus.

Another well-characterized group with a possible role in plant defence are the lectins, many of which are secretory proteins that accumulate in cell walls or vacuoles. Lectins bind to different combinations of

(a)

(b)

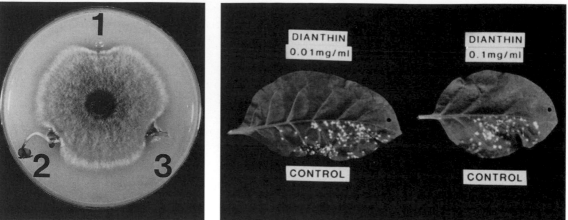

Fig. 9.4 (a) Inhibition of growth of a fungal culture on an agar medium by exudate from germinating radish seeds (2 and 3) and a protein extract purified from the seeds (1). (Courtesy of Willem Broekaert.) (b) Effect of the antiviral protein dianthin on local lesion formation by tobacco mosaic virus (TMV). The upper half of each tobacco leaf was treated with the virus and dianthin at the concentrations shown, while the lower half was treated with the virus alone. (From Taylor *et al.* 1994.)

Table 9.1 Some plant proteins that confer resistance to pests and pathogens.

Type	Examples	Biological activity
Hydrolases	Chitinase / Glucanase	Digest fungal cell walls
	Lysozyme	Lyses bacterial cells
Enzyme inhibitors	Polygalacturonase-inhibiting proteins (PGIPs)	Inhibit fungal pectinases
	Protease inhibitors / Amylase inhibitors	Inhibit insect digestive enzymes
Chitin-binding proteins	Hevein (in latex from rubber)	Inhibits growth of fungi
	Lectins	Inhibit growth of insects and fungi
Antifungal peptides (AFPs)	Plant defensins	Inhibit growth of fungi
Thionins	Hordothionin (from barley)	Inhibits growth of fungi and bacteria
Ribosome-inactivating proteins (RIPs)	Pokeweed antiviral protein (PAP)	Inhibits mechanical transmission of viruses

sugars present in polysaccharides, glycoproteins and glycolipids. They attach to the surface of cells containing particular sugar residues, and may thus form part of a recognition system. For instance, one series (or family) of lectins binds to molecules containing the amino sugar *N*-acetyl glucosamine, the basic building block of chitin. Chitin-binding lectins are commonly found in cereal grains, as well as unrelated plants such as tomato and potato, and in the latex of rubber trees. These proteins, while widely distributed in the plant kingdom, share some sequence homology, suggesting that they may all have evolved from an ancestral gene with a common function. Chitin is found in the insect exoskeleton, and is also a component of the cell walls of most fungi; these lectins may thus function in defence against both herbivorous insects and fungi. Binding of the protein to the fungus may slow hyphal growth, thereby creating time for other defence systems to act. If eaten by animals, other lectins bind to the digestive tract, preventing effective absorption of nutrients. Together these proteins may protect plant tissues, especially nutrient-rich seeds, against infection or predation.

Antiviral proteins

It has been known for many years that protein extracts from certain plant species reduce the infectivity of plant viruses. For instance, when proteins from carnation and related *Dianthus* species are mixed with tobacco mosaic virus (TMV) prior to inoculation, local lesions are reduced or completely suppressed (Fig. 9.4b). Two theories have been proposed to explain this effect. One is that the proteins involved directly inactivate the virus; the second is that these inhibitors interfere with virus replication via effects on the host plant. Recent evidence supports the latter view, as the molecules concerned appear to inhibit protein synthesis by affecting ribosomes in host cells, and hence block virus replication. Consequently these inhibitors are described as **ribosome-inactivating proteins** (RIPs).

The examples above illustrate some of the biochemical diversity of plant proteins with a proposed role in constitutive resistance to pathogens and pests. In fact, some of these inhibitors may also be produced in response to attempted infection or other forms of stress. Synthesis of antifungal thionins in cereal leaves, for instance, is triggered following inoculation with foliar pathogens such as powdery mildews. The distinction between constitutive and inducible resistance mechanisms is therefore not absolute, and the regulation of these defence proteins may vary from tissue to tissue.

There is currently considerable interest in plant defence proteins because many of them have been sequenced and the genes encoding them isolated, providing the opportunity to engineer transgenic plants expressing such proteins to improve resistance to both pathogens and pests. This novel breeding strategy will be discussed further in Chapter 12.

Active defence mechanisms

The various structural and chemical components described in the previous section serve as a first line of defence against microbial attack. These mechanisms are probably highly effective against the large majority of potential colonists. Nevertheless, in the constant evolutionary arms race between plant and pathogen, a proportion of microorganisms have developed pathogenicity factors able to breach or inactivate preformed plant defences. Such pathogens also face a series of active defence responses induced once penetration of the host begins.

Host reactions to penetration

The initial events during host–pathogen interaction appear to be similar irrespective of whether the combination is compatible or incompatible (see Fig. 2.5). Thus, in the case of a fungal pathogen penetrating a highly resistant host, spore germination, growth of the germ tube, appressorium formation and subsequent entry of the infection hypha can take place on a time scale comparable to these events in a host lacking resistance. At this early stage in the interaction it is often impossible to distinguish between the two host reaction types. Once the pathogen begins to breach the cell wall, however, events take a very different course depending upon the degree of host resistance. If the combination is compatible, further hyphal growth and invasion of host tissues continues unrestricted. In resistant hosts, though, a number of changes take place in penetrated cells and adjacent tissues which ultimately halt the advance of the pathogen.

Changes in host cell walls

The first detectable response of a plant cell to an invading microorganism is often an alteration in the appearance or properties of the cell wall. For instance, attempted penetration of cereal leaves by non-pathogenic or avirulent fungi is accompanied by the deposition of a plug of material, known as a papilla, directly beneath the penetration site (Fig. 9.5). The epidermal cell wall surrounding the papilla may be changed to leave a disc-shaped zone or halo. Similar thickening and modification of host cell walls has been observed in other plant–fungus interactions (see Fig. 9.16); root cortical cells penetrated by root-infecting fungi may form characteristic protrusions of the wall known as lignitubers. Collectively, all these

(a)

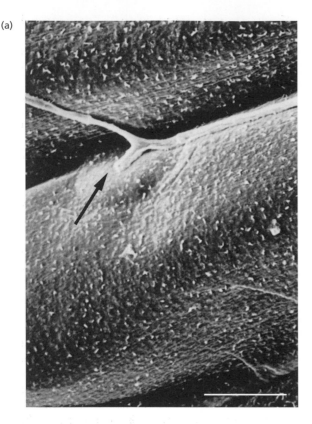

(b)

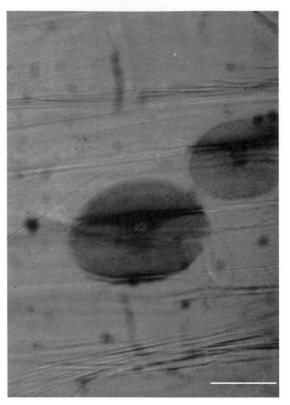

Fig. 9.5 Attempted penetration of a resistant cereal leaf by *Septoria (Stagonospora) nodorum*. (a) Scanning electron micrograph showing surface hypha and apparent swelling in epidermal cell wall at penetration site (arrow). Scale bar = 10 µm. (b) Stained epidermal strip with halos of phenolic material around penetration sites. Scale bar = 20 µm. (Courtesy of J.A. Hargreaves.)

structures involving the accretion of new wall material are described as **wall appositions**.

Morphologically, wall deposits of various kinds appear similar, but what are they made of? Studies using histochemical stains, fluorescence microscopy and X-ray microanalysis have shown that quite different types of material accumulate, ranging from minerals such as silica to complex organic polymers like lignin. Very recently it has been shown that resistant cocoa (*Theobroma cacao*) varieties reacting to invasion by the vascular wilt fungus, *Verticillium dahliae,* accumulate elemental sulphur, a well-known fungicide, in cell walls adjacent to infected vessels. Some of the changes known to take

place in cell walls in response to infection are listed in Table 9.2.

Although the accumulation of reaction material in cell walls is one of the earliest and most regular plant responses to challenge by a pathogen, there is debate about the importance of these changes in host resistance. The most obvious suggestion is that wall appositions impede penetration by the pathogen, cither by increasing the strength and thickness of the cell wall or by enhancing its resistance to enzymic attack. Alternatively, successful parasitism may be impaired through changes in the permeability of cell walls. Finally it should be noted that the monomeric precursors of wall polymers such as lignin may themselves be toxic to microorganisms. These various possibilities are listed in Table 9.3. What is not clear, however, is the extent to which any or all of these functions actually determine the outcome of a particular host–pathogen confrontation. Wall appositions are also produced in response to other types of injury, and it is possible that they may be a non-specific wound response and therefore of secondary importance.

Table 9.2 Some cell-wall changes occurring in response to infection.

Deposition of callose
Deposition of suberin
Impregnation with oxidized phenols, e.g. melanins
Accumulation of calcium, silicon or sulphur
Lignification
Changes in amount or type of cell-wall proteins
Oxidative cross-linking of cell-wall proteins

Table 9.3 Possible functions of cell-wall changes in host resistance.

Mechanical barrier
Increased resistance to cell-wall-degrading enzymes
Reduced diffusion of compounds from host to
  pathogen or vice versa
Direct toxicity of wall precursors, e.g. phenols, to
  pathogen

The hypersensitive response (HR)

One hundred years ago, the US plant pathologist Marshall Ward, working on rust fungi infecting grasses, noted that in resistant hosts the cells adjacent to the infection site rapidly became discoloured, granular and necrotic. A few years later the term 'hypersensitive' was introduced to describe this type of reaction; host cells are apparently so sensitive to the pathogen that they collapse and die as soon as they are penetrated (Fig. 9.6).

The hypersensitive response (HR) has been the subject of intense study, partly because it is a clear example of the dynamic role of the host in the early stages of pathogen attack, but also because it confers a high degree of resistance on the host and has therefore tended to be the type of resistance selected for by the plant breeder. For instance, the resistance genes (*R* genes) introduced into the cultivated potato (*Solanum tuberosum*) through hybridization with the wild South American species *S. demissum,* give good protection against specific races of *Phytophthora infestans*. Resistance controlled by these *R* genes is expressed as a hypersensitive response against incompatible races of the fungus. The high degree of specificity has proved in this case to be a stumbling block, as new pathogen races capable of overcoming the effects of these genes have arisen (see Chapter

12). Nevertheless, hypersensitivity has continued to attract interest, particularly now it is realized that this type of defence reaction is widespread amongst plants and effective against a variety of pathogens.

Initially, hypersensitivity was thought of as being significant only in interactions involving biotrophic fungi. Immediate death of living host cells would alone be sufficient to prevent the fungus from establishing an effective relationship. For a while this simple but attractive hypothesis seemed to be an adequate explanation for the basis of hypersensitive resistance. However, superficially similar and equally effective host reactions were found to occur with necrotrophic fungi and bacteria (Fig. 9.6c). In many respects the local lesion reaction of plants to viruses (Plate 3b, facing p. 12) also bears similarities to a hypersensitive response; necrosis of cells around the inoculation site prevents the virus from spreading systemically throughout the host.

The fact that HR operates against both necrotrophs and biotrophs indicates that programmed cell death may not in itself determine the fate of the pathogen. Instead it appears that host cell necrosis is accompanied by a whole series of events, including changes in oxidative metabolism, the accumulation of toxic compounds, and lignification of cell walls. The breakdown of membranes and the cellular disorganization that accompanies necrosis may in fact trigger metabolic changes in adjacent living cells. The sum total of these changes is the creation of an inhibitory environment which restricts further growth of the pathogen, either by starving it or poisoning it or physically walling it in, or a combination of all three.

Is HR a primary determinant of resistance?

The original concept of hypersensitivity envisaged a straightforward sequence of events. Upon contact with the pathogen, host cells swiftly die and the pathogen then ceases to grow. In effect, the host 'sacrifices' several cells so that the rest can survive. For example, in potato containing a specific *R* gene, cells remain viable for up to 48 h after initial contact with a compatible race of the pathogen. Similar cells penetrated by an incompatible race begin to degenerate within 1 h of contact and die within 2–3 h. Comparison of growth rates of compatible and incompatible races of the pathogen in cells of one

(a)

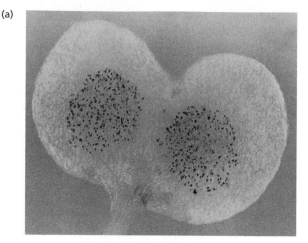

(b)

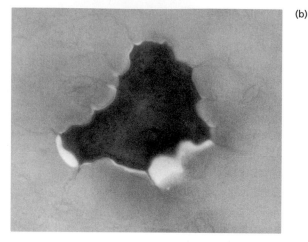

(c)

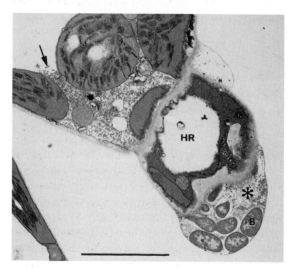

Fig. 9.6 The hypersensitive response. (a) Light micrograph of cabbage cotyledon inoculated with two droplets containing spores of an incompatible isolate of the downy mildew pathogen *Peronospora parasitica*, showing small necrotic spots at penetration sites. (b) Higher power view of host epidermis at penetration site, showing single hypersensitive cell with dark pigmentation and deposition of fluorescent material on cell wall (×350). (c) Electron micrograph of bean leaf tissues inoculated with an incompatible race of the bacterium *Pseudomonas syringae* pv. *phaseolicola*. A mesophyll cell has collapsed due to a hypersensitive response (HR) to adjacent bacterial cells (B) present as a colony with fibrillar material (∗). The tonoplast has ruptured (arrow) in the next host cell. Scale bar = 5 μm. (c, From Brown & Mansfield 1988.)

cultivar has shown that there is very little difference between the two until at least 8 h after penetration (Fig. 9.7). In other words, the death of host cells precedes inhibition of pathogen growth by several hours.

Not all studies support the idea that host-cell death is a primary event. If one treats a compatible potato–*Phytophthora* combination with an antibiotic such as streptomycin, it is possible to induce a hypersensitive response; this conversion of a normally susceptible reaction type into HR can be interpreted as evidence that inhibition of pathogen growth can itself trigger host-cell death. In a naturally occurring HR some factor might initially slow down or damage the pathogen, thereby initiating a cascade of events in

which host-cell necrosis is a consequence, rather than a cause of resistance.

Debate over the relationship between host-cell death and resistance to some extent misses the central point—what actually switches on the whole process? When cells of saprophytic (i.e. non-pathogenic) bacteria are inoculated into plant leaves, the bacteria fail to multiply to any significant extent, and there is no apparent host reaction. If, however, pathogenic bacteria are introduced into the leaf, the population multiplies, leading in a susceptible host to a progressively spreading, water-soaked lesion. In a resistant host (or a non-host plant), the initial multiplication phase is accompanied by rapid changes in host-cell permeability, perhaps reflecting membrane damage,

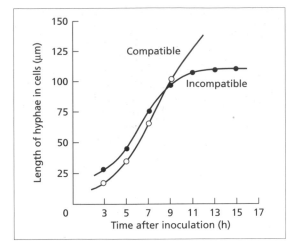

Fig. 9.7 Total length of hyphae of compatible and incompatible races of *Phytophthora infestans* in potato cells. Incompatible cells die within 3 h while compatible cells are alive after 48 h. (Data from Tomiyama 1971.)

followed by a visible HR. No further multiplication of the bacteria takes place. Interestingly, induction of this HR requires the presence of living bacteria, suggesting that some product of bacterial growth and metabolism triggers the host-cell response. A similar interpretation can be drawn from many incompatible, fungal–plant interactions, in which a period of pathogen growth occurs prior to a visible host reaction. Thus induction of active host defence may depend upon recognition of signal molecules released by the growing pathogen, or exposed on the surface of pathogen cells or hyphae. Further discussion of the possible identity of such signal molecules will be delayed until Chapter 10, after we have

considered some other aspects of active host defence (see p. 169).

Phytoalexins

The fact that necrotrophic pathogens, which normally thrive in necrotic tissues, are effectively contained by the hypersensitive response, indicates that other inhibitory factors must be involved. The first evidence for such factors was obtained more than 50 years ago, in experiments by Müller and Borger on the interaction of different isolates of *Phytophthora infestans* with potato tuber tissues. If a compatible isolate of the blight fungus is applied to the cut surface of tubers it invades the underlying tissues and also forms a fluffy superficial mycelium. If an incompatible isolate is applied, the host tissues become discoloured and necrotic and fungal growth is restricted. Müller and Borger first inoculated tuber tissues with an incompatible isolate of the fungus, followed after one or two days by a compatible isolate. This second inoculation, which would normally lead to infection, failed to invade tissues and no mycelial growth was observed. Müller and Borger explained this result by postulating the production of inhibitors, termed phytoalexins, in response to the first inoculation, which subsequently prevented growth of the virulent isolate.

Since this pioneering work a large number of phytoalexins have been isolated from different plant tissues and chemically characterized (Fig. 9.8). Phytoalexins are structurally diverse, but share certain features in common (Table 9.4). A convenient working definition is 'low-molecular-weight antibiotic compounds which are synthesized by and accumulate in plant tissues in response to microbial

Table 9.4 Some features of phytoalexins.

Low-molecular-weight antimicrobial compounds synthesized from remote precursors
Broad-spectrum inhibitors active against a wide range of organisms (low specificity)
Induced by microbial challenge, chemical or physical agents (low specificity)
Chemically diverse, with different phytoalexins characteristic of different plant species, genera or families, i.e. host-specific rather than pathogen-specific
Induction is associated with *de novo* synthesis of phytoalexin biosynthetic enzymes

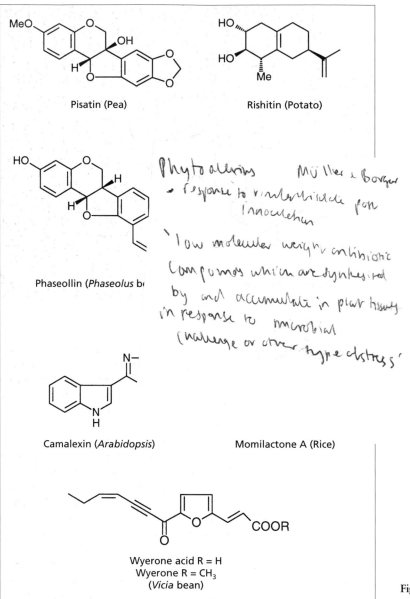

Pisatin (Pea)

Rishitin (Potato)

Phaseollin (*Phaseolus b*

Camalexin (*Arabidopsis*)

Momilactone A (Rice)

Wyerone acid R = H
Wyerone R = CH₃
(*Vicia* bean)

Handwritten note:
Phytoalexins   Müller & Borger
→ response to v. inter... per
        Inoculation

'low molecular weight antibiotic
Compounds which are synthesised
by and accumulate in plant tissues
in response to microbial
challenge or other type distress'

Fig. 9.8 Origin and structure of some phytoalexins.

challenge or other types of stress'. Unlike preformed inhibitors, phytoalexins are not present in healthy tissues; instead their production is induced by infection or injury. Detection of phytoalexins relies upon procedures for extracting and purifying the compounds, coupled with an appropriate bioassay to determine their antimicrobial activity. An example is shown in Fig. 9.9, where extracts from broad bean (*Vicia faba*) leaves inoculated with *Botrytis cinerea* have been separated by chromatography, and antifungal compounds visualized by spraying the plate with a coloured fungus. Inhibitory zones show up as clear white areas of silica gel. This experiment demonstrates a further significant point; plants challenged with microorganisms often produce several phytoalexins which may be structurally related—

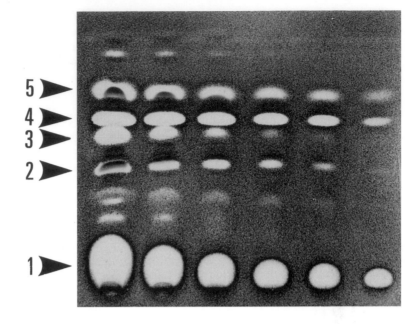

Fig. 9.9 Bioassay of phytoalexins produced by broad bean leaves in response to inoculation with the fungus *Botrytis cinerea*. Extracts from 0.4, 0.2, 0.1, 0.05, 0.025 and 0.0125 g of tissue were collected 3 days after inoculation and separated by thin-layer chromatography. Antifungal compounds were then located by spraying the plate with a spore suspension of the dark-coloured fungus *Cladosporium herbarum*. Phytoalexins identified are (1) wyerone acid, (2) medicarpin, (3) wyerol, (4) wyerone epoxide and (5) wyerone. (From Hargreaves *et al*. 1977.)

even isomeric forms — or alternatively produced by quite different biosynthetic pathways.

Phytoalexins and resistance

The role played by phytoalexins in defence has proved difficult to evaluate critically for several reasons. Firstly, phytoalexin induction is non-specific. As well as microorganisms, treatment with a wide variety of chemical and physical agents can trigger phytoalexin synthesis. Anything which injures plant cells is likely to induce phytoalexins, and they can therefore be regarded as stress metabolites, rather than specific defence molecules analogous to animal antibodies.

The second problem is to establish that phytoalexins accumulate in the right place at the right time in sufficient concentrations to account for inhibition of pathogen growth. Bulk extracts of tissues may show the presence of inhibitory compounds, but often at low levels. Also, the timing of phytoalexin production may be critical. The relative rate at which a phytoalexin is produced in a particular host–pathogen combination is probably more important than the final concentration of compound present. Figure 9.10 shows that the French bean phytoalexin phaseollin (Fig. 9.8) accumulates earlier in plants reacting hypersensitively to the fungus *Col-*

letotrichum lindemuthianum than in beans infected by a compatible isolate of the same pathogen. Although higher levels of phytoalexin eventually accumulate in the susceptible combination the lesions are already well established and the fungus has, in effect, escaped.

Detailed correlations between the growth rate of a pathogen and the amount of phytoalexin present at particular tissue sites have been determined for relatively few host–pathogen systems; one of the most intensively studied is the interaction of the fungus *Phytophthora megasperma* var. *sojae* (Pms) with soybean. For the first 8 h following inoculation the growth rate of the fungus is similar in both a susceptible cultivar and one containing a single gene for resistance. After this time growth is arrested in the resistant cultivar, but continues unchecked in the susceptible cultivar. Analysis of thin slices of hypocotyl tissue taken directly below the inoculation site showed that the concentration of the major soybean phytoalexin, glyceollin (see Fig. 9.8) accumulating in the resistant cultivar was more than sufficient to inhibit the fungus by the time growth was checked. While some glyceollin was detected in susceptible tissues, it never reached the threshold concentration necessary to affect the pathogen.

Further evidence that phytoalexins play an important role in resistance has come from experiments

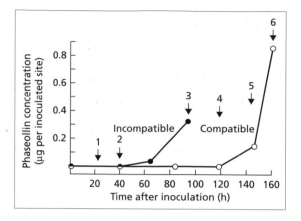

Fig. 9.10 Accumulation of phaseollin in beans inoculated with compatible and incompatible races of the fungus *Colletotrichum lindemuthianum*. Stage 1, appressorium formation; stage 2, hypersensitive response visible; stage 3, hypersensitive response complete; stage 4, 1% lesions in compatible combination; stage 5, 80% lesions; stage 6, 100% lesions. (Data from Bailey & Deverall 1971.)

using inhibitors of the biosynthetic pathways leading to phytoalexin production. When phytoalexin synthesis is blocked, the resistance of tissues to pathogen attack is reduced. Even more convincing data have been obtained in genetic studies on the degradation of

phytoalexins by pathogenic fungi. *Nectria haematococca* is a pathogen of peas that produces an enzyme, pisatin demethylase, which inactivates the major pea phytoalexin, pisatin (Fig. 9.8). In crosses between fungus strains producing high and low amounts of the enzyme, only progeny strains producing high amounts of the enzyme were able to cause significant lesions on pea plants (Fig. 9.11). Pathogenicity towards the host was therefore correlated with ability to inactivate the phytoalexin. More recently the gene encoding pisatin demethylase has been cloned and introduced into another fungus, *Cochliobolus heterostrophus*, which is normally a pathogen of maize. Surprisingly, transformed strains expressing the gene were able to grow in pea plants, suggesting that ability to degrade a host phytoalexin is a specificity determinant. Yet even in this genetically well-characterized example, there is still controversy over the real importance of phytoalexin degradation, as some mutant strains of *N. haematococca*, in which enzyme production has been abolished through disruption of the gene, apparently retain pathogenicity to peas. The current consensus appears to be that ability to detoxify phytoalexins can affect the extent of lesion development, but that additional factors are required for full pathogenicity. Table 9.5 summarizes the evidence that phytoalexins play a role in active defence.

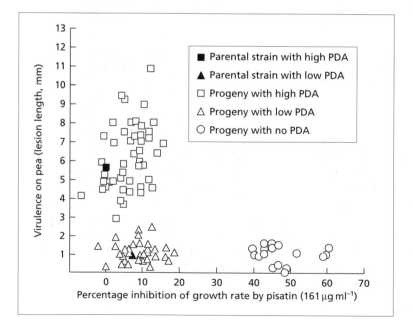

Fig. 9.11 Segregation of virulence towards pea, tolerance of the phytoalexin pisatin, and pisatin demethylase (PDA) activity in progeny from a cross between two strains of *Nectria haematococca*. The parent and progeny with high PDA activity are tolerant of pisatin in culture and cause extensive lesions on peas, whereas the parent and progeny with low PDA activity are tolerant in culture but unable to cause significant lesions. A third group of progeny with no PDA activity are inhibited in culture and also lack virulence. (After Kistler & VanEtten 1984.)

Table 9.5 Evidence that phytoalexins play a role in plant defence.

Accumulate *de novo* in challenged tissues

Accumulate earlier in resistant than in susceptible host–pathogen combinations

Local concentrations in tissues are sufficient to inhibit microbial growth

Inhibitors of phytoalexin biosynthesis increase the susceptibility of plant tissues to infection

Mutant plants deficient in phytoalexin production allow more extensive microbial colonization than wild-type plants

Fungal strains which degrade phytoalexins are more virulent than strains unable to detoxify these compounds

Molecular biology of induced plant defence

The previous section described some of the responses observed in plant tissues following microbial challenge, including changes in cell walls, programmed cell death, and the accumulation of antibiotic compounds such as phytoalexins. The next part of this chapter considers the molecular events underlying these changes, and in particular the regulation of induced defence responses.

Pathogenesis-related proteins

If one prepares protein extracts from healthy plants and compares them with extracts from plants responding actively to microbial attack it is clear that the pattern of protein synthesis is altered following challenge. Some of the new proteins detected are likely to be enzymes involved in the synthesis of defence compounds such as phytoalexins, but others have quite different functions.

Inoculation of tobacco leaves with tobacco mosaic virus (TMV) can lead either to systemic infection, or to the formation of local necrotic lesions, depending upon the presence of a dominant resistance gene (Plate 3b, facing p. 12). This hypersensitive-type reaction is accompanied by the production of new, host-encoded proteins described as pathogenesis-related proteins (usually abbreviated to PRs). PRs are relatively small, stable proteins which accumulate predominantly in the intercellular spaces of plant tissues. Their appearance is easily demonstrated by gel electrophoresis of tissue extracts (Fig. 9.12). Similar proteins have now been detected in many other plant species responding to virus infection or challenge by

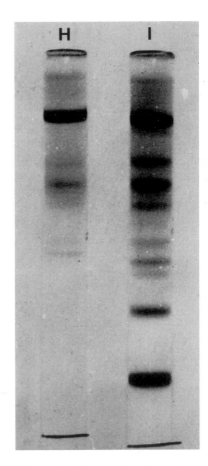

Fig. 9.12 Electrophoretic separation of proteins extracted from healthy tobacco leaves (H) and leaves forming local lesions in response to inoculation (I) with tobacco mosaic virus (TMV). Note appearance of novel, PR proteins in infected leaves. (Courtesy of Jon Antoniw.)

microorganisms; PR production is also induced by various chemicals, including hormones such as ethylene, as well as other types of stress.

PRs were initially classified on the basis of their physical properties, such as electrophoretic mobility, size, acidity and so on, and grouped into 'families' of related proteins (Table 9.6). For several years after their discovery the biological function of PRs was unknown, although their appearance in infected or stressed plants suggested some role in defence or protection from tissue damage. Biological functions have now been assigned to most groups of PRs, although the activity of some, for instance the PR1 family, was not demonstrated until very recently (Table 9.6). PR1 proteins from tobacco and tomato inhibit the growth of oomycete pathogens such as *Phytophthora infestans*. PR2 and PR3 turned out to be enzymes active against polysaccharides containing β-1,3-glucans, or another polymer, chitin. Interestingly, neither of these substrates occurs naturally in plants. Instead they are the main polymers found in the cell walls of many fungi, while chitin is also a key component of the insect exoskeleton. A direct role in defence against pathogens and pests therefore seems likely.

Purified chitinases from plants are able to inhibit the growth of at least some fungi *in vitro*, while mixtures of chitinase and β-1,3-glucanase are much more active. A more direct test of the possible role of these enzymes in defence is to assess their effects in transgenic plants; when a chitinase gene was transferred into tobacco or oilseed rape plants (see p. 229) production of the enzyme enhanced resistance to certain fungal pathogens. Current evidence suggests that such enzymes may have a direct role in slowing down fungal growth, thereby delaying invasion of the plant and allowing other host defence systems to operate, and an indirect role in releasing breakdown products from fungal cell walls, which act as signal molecules

which the plant can recognize. Plants may therefore have evolved these enzymes both as a general type of resistance against fungi, as well as an early warning system.

Other defence-related enzymes

The participation of active host metabolism in the expression of induced defence has been known since early experiments showed that treatment of plant tissues with metabolic inhibitors increased their susceptibility to microbial infection. More recent work has defined the metabolic pathways involved, and identified some of the key biosynthetic enzymes producing defence compounds such as phytoalexins.

An important precursor for the synthesis of phenolic compounds is the aromatic amino acid phenylalanine; in healthy tissues this and other amino acids are usually incorporated into proteins. Following infection or injury, however, phenylalanine is converted to *trans*-cinnamic acid and enters the pathway for biosynthesis of phenylpropanoid compounds (Fig. 9.13). These include various groups of phytoalexins, as well as precursors of structural defence molecules such as lignin.

If one assays enzymes such as phenylalanine ammonia lyase (PAL), catalysing the first step in the pathway, and chalcone synthase (CHS), catalysing the branch point to isoflavonoid compounds (Fig. 9.13), activity is usually found to be increased in infected plant tissues. This suggests that phenylpropanoid biosynthesis is accelerated as part of the plant response. Assays of activity alone, however, provide only limited information. It is not possible to decide, for instance, whether any increase is due to synthesis of new enzyme, or activation of preformed enzyme. A further problem is that the induction of enzymes such as PAL seems to be nonspecific, and

---

Table 9.6 Pathogenesis-related (PR) proteins of tobacco.

Family	Function	Size
1	Antifungal	16 kDa
2	β-1,3-Glucanase	30–40 kDa
3	Chitinase	20–28 kDa
4	Antifungal	13–15 kDa
5	Possible protease inhibitor	23 kDa

Fig. 9.13 (*Facing page*) Pathway for biosynthesis of phenylpropanoid defence metabolites in legumes. PAL, phenylalanine ammonia lyase; CA4H, cinnamic acid 4-hydroxylase; 4CL, 4-coumarate : CoA ligase; SS, stilbene synthase; CAD, coniferyl alcohol dehydrogenase; CHS, chalcone synthase; CHI, chalcone isomerase; IFS, isoflavone synthase; IFOMT, isoflavone O-methyl transferase; IF2′OHase, isoflavone 2′hydroxylase; IFR, isoflavone reductase; PS, pterocarpan synthase. (After Dixon & Harrison 1990.)

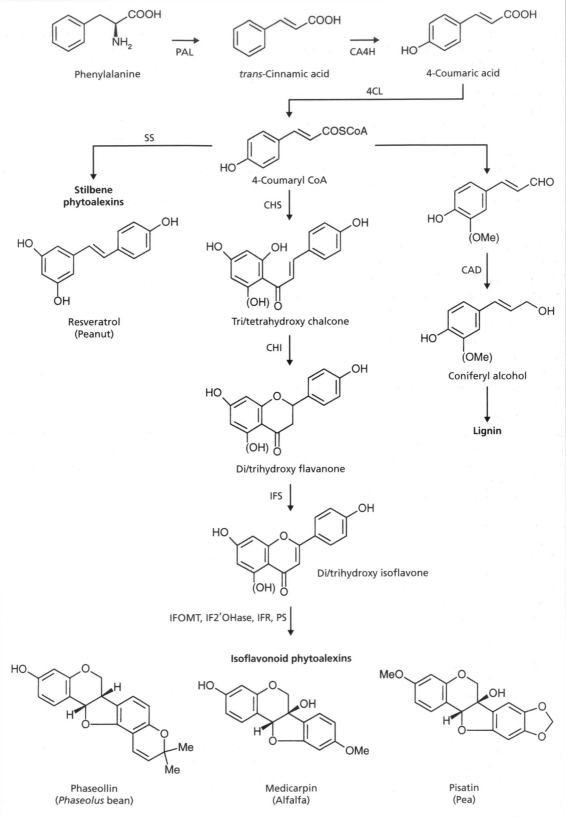

activity increases in response to a number of stimuli, including light and mechanical injury as well as infection by fungi, bacteria, and virus strains which cause local lesions. Much more precise molecular approaches are required to determine whether the mechanism of increase is the same in all cases.

Many of the enzymes involved in this pathway have now been purified, which opens up further ways of dissecting the defence response. For a start the pure protein can be used to raise antibodies specific for the enzyme; alternative assays using such antibodies show that the increase in enzyme activity is accompanied by increases in the amount of enzyme protein present. This suggests that the response involves enzyme synthesis and most likely therefore activation of host genes. Deciphering some or all of the amino acid sequence of the purified enzyme also permits prediction of the nucleotide sequence of the corresponding genetic message. Hence one can design an oligonucleotide probe capable of hybridizing with the message. This allows detection of any mRNA produced and ultimately detection of the structural gene itself. A different approach to the problem is to isolate cell fractions, typically nuclei, producing the message, and express the protein products through an *in vitro* translation system. A combination of all these molecular tricks has been used successfully to analyse the early induction of phenylpropanoid biosynthetic enzymes such as PAL in great detail.

Figure 9.14 shows the pattern of increase in PAL activity in beans undergoing a hypersensitive response to the bacterium *Pseudomonas syringae* pv. *phaseolicola*. Activity peaks around 20 h after inoculation and then declines; the corresponding mRNA for PAL, detected using a copy (cDNA) probe, peaks around 12 h. Isoflavonoid phytoalexins begin to appear around 24 h. This experiment suggests that following challenge with an incompatible bacterial isolate the PAL gene is switched on, new message is transcribed for several hours, new enzyme protein is synthesized, and antibiotic secondary metabolites produced via the phenylpropanoid pathway accumulate. In the compatible (i.e. susceptible) bean plus bacterium combination, none of these changes occurs within the first 24 h after inoculation.

Even more rapid responses to microbial challenge can be demonstrated using cell cultures rather than tissues or intact plants. This approach permits simultaneous exposure of a whole population of cells to a pathogen, or more usually some biologically active

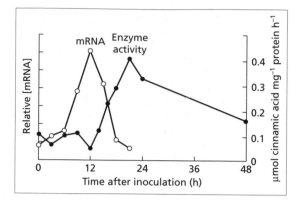

Fig. 9.14 Timing of induction of phenylalanine ammonia lyase (PAL) mRNA and enzyme activity in the hypersensitive response of *Phaseolus* bean to an incompatible race of *Pseudomonas syringae* pv. *phaseolicola*. Relative mRNA concentration was determined by probing dot blots and northern blots with a cDNA probe for bean PAL. Chalcone synthase mRNA showed a similar pattern of induction. PAL enzyme activity increased in parallel with mRNA concentration but about 6 h later. Antibacterial isoflavonoids were detectable by 24 h after inoculation. (After Slusarenko *et al.* 1991; by permission of Oxford University Press.)

fraction (or **elicitor**; see p. 169) derived from the pathogen. Such studies show that defence genes encoding enzymes such as PAL, CHS and cinnamyl-alcohol dehydrogenase, an enzyme on the branch to lignin synthesis, are activated within minutes of exposure to challenge. Figure 9.15 shows the rapid appearance of mRNA transcripts for several defence-related enzymes in bean suspension cells treated with an extract from fungal cell walls. This represents one of the most rapid gene activation responses known in plant cells. What is more, coordinated induction of these genes takes place; thus, in response to a microbial signal, genes specifying a series of enzymes in a biosynthetic pathway not operating in healthy plant cells are simultaneously activated, and the whole pattern of secondary metabolite production by the host is altered.

Cell-wall proteins

A further group of plant proteins which change during active defence are located in the cell wall. Because these structural proteins contain a high proportion of particular amino acids, especially hydrox-

Fig. 9.15 Induction of transcription of several defence-related enzymes in bean cells treated with a fungal elicitor. Blot shows appearance of mRNA for phenylalanine ammonia lyase (PAL), chalcone synthase (CHS), chalcone isomerase (CHI) and chitinase within minutes of exposure to a fungal cell-wall preparation. H1 is a constitutively expressed gene unaffected by microbial challenge. (From Hedrick *et al.* 1988. © American Society of Plant Physiologists. Reprinted by permission of the publishers.)

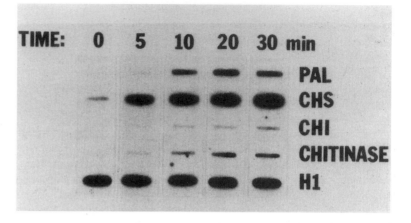

yproline, they are termed **hydroxyproline-rich glycoproteins**, usually abbreviated to **HRGPs**. Following infection, HRGPs accumulate in host cell walls and may contribute to resistance by trapping pathogen cells, or acting as structural barriers and sites for lignin deposition (Fig. 9.16).

Molecular analysis of this response has revealed a rather similar pattern of events to that observed with enzymes involved in phytoalexin synthesis. Following challenge, mRNA encoding HRGPs appears in host cells, and new protein is synthesized. This response occurs earlier in resistant host–pathogen combinations than in susceptible combinations (Fig. 9.17). One subtle difference, however, is in the timing of the response. Experiments with cell cultures exposed to challenge suggest that message for HRGPs appears later than that for PAL and related enzymes; hence, there is a slower, more sustained activation of gene expression, rather than a very rapid, transient increase. A further difference which may be significant has been observed in intact tissues. Typically, phytoalexin synthesis is localized to tissues directly challenged by the pathogen, while HRGP gene activation can be detected in cells more distant from the site of infection (see Fig. 9.17).

Recently, another change in the walls of plant cells exposed to challenge or wounding has been detected. Analysis of protein extracts from such cells showed that certain types of HRGPs rapidly disappear following treatment of the cells with pathogen elicitors. At first sight this observation seems to conflict with the notion that cell walls are reinforced by the addition of new protein and other polymers following infection. The mystery was solved when it was shown

that the apparent disappearance of these proteins was in fact due to a sudden change in their properties. Cross-linking of amino acid chains in the proteins alters their solubility so that they remain bound to the wall during the extraction procedure, and hence are no longer detected. Insolubilization of these proteins appears to toughen the cell wall and also increases its resistance to digestive enzymes. The cross-linking reaction is extremely rapid, taking place within a few minutes of challenge, and hence is complete before any of the molecular changes based on gene activation. The reaction appears to be due to sudden changes in oxidative metabolism, and in particular the generation of hydrogen peroxide at the cell surface (see below). Thus, transcription of host genes is not involved in this response. Similarly, the synthesis of callose (Table 9.2) appears to rely on enzyme activation rather than gene transcription.

The trigger for active plant defence

It should be clear from the previous sections that active defence in plants is a multicomponent process. Expression of resistance includes early, irreversible membrane damage, programmed cell death, and a cascade of molecular responses such as activation of genes for phytoalexin synthesis, HRGPs, chitinase and other PR proteins. The precise timing of these responses may vary from system to system. One central question concerns the initial trigger which activates the whole process. The basic model envisages a receptor which senses the presence of a pathogen, coupled to a signal transduction

(a)

(c)

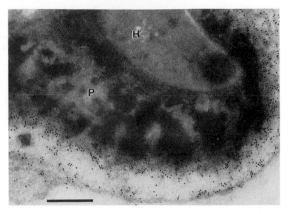

(b)

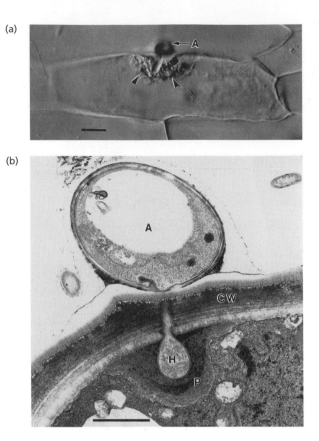

Fig. 9.16 Penetration of epidermal cell of bean by the fungus *Colletotrichum lindemuthianum*. (a) Light micrograph showing papilla (arrow heads) formed beneath fungal appressorium (A). Scale bar = 10 μm. (b) Electron micrograph of section through penetrated cell showing cell wall (CW) and a small infection hypha (H) encased by a papilla (P). Scale bar = 1 μm. (c) High-power view of penetrating hypha (H) and surrounding papilla (P) labelled with a gold-conjugated antibody specific for hydroxyproline-rich glycoproteins. Small black particles locate the protein mainly in the outer layers of the papilla. Scale bar = 0.5 μm. (All parts courtesy of Richard O'Connell.)

pathway which initiates the host-cell response. This is discussed in more detail in Chapter 10. For the moment we will concentrate on the earliest detectable responses by plant cells, which may provide clues to the identity of the recognition system itself.

It seems likely that the first interactions take place at the plasma membrane, and that changes in membrane properties may initiate other cell responses. Studies on the electrical membrane potential of cells undergoing HR suggest that there are early changes in the lipid matrix or, possibly, protein channels. There is also a burst of oxidative activity with the release of highly reactive superoxide anions and hydrogen peroxide. Such active oxygen species may have diverse effects, including oxidizing membrane lipids, catalysing protein cross-linking, and causing cell damage. Hydrogen peroxide can induce defence genes and trigger localized cell death and hence is implicated in the overall orchestration of the hypersensitive response. Interestingly, a similar early oxidative burst is seen in animal defence cells such as phagocytes responding to bacterial infection.

The oxidative burst is associated with increases in the activity of several enzymes, including lipoxygenase, which acts on membrane lipids to generate products which are active as signal molecules switching on plant defence (see below). The exact temporal relationship between all these events is not yet clear. Programmed cell death may itself be a trigger for active defence systems in surrounding cells, thereby amplifying the response.

Responses to wounding

When plant tissues are damaged, for instance by mechanical wounding or attack by insects, defence genes are switched on in cells adjacent to the wound site. These wound-inducible genes encode products such as proteinase inhibitors which are active against digestive enzymes and hence interfere with the gut function of herbivorous insects. Similar inhibitor pro-

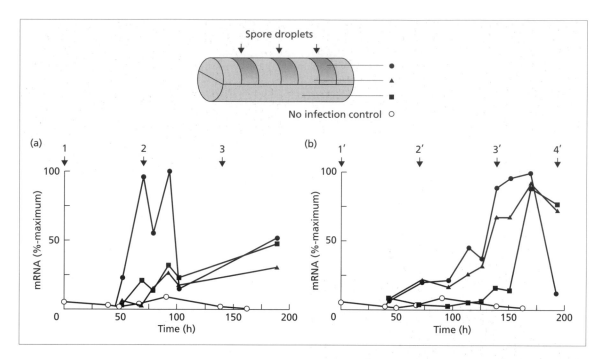

Fig. 9.17 Time course of accumulation of mRNA for hydroxyproline-rich glycoprotein in *Phaseolus* bean hypocotyl tissues inoculated with different races of *Colletotrichum lindemuthianum*. The diagram at the top indicates the different tissue samples taken for analysis shown in the corresponding graphs below. (a) Incompatible race. Time points: 1, spore inoculation; 2, onset of hypersensitive flecks; 3, intense hypersensitive response at all inoculated sites. (b) Compatible race. Time points: 1′, spore inoculation; 2′, no visible symptoms; 3′, onset of symptom development; 4′, extensive water-soaked lesions spreading from inoculation site. (After Showalter *et al.* 1985.)

teins are also synthesized in tissues remote from the initial injury. This observation suggests that plants can respond to an initial attack by activating defence processes in other parts of the plant, as if in anticipation of further challenge. Presumably a molecular signal generated by the initial wound process must move both locally and also to distant cells to activate these defence responses. How might such a signal pathway operate?

Several possible chemical inducers of the wound response have now been identified. One active class of compounds are cell-wall fragments, known as oligosaccharins (see p. 171), which may be released as a result of cell damage. Others are the hormone abscisic acid, lipid-derived compounds such as jasmonic acid, and a small peptide molecule comprising 18 amino acids, christened systemin (Fig. 9.18). The latter is highly active at low concentrations, is mobile in the plant, and hence is a good candidate for the systemic signal itself. Systemin is derived from a much larger 200 amino acid precursor, prosystemin, from which it is released upon wounding. In this respect systemin behaves much like certain polypeptide hormones found in animals. Based on these observations, a model of the wound-inducible signal pathway in plants has now been proposed (Fig. 9.19). Following damage, localized signals such as oligosaccharins are generated from cell walls. Systemin is released from the precursor protein, prosystemin, and translocated throughout the plant, inducing defence functions in remote tissues. Binding of oligosaccharins or systemin to receptors in the plasma membrane generates intracellular signals derived from membrane lipids. The enzyme lipoxygenase is involved in the conversion of lipid precursors to jasmonic acid which switches on the genes encoding defence proteins, such as the proteinase inhibitors.

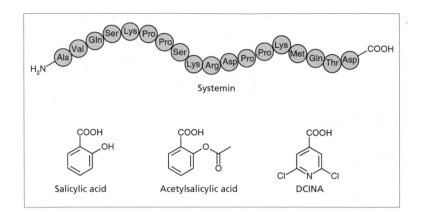

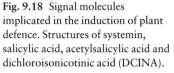

Fig. 9.18 Signal molecules implicated in the induction of plant defence. Structures of systemin, salicylic acid, acetylsalicylic acid and dichloroisonicotinic acid (DCINA).

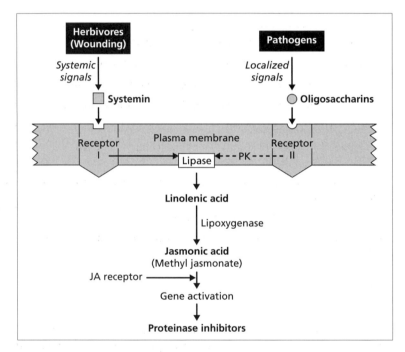

Fig. 9.19 Proposed model for the signal pathway leading to expression of proteinase inhibitor genes in plants. JA, jasmonic acid; PK, protein kinase. (After Farmer & Ryan 1992.)

Can plants be immunized?

Active defence in plants bears some similarity to the immune response of animals (Fig. 9.20). Like animals, plants can recognize pathogens and respond by activating a series of defence systems, but can they remember the encounter so that they are protected against later exposure to a pathogen?

During the early years of this century there were repeated attempts to demonstrate the existence in plants of an immune system capable of producing specific antibodies against microorganisms. These experiments found no reliable evidence for plant antibodies, or indeed defence cells similar to phagocytes. Nevertheless the belief in plant 'immunity' persisted, typified by K. Chester, who in 1933 suggested that acquired resistance plays 'an important role in the preservation of plants in nature'. More recent work has to some extent validated this claim, as it has proved possible to protect certain plants against pathogens using immunization techniques similar to those employed so successfully in medicine.

Fig. 9.20 Animal and plant immune systems compared.

Table 9.7 Some characteristics of systemic acquired resistance (SAR).

Induced by agents/pathogens causing necrosis, e.g. local lesions
Delay of several days between induction and full expression
Protection conferred on tissues not exposed to inducer inoculation
Expressed as reduction in lesion number, size, spore production, etc.
Protection is long-lasting, often for weeks or even months
Protection is non-specific, i.e. effective against pathogens unrelated to inducing agent
Development of SAR associated with expression of several gene families, e.g. PR proteins
Signal for SAR is translocated and graft-transmissible
Protection not passed on to seed progeny; transmission to clonal tissues is unresolved

PR, pathogenesis-related.

Figure 9.21a diagrammatically shows a typical 'immunization' experiment. Prior inoculation of a leaf with an incompatible pathogen, or elicitor extracted from a pathogen, protects against infection by a live, virulent pathogen administered several days later. This phenomenon has been demonstrated with numerous host–pathogen combinations, and was instrumental in the discovery of phytoalexins. Such protection is, however, short-lived and localized to the treated tissues; developing leaves remain susceptible to subsequent challenge.

A more intriguing experiment is shown in Fig. 9.21b. In this case exposure of leaf 1 to a pathogen leads to long-lasting protection of all subsequently developing tissues. Such **systemic acquired resistance** (often abbreviated to SAR) has been described in several host–pathogen systems, notably members of the family Cucurbitaceae inoculated with various fungi, bacteria or viruses, and tobacco plants exposed to local-lesion-inducing strains of TMV or the blue mould fungus *Peronospora tabacina*. More recently SAR has also been demonstrated in the experimental model plant *Arabidopsis*. Unlike animal immunization this form of protection is nonspecific; the acquired resistance operates against a broad range of pathogens and can last for several weeks to months after the initial, inducing infection. The degree of protection is often high, and the resistance has been shown to be effective under field conditions. In the case of tobacco blue mould, protection compares

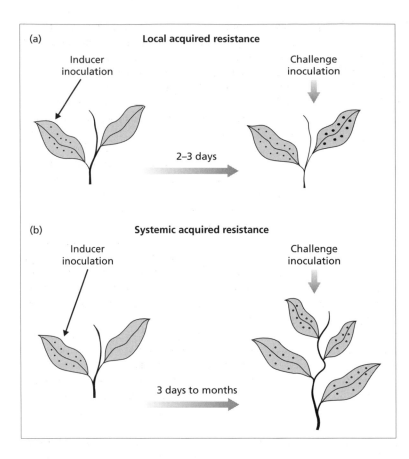

(a)

Local acquired resistance

Inducer
inoculation

Challenge
inoculation

2–3 days

(b)

Systemic acquired resistance

Inducer
inoculation

Challenge
inoculation

3 days to months

Fig. 9.21 (a) Local and (b) systemic acquired resistance.

favourably with that achieved using the best available fungicides. Table 9.7 summarizes the main characteristics of SAR.

Due to the potential practical significance of SAR for disease control it is important to understand how it works. There are, in fact, several ways in which prior treatment with a pathogen can affect subsequent challenge inoculations. Some do not involve any host response, and instead are due to direct interactions between the two inoculants. Such suppression or antagonism is the basis for some forms of biological control (see p. 242). Also, SAR should not be confused with cross-protection, the phenomenon whereby infection with an attenuated or weakly virulent virus protects the plant against subsequent infection with a related, virulent virus (see p. 230).

Typically, in SAR the primary inoculation involves a chemical or biological agent which induces necrosis in the treated tissues. This triggers active defence systems in the surrounding cells. In addition, a signal must be produced and translocated to other parts of the plant, where it primes unexposed tissues against subsequent challenge (Fig. 9.22). The changes induced in distant tissues are not fully understood, although alterations in the structure and chemistry of the cuticle and cell wall have been detected, and defence-related genes such as those encoding PR proteins are expressed. Upon challenge, it appears that the full repertoire of active defence systems is activated more quickly then in unprotected plants. Thus SAR seems to sensitize the plant to subsequent exposure to pathogens.

Most current interest in SAR concerns the possible identity of the translocated signal. It has been known for some time that the signal is graft-transmissible and presumably moves in the vascular system. As the defence genes switched on differ from those induced by damage, the signal pathway is presumed to be distinct from the wound-inducible process shown in Fig. 9.19. Work on SAR in tobacco and cucumber plants has detected changes in the composition of phloem

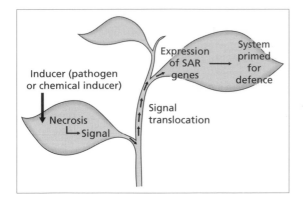

Fig. 9.22 Model for systemic acquired resistance (SAR).

sap following inducer treatments; one compound which increases on a time scale correlated with the expression of induced resistance is salicylic acid, a chemical relative of aspirin (see Fig. 9.18). When applied to tobacco plants this simple compound is also an active chemical inducer of SAR. Further evidence that salicylic acid plays an important part in the signal pathway has come from work on *Arabidopsis* mutants altered in SAR expression, and transgenic plants engineered to break down salicylic acid. In the latter experiments, a bacterial gene for salicylate hydroxylase, an enzyme which degrades salicylic acid, was introduced into tobacco. These trans-

formed plants were unable to accumulate salicylic acid to any significant extent, and at the same time lost their ability to express SAR. Hence there is a direct correlation between the accumulation of salicylic acid and induction of SAR. However, there is still a question mark over the identity of the long-distance translocated signal. Grafting experiments in which the transgenic tobacco plants described above were used as rootstocks and normal tobacco as the shoot (scion) have shown that such grafted plants can still express SAR in their leaves following an inducer treatment of the rootstock. As any salicylic acid produced in the rootstock tissues is quickly degraded, this result suggests that the signal moving from root to shoot cannot be salicylic acid. The observation that some chemical agents (see below) are able to activate SAR both in *Arabidopsis* mutants, and in transgenic tobacco plants unable to accumulate salicyclic acid, also supports the idea that a further signal molecule must be involved.

Another intriguing possibility is that plants under threat from insect pests or pathogens might be able to alert other plants in the vicinity. This initially fanciful idea has now received experimental support from work on tobacco reacting to inoculation with TMV. During local lesion formation these plants evolved significant amounts of methyl salicylate, a volatile derivative of salicylic acid. The airborne signal was sufficient to induce PR protein expression in neigh-

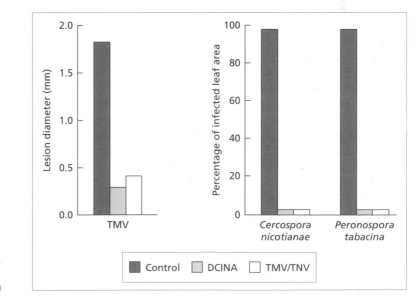

Fig. 9.23 Protection of tobacco against tobacco mosaic virus (TMV), frog-eye disease (*Cercospora nicotianae*) or blue mould (*Peronospora tabacina*) by pretreatment with dichloroisonicotinic acid (DCINA), TMV or tobacco necrosis virus (TNV). (After Vernooij *et al.* 1995.)

bouring healthy plants, and to boost their resistance to the virus. It is not yet known whether this type of communication occurs in the field, but it has been suggested that among crowded populations of crop plants methyl salicylate might accumulate to levels able to exert a physiological effect. This discovery adds a whole new dimension to the phenomenon of systemic acquired resistance. Generation of a volatile signal could serve as an early warning system priming the defence pathways of plants not yet under attack.

One of the most significant practical aspects of SAR is the recent discovery of several chemical inducers of the response. For instance, treatment with the compound dichloroisonicotinic acid (DCINA; see Fig. 9.18) provides systemic protection as effective as that induced by biotic agents such as viruses (Fig. 9.23). Screening of chemicals for their ability to induce SAR has now led to the development of a new generation of fungicides which act via plant defence systems, rather than through direct effects on the pathogen itself (see Chapter 11).

Conclusions

The elegant and often ingenious molecular analyses described in the previous section have confirmed that plants possess a sophisticated battery of inducible defence systems, activated upon challenge by potential pathogens. Early, rapid changes in cells appear to slow the progress of an invader, buying time for the expression of systems which first confine and then kill the pathogen. Coordinated induction of defence genes underlies most of the observed changes.

There are still important questions to be answered, however, concerning the regulation and specificity of such systems. Many of the molecular events accompanying active defence against pathogens are also seen in plant tissues subjected to wounding or other abiotic treatments. Undoubtedly, a proportion of the observed changes are non-specific stress responses occurring in damaged cells; others may be more directly linked to specific resistance mechanisms. Work on some defence proteins, for instance HRGPs, now suggests that more than one gene may encode similar proteins, and that selective activation of different members of a multigene family may occur in response to different types of stress. Such molecular specialization may permit flexible deployment of plant defence systems depending upon the precise nature of the threat encountered. It also suggests that subtle differences exist in the signal transduction pathways operating in response to different microorganisms, chemicals or injury.

Ultimately such differences in the regulation of defence systems might explain, at least in part, why certain microbial invaders successfully colonize the host, while the majority fail. Most of the vital clues to this question of specificity concern the initial recognition event, which is the subject of the next chapter.

Further reading

Books

Bailey, J.A. & Mansfield, J.W. (eds) (1982) *Phytoalexins.* Blackie, Glasgow.

Fritig, B. & Legrand, M. (eds) (1993) *Mechanisms of Plant Defense Responses.* Kluwer Academic Publishers, Dordrecht.

Goodman, R.N. & Novacky, A.J. (eds) (1994) *The Hypersensitive Reaction in Plants to Pathogens. A Resistance Phenomenon.* APS Press, St Paul, Minn.

Reviews and papers

Bowyer, P., Clarke, B.R., Lunness, P., Daniels, M.J. & Osbourn, A.E. (1995) Host range of a plant pathogenic fungus determined by a saponin detoxifying enzyme. *Science* **267**, 371–374.

Bradley, D.J., Kjellbom, P. & Lamb, C.J. (1992) Elicitor- and wound-induced oxidative cross-linking of a proline-rich plant cell wall protein: A novel, rapid defense response. *Cell* **70**, 21–30.

Cao, H., Bowling, A., Gordon, S. & Dong, X. (1994) Characterization of an *Arabidopsis* mutant that is nonresponsive to inducers of systemic acquired resistance. *The Plant Cell* **6**, 1583–1592.

Collinge, D.B., Kragh, K.M., Mikkelsen, J.D., Nielsen, K.K., Rasmussen, U. & Vad, K. (1993) Plant chitinases. *The Plant Journal* **3**, 31–40.

Delaney, T.P. (1997) Genetic dissection of acquired resistance to disease. *Plant Physiology* **113**, 5–12.

Dixon, R.A. (1986) The phytoalexin response: elicitation, signalling and control of host gene expression. *Biological Reviews* **61**, 239–291.

Hedrick, S.A., Bell, J.N., Boller, T. & Lamb, C.J. (1988) Chitinase cDNA cloning and mRNA induction by fungal elicitor, wounding and infection. *Plant Physiology* **86**, 182–186.

Hunt, M.D. & Ryals, J.A. (1996) Systemic acquired resistance signal transduction. *Critical Reviews in Plant Sciences* **15**, 583–606.

Lamb, C.J., Lawton, M.A., Dron, M. & Dixon, R.A. (1989) Signals and transduction mechanisms for activation of plant defenses against microbial attack. *Cell* 56, 215–224.

Malamy, J. & Klessig, D.F. (1992) Salicylic acid and plant disease resistance. *The Plant Journal* 2, 643–654.

Métraux, J.P., Signer, H., Ryals, J. *et al.* (1990) Increase in salicylic acid at the onset of systemic acquired resistance in cucumber. *Science* 250, 1004–1006.

Niderman, T., Genetet, I., Bruyère, T. *et al.* (1995) Pathogenesis-related PR-1 proteins are antifungal: Isolation and characterisation of three 14-kilodalton proteins of tomato and of a basic PR-1 of tobacco with inhibitory activity against *Phytophthora infestans*. *Plant Physiology* 108, 17–27.

O'Connell, R.J., Brown, I.R., Mansfield, J.W. *et al.* (1990) Immunocytochemical localization of hydroxyproline-rich glycoproteins accumulating in melon and bean at sites of resistance to bacteria and fungi. *Molecular Plant–Microbe Interactions* 3, 33–40.

Osbourn, A., Bowyer, P., Lunness, P., Clarke, B. & Daniels, M. (1995) Fungal pathogens of oat roots and tomato leaves employ closely related enzymes to detoxify different host plant saponins. *Molecular Plant–Microbe Interactions* 8, 971–978.

Peumans, W.J. & Van Damme, E.J.M. (1995) Lectins as plant defense proteins. *Plant Physiology* 109, 347–352.

Rogers, E.E., Glazebrook, J. & Ausubel, F.N. (1996) Mode of action of the *Arabidopsis thaliana* phytoalexin camalexin and its role in *Arabidopsis*–pathogen interactions. *Molecular Plant–Microbe Interactions* 9, 748–757.

Schäfer, W., Straney, D., Ciuffetti, L., Van Etten, H.D. & Yoder, O.C. (1989) One enzyme makes a fungal pathogen, but not a saprophyte, virulent on a new host plant. *Science* 246, 247–248.

Shewry, P.R. & Lucas, J.A. (1997) Plant proteins that confer resistance to pests and pathogens. *Advances in Botanical Research* 26, 136–192.

Showalter, A.M., Bell, J.N., Cramer, C.L., Bailey, J.A.,

Varner, J.A. & Lamb, C.J. (1985) Accumulation of hydroxyproline-rich glycoprotein mRNAs in response to fungal elicitor and infection. *Proceedings of the National Academy of Sciences of the USA* 82, 6551–6555.

Shulaev, V., León, J. & Raskin, I. (1995) Is salicylic acid a translocated signal of systemic acquired resistance in tobacco? *The Plant Cell* 7, 1691–1701.

Shulaev, V., Silverman, P. & Raskin, I. (1997) Airborne signalling by methyl salicylate in plant pathogen resistance. *Nature* 385, 718–721.

Taylor, S., Massiah, A., Lomonossoff, G., Roberts, L.M., Lord, J.M. & Hartley, M. (1994) Correlation between the activities of five ribosome-inactivating proteins in depurination of tobacco ribosomes and inhibition of tobacco mosaic virus infection. *The Plant Journal* 5, 827–835.

Tenhaken, R., Levine, A., Brisson, L.F., Dixon, R.A. & Lamb, C. (1995) Function of the oxidative burst in hypersensitive disease resistance. *Proceedings of the National Academy of Sciences of the USA* 92, 4158–4163.

Terras, F.R.G., Eggermont, K., Kovaleva, V. *et al.* (1995) Small cysteine-rich antifungal proteins from radish: their role in host defense. *The Plant Cell* 7, 573–588.

Uknes, S., Mauch-Mani, B., Moyer, M. *et al.* (1992) Acquired resistance in *Arabidopsis*. *The Plant Cell* 4, 645–656.

Vernooij, B., Friedrich, L., Ahl Goy, P. *et al.* (1995) 2,6-Dichloroisonicotinic acid-induced resistance to pathogens without the accumulation of salicylic acid. *Molecular Plant–Microbe Interactions* 8, 228–234.

Ward, E.R., Uknes, S.J., Williams, S.C. *et al.* (1991) Coordinate gene activity in response to agents that induce systemic acquired resistance. *The Plant Cell* 3, 1085–1094.

Wasmann, C.C. & VanEtten, H.D. (1996) Transformation-mediated chromosome loss and disruption of a gene for pisatin demethylase decrease the virulence of *Nectria haematococca* on pea. *Molecular Plant–Microbe Interactions* 9, 793–803.

10   Host–Pathogen Specificity

'The nature of specific or major gene resistance . . . is one of the most challenging yet elusive problems in plant pathology. What are the mechanisms whereby the genes in the host and the parasite interact to control the development of the disease?' [H. Flor, 1900–1991]

Why certain microorganisms are able to cause disease, whilst the majority cannot, is a crucial question in both medical microbiology and plant pathology. Some of the properties distinguishing pathogens from non-pathogens have already been described in Chapter 8. Even where a microorganism has the ability to cause disease, it does not attack all plants equally. For instance, the bean rust fungus, *Uromyces appendiculatus,* infects bean plants but not coffee trees, while the coffee rust fungus, *Hemiliea vastatrix,* behaves vice versa. Hence each pathogen is restricted to particular plant types and has a characteristic **host range**.

Host–pathogen specificity involves not only those factors determining virulence in the pathogen, but also those conferring resistance on the host. Experimental analysis is therefore complex. Understanding specificity is, however, important for both practical and conceptual reasons; a complete analysis should suggest more precise and reliable ways of intervening in the disease process, either through rationally designed chemicals which shift the balance of the interaction towards host resistance, or by genetic manipulation to produce novel types of resistant crops. Furthermore, insights into the nature of host–pathogen specificity are likely to prove relevant to other biological systems involving compatibility, such as the pollen–stigma interaction in plant fertilization, or recognition between mutualistic microorganisms and their hosts.

Types of specificity

Host–pathogen specificity involves several separate and probably different phenomena (Table 10.1). First there is the distinction between pathogenic organisms and those which are unable to attack living hosts under any circumstances. In the example shown in Table 10.1, *Cladosporium herbarum* is a saprophytic fungus which lives on leaf surfaces without causing disease, while the related species *C. fulvum* can invade leaves and grow in intercellular spaces, eventually causing necrotic lesions. Next, a particular pathogen has host species, which are infected, and **non-hosts**, which are never infected. *Cladosporium fulvum,* which is a tomato leaf mould pathogen, attacks tomatoes but not other plants such as bean. Some pathogens have very wide host ranges, attacking many different host species, while others are highly specific, attacking only a few closely related plants. An example of such specialization is seen with the rust and powdery mildew fungi, where distinct form species (abbreviation f.sp.) occur on particular cereal hosts. For instance, the rust fungus *Puccinia graminis* occurs as forms attacking wheat (*Triticum*; *P.g.* f.sp. *tritici*), barley (*Hordeum*; *P.g.* f.sp. *hordei*) and other cereals. Even greater specificity may be detected in the interaction of particular pathogen isolates (often described as races) and specific host lines (cultivars) (Table 10.1b). Thus a single genotype of a pathogen may infect only certain susceptible genotypes of the host. This race–cultivar specificity is described in genetic terms by the gene-for-gene model (see p. 27).

A further form of specificity, often overlooked, is where a parasitic microorganism is restricted to particular host tissues or organs. Such **tissue specificity** is suggested in many of the common names for plant diseases, such as root rot, flower blight and vascular wilt. Table 10.1c shows three fungal pathogens which

Table 10.1 Some examples of host–pathogen specificity.

(a) Species specificity.

	Pathogen (*Cladosporium fulvum*)	Non-pathogen (*Cladosporium herbarum*)
Non-host (bean)	–	–
Host (tomato)	+	–

(b) Race × cultivar specificity.

Tomato cultivar genotype	*Cladosporium fulvum* race (genotype)					
	0 (A1, A2, A5, A9)	1 (a1, A2, A5, A9)	2 (A1, a2, A5, A9)	1,2 (a1, a2, A5, A9)	5 (A1, A2, a5, A9)	9 (A1, A2, A5, a9)
Moneymaker (no *R* genes)	+	+	+	+	+	+
Cf1	–	+	–	+	–	–
Cf2	–	–	+	+	–	–
Cf5	–	–	–	–	+	–
Cf9	–	–	–	–	–	+

Cf, host resistance gene; *A*, pathogen avirulence gene (dominant allele); *a*, pathogen avirulence gene (recessive allele); +, susceptible interaction; –, resistant interaction.

(c) Tissue specificity.

Pathogen	Host tissue		
	Leaf	Root	Vascular tissue
Cladosporium fulvum	+	–	–
Pyrenochaeta lycopersici	–	+	–
Fusarium oxysporum f.sp. *lycopersici*	–	–	+
Tomato mosaic virus (ToMV)	+	+	+

colonize different types of tomato tissue, and a virus which can spread systemically throughout the plant. The factors underlying these patterns of colonization are poorly understood, although the entry route exploited by the pathogen is obviously important (see p. 104).

No single model of host–pathogen specificity is likely to explain all these phenomena. Nevertheless, concerted efforts are being made to identify the mechanisms involved, especially in cases where single genes of major effect appear to be present in the host, or pathogen, or both. Once again, this allows the powerful tools of molecular biology to be applied to the problem.

Basic compatibility

To successfully invade a host plant, to survive in the host, and to secure sufficient nutrients to grow and reproduce, a parasite must possess particular properties. One idea is that during co-evolution with plants, some microorganisms acquired pathogenicity factors enabling them to overcome host defences and establish a state of **basic compatibility**. During subsequent co-evolution there emerged genetic variants of the susceptible host that were able to recognize the pathogen and resist infection by activating host defence systems. These in turn were matched by the evolution of pathogen variants able to avoid or counter the recognition process, equivalent to a type of 'arms race'. According to this theory, specificity must include factors conferring basic compatibility, and a further set of interactions involving single complementary genes in the host and pathogen (Table 10.1b). The hypothesis thus provides an evolutionary explanation for gene-for-gene complementarity.

Such ideas have spawned a series of hypothetical models, usually based on incomplete experimental evidence. One current debate, for instance, concerns the extent to which specific gene-for-gene interactions determine the host range of pathogens in wild plant species. It has been argued that such specificity might be an artefact of plant breeding, where single resis-

tance genes are introduced into crop cultivars, and matching virulence is eventually selected in the pathogen population (see p. 220). However, recent analysis of host–pathogen relationships in wild plants such as *Arabidopsis* has shown that gene-for-gene interactions also regulate the host range of bacterial and fungal pathogens in these natural systems. The model may therefore be of universal relevance to plant–pathogen relationships. This chapter will focus on the genetic control of host–pathogen specificity, and review the extent to which recent experimental data support the basic model.

Genes determining host–pathogen specificity

In a gene-for-gene system (Fig. 10.1), specific recognition occurs when a resistance (*R*) gene in the host plant is matched by a corresponding avirulence (*avr*) gene in the pathogen. Almost always, plant resistance genes and pathogen avirulence genes segregate as single dominant characters. The simplest interpretation is that the product of the *avr* gene interacts directly with the product of the *R* gene; models based

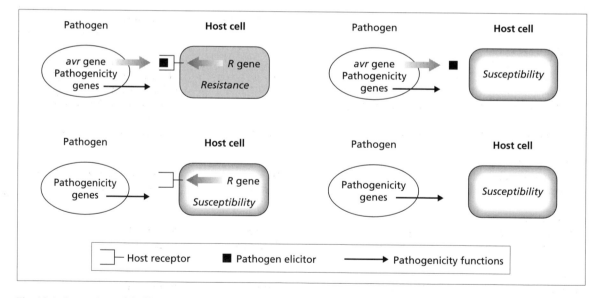

Fig. 10.1 A genetic model of host–pathogen specificity. The pathogen possesses functions, encoded by pathogenicity genes, enabling it to infect the host and cause disease. A gene-for-gene system, based on the interaction of a pathogen avirulence (*avr*) gene with a plant resistance (*R*) gene, determines specific recognition. This occurs when the

product of the *avr* gene (an elicitor) binds to the product of the host *R* gene (a receptor), leading to activation of host resistance mechanisms. Absence of either the elicitor or the receptor leads to susceptibility. (Redrawn with permission from Staskawicz *et al.* 1995. Copyright 1995 American Association for the Advancement of Science.)

on this predict that the *R* gene product is likely to be a receptor located on the host cell membrane, while the *avr* gene encodes a secreted, or surface-located, signal molecule which binds to the receptor. This molecular interaction then triggers the complex cascade of responses involved in active plant defence.

Two experimental approaches have been used to try to find the molecules and genes involved in this specific interaction. The first is to attempt to identify directly the piece of host or pathogen DNA carrying the host *R* or pathogen *avr* gene. Some ways of doing this are described later (see p. 174). Alternatively, one can try to isolate the pathogen signal molecule. This should bind specifically to the host receptor (Fig. 10.1) and can therefore be used as a molecular probe able to locate the corresponding host molecule. Assuming the receptor is a protein, information on its amino acid sequence can then be used to predict the DNA sequence and to construct an oligonucleotide probe complementary to the *R* gene itself. If the signal molecule also turns out to be a protein, a similar approach can be adopted to predict the sequence of the *avr* gene. The first fungal *avr* gene cloned was in fact isolated by this procedure (see p. 171).

Cloning *avr* genes

Nature of the signal molecule

It has been known for many years that microbial culture filtrates, or extracts from microbial cells, can act as potent inducers of plant defence responses. For instance, extracts from fungal cell walls, when applied to plant tissues, induce the synthesis and accumulation of phytoalexins (see p. 149). The active components in such extracts are referred to as **elicitors**; this term is now used in a more general sense to denote molecules which induce any plant defence response, including the accumulation of pathogenesis-related (PR) proteins, cell-wall changes, and hypersensitive cell death.

Evaluating the role played by elicitors in host–pathogen interactions is difficult for several reasons. Many of the early experiments used crude

Table 10.2 Some elicitors of plant defence responses.

Source	Common name	Chemical nature	Biological activity
Fungal pathogens			
Monilinia fructicola	Monilicolin	Peptide	Induces phytoalexins in bean
Phytophthora megasperma	–	Glucan	Induces phytoalexins in soybean
Phytophthora parasitica	Elicitin	Peptide	Induces necrosis
Rhynchosporium secalis	Necrosis-inducing peptides	Peptide	Induces necrosis in resistant barley
Cladosporium fulvum	avr elicitor	Peptide	Induces HR in resistant tomato
Fungal non-pathogens			
Saccharomyces cerevisiae	–	Glucan	Induces phytoalexins
Fungal cell walls	–	Chitosan	Induces phytoalexins
Bacterial pathogens			
Erwinia amylovora	Harpin	Protein	Induces HR
Pseudomonas syringae	Syringolide	Glycoside	Induces HR
Viral pathogens			
Tobacco mosaic virus	Coat protein	Protein	Induces HR
Potato virus X	Coat protein	Protein	Induces resistance in protoplasts
Plant elicitors			
Endogenous	Oligosaccharins	Oligosaccharides	Induce host defence

HR, hypersensitive response.

extracts which contained complex mixtures of chemicals in which the biologically active fractions were not clearly defined. Attempts to purify elicitors have since shown that they are chemically diverse, including polysaccharides, glycoproteins and proteins (Table 10.2). It is unlikely that these different types all function in the same way. Some, described as **endogenous elicitors**, are actually molecules released from plant cell walls, rather than products of the pathogen. The specificity of elicitors also varies widely. To fulfil the predictions of the gene-for-gene model, an elicitor must be the specific product of a pathogen avirulence gene which is recognized by the product of the appropriate host resistance gene to activate defence. In fact, most of the elicitors described to date are non-specific in action, inducing defence responses in both resistant and susceptible hosts. Potent elicitors have also been isolated from non-pathogens, such as the yeast *Saccharomyces cerevisiae*. Some of these molecules might therefore be important as part of a non-specific surveillance system enabling plant cells to recognize foreign organisms, and most seem unlikely to be the products of *avr* genes. Investigation of their properties has nevertheless provided valuable insights into how plant cells perceive and respond to microbial signals.

Oligosaccharide elicitors

Partial hydrolysis of fungal cell walls releases a mixture of breakdown products, some of which are highly active in switching on host defence. During infection, enzymes produced by the plant, such as glucanase and chitinase (p. 154), may serve to release similar fragments from invading hyphae. Analysis of the glucans extracted from *Phytophthora megasperma* f.sp. *glycinea*, a pathogen of soybean, has identified the most active fraction as a branched oligosaccharide containing seven sugar residues (Fig. 10.2). This heptaglucoside is the most potent elicitor of phytoalexins known, inducing glyceollin synthesis in soybean at concentrations as low as 10^{-8} M. Using radioactively labelled elicitor as a probe, a high-affinity binding site has been detected in soybean cell membranes. Hence the first step in signal transduction appears to be specific binding of the elicitor to a membrane receptor, most likely a protein.

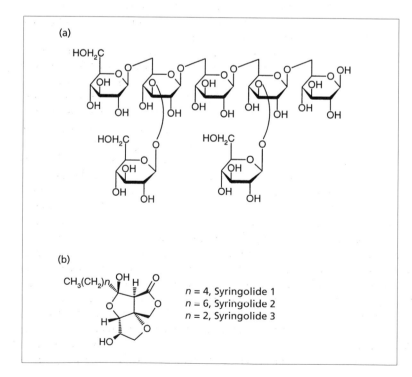

Fig. 10.2 Some characterized elicitors of plant defence. (a) A hepta-glucoside from *Phytophthora*. (b) Syringolides from *Pseudomonas syringae*.

n = 4, Syringolide 1
n = 6, Syringolide 2
n = 2, Syringolide 3

Breakdown products from chitin, another component of the cell walls of many fungi, are also effective elicitors; in this case the structural requirements for biological activity are less well-defined, although a minimum chain length of four sugar residues seems to be critical.

A further class of oligosaccharide elicitors originate from the host cell wall, rather than from the pathogen. It is well known that many plant-pathogenic bacteria and fungi produce cell-wall-degrading enzymes during tissue colonization (see p. 125), especially pectolytic enzymes which digest the middle lamella. Fragments of pectic polymers, ranging from 10 to 13 sugar residues in size, turn out to be very effective elicitors of plant defence. The products of a key process in pathogenesis might therefore alert the plant to the presence of an invading pathogen. An additional intriguing observation is that plant proteins which inhibit the activity of pectic enzymes such as polygalacturonase may also play a role by preventing complete hydrolysis of the polymer, thereby prolonging the half-life of elicitor-active fragments.

These types of elicitors, known as **oligosaccharins**, may in fact be of much wider significance in plant biology, acting as regulatory signals affecting growth and development. Similar wall fragments can, for instance, induce flowering or inhibit root formation. In animal cells, complex carbohydrates are known to act as cell surface receptors for hormones or toxins, and are also involved in tissue specificity. Hence, oligosaccharides play diverse roles in cell signalling.

Protein and peptide elicitors

For an elicitor to be a candidate *avr* gene product it should display activity on plants containing the complementary *R* gene, and be inactive on plants lacking this gene. There have been numerous attempts to detect this type of activity in culture filtrates or extracts from pathogens, but most have ended in failure. This is not entirely surprising as such extracts almost invariably contain potent non-specific elicitors, such as the oligosaccharides described above, and any other activity is masked. Finding specific elicitors therefore requires a more subtle approach.

The fungus *Cladosporium fulvum* causes leaf mould disease of tomato; this pathogen enters via stomata and grows in the intercellular spaces of the leaf, eventually sporulating on the surface. The fungus occurs as individual races which interact with different tomato cultivars according to a well-defined gene-for-gene system (Table 10.1b). When a race containing a specific *avr* gene is inoculated into a tomato leaf containing the corresponding *R* gene, resistance is expressed as a hypersensitive response (HR).

Early work on this system suggested that culture filtrates from avirulent races might contain factors which induced HR when infiltrated into leaves with the appropriate *R* gene. However, attempts to repeat these experiments proved inconclusive. Instead, Pierre de Wit and colleagues at Wageningen in the Netherlands adopted a different approach. They reasoned that *avr* genes might only be expressed within the host plant; if so the gene product(s) should, in theory, be secreted into the intercellular spaces. They therefore developed a procedure for extracting fluids from leaves inoculated with different races of the fungus (Fig. 10.3). Infected leaves of a susceptible cultivar were infiltrated with water under vacuum, and the fluids recovered by centrifugation. These extracts were then tested by injecting them into the leaves of tomato cultivars containing different *R* genes. If a specific elicitor is present, a necrotic reaction should occur only in tomato cultivars containing the complementary *R* gene.

Initial experiments of this type concentrated on the interaction between the tomato gene *Cf9* and races of the fungus possessing the putative avirulence gene *avr9* (*A9*, Table 10.1b). First, the pathogen was inoculated onto a susceptible cultivar lacking the *Cf9* gene (see Fig. 10.3). Recovered fluids were then injected into leaves containing *Cf9*. As predicted, a necrotic reaction typical of a positive HR was observed. Fluids recovered from compatible races, i.e. those able to infect *Cf9* cultivars, did not induce necrosis. Furthermore, the fluids inducing a positive HR on *Cf9* tomatoes were inactive on cultivars lacking this gene, or containing alternative *R* genes. Altogether this was encouraging evidence that races of the fungus incompatible with *Cf9* cultivars produced a specific elicitor recognized by this host gene.

The next step was to purify the active molecule from the leaf extracts. The specific elicitor turned out to be a peptide containing 28 amino acids. Once the sequence of this peptide was known it could then be used to design an oligonucleotide probe complemen-

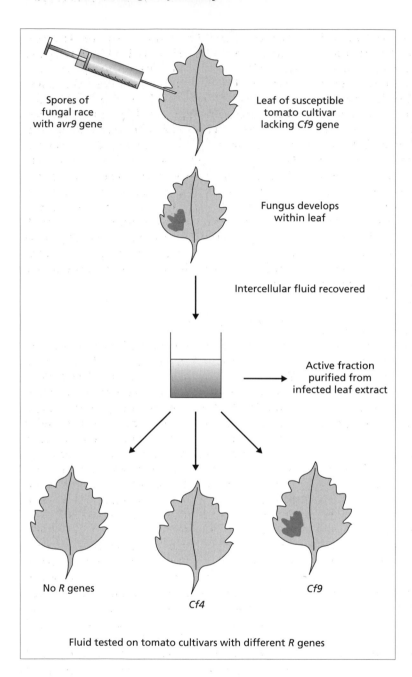

Fluid tested on tomato cultivars with different *R* genes

Fig. 10.3 Procedure used to isolate a specific elicitor of host resistance produced by the fungus *Cladosporium fulvum*.

tary to the DNA sequence coding for the peptide, i.e. the putative *avr9* gene. In fact, de Wit and colleagues used this probe initially to identify a DNA copy (cDNA) clone prepared from messenger RNA (mRNA) isolated from leaves infected with a pathogen race containing *avr9*. As the fungus grew within the host it produced the message encoding the *avr9* elicitor. The cDNA could then be used to hybridize with samples of genomic DNA extracted from the pathogen to identify the *avr9* sequence. This elegant experiment led to the first successful cloning of an avirulence gene from a fungal pathogen.

To confirm that *avr9* functioned according to the predictions of the gene-for-gene model, several further experiments were necessary. Tests showed that races of *C. fulvum* virulent on *Cf9* tomato genotypes lacked the *avr9* gene; introducing the gene into such races by transformation converted them into avirulent forms. Conversely, when the *avr9* gene was inactivated by a gene disruption procedure, a race previously avirulent on *Cf9* genotypes was converted into a virulent form. Hence *C. fulvum* can evade recognition on *Cf9* tomatoes either if it lacks the specific avirulence gene, or if the gene is unable to direct synthesis of the peptide elicitor. Subsequent work on another avirulence gene, *avr4*, has in fact suggested that non-functional versions of this gene may be present in pathogen races virulent on tomato cultivars containing the resistance gene *Cf4*. Recent studies have shown that a single base-pair change is sufficient to abolish recognition on tomato cultivars containing the corresponding resistance gene.

Some further features of the *avr9* gene merit comment. The primary product is a 63 amino-acid precursor peptide, which is cleaved to yield the smaller 28 amino-acid elicitor found in the intercellular fluids of leaves. Experiments on regulation of the *avr9* gene show that it is expressed in the host plant, but not in nutrient-rich culture media. This appears to be typical of pathogen genes which play a role in specificity; they are only switched on in nutrient-poor environments such as those prevailing in plant tissues. Only a single copy of the avirulence gene is present, which provides an explanation for the frequent changes in the virulence of fungal pathogens encountered in the field. A single mutation, or deletion, of the avirulence gene will result in a gain of virulence on a previously resistant host containing the matching *R* gene.

Further avirulence genes from fungi have now been cloned, including from *Magnaporthe grisea*, causal agent of the destructive blast disease of rice. Mapping of one of these rice blast genes, *avr2*, shows it to be located very close to the end of a chromosome, within 50 base pairs of the telomere. This region of the genome is known to be relatively unstable, and deletion or inactivation of the gene can therefore take place. Such changes seem likely to account for the variability in host range often observed in this fungus. The *avr2* gene sequence shows some homology with other genes encoding a type of protease enzyme; in this case the gene product may therefore be a secreted enzyme which acts to release peptide elicitors, rather

than interacting directly with a host receptor. Interestingly, a further gene identified in *M. grisea* (known as *PWL2*) determines pathogenicity or non-pathogenicity to a wild relative of rice, weeping love grass. This is an example of a gene determining host species-specificity, rather than host range on different genotypes of the same species, which is the usual pattern with *avr* genes. The gene product appears to be a secreted protein composed of 145 amino acids. Expression of this gene is again unstable, with frequent mutations converting forms of the fungus non-pathogenic to love grass to pathogenic forms. Investigation of a collection of different isolates of *M. grisea* has shown that variants of the *PWL2* gene are widely distributed in the pathogen population, but only some are actually functional and influence host range.

Other protein elicitors

Small extracellular proteins which elicit necrosis and defence responses in plants have also been found in filtrates from cultures of several *Phytophthora* species, as well as from the fungus *Rhynchosporium secalis* (Table 10.2). The first *Phytophthora* proteins characterized, known as **elicitins**, are all remarkably similar to one another, consisting of 98 amino acids and sharing a high degree of sequence homology. Some evidence suggests that strains of *P. parasitica* able to attack host plants such as tobacco do not produce elicitins, and that lack of elicitin production therefore correlates with virulence to specific hosts. This is analogous to the situation with *avr* gene products, but genetic evidence to confirm the role played by these proteins is not yet complete. A larger glycoprotein which induces a hypersensitive response and systemic acquired resistance in tobacco has now been identified in *Phytophthora*, so this pathogen can potentially produce several different types of signal molecules capable of eliciting host defence.

Rhynchosporium secalis is a foliar pathogen of barley, causing necrotic lesions known as leaf scald disease. The fungus grows initially just beneath the cuticle without penetrating epidermal cells, but induces disease by secreting factors active in killing host tissues. Several small, toxic proteins, known as necrosis-inducing peptides (NIP), have been purified from culture filtrates of the fungus. These proteins are toxic to both resistant and susceptible barley cultivars, and also to some other plant species. Recent work has shown, however, that resistant barley cultivars may be able to recognize specific NIP products of the fungus. For instance, the protein NIP1 induces defence responses in barley plants containing a particular resistance gene *Rrs1*. Thus in this system involving a necrotrophic pathogen, a key factor determining pathogenicity can also act as a specific elicitor. This is of particular interest as it provides an explanation both for the ability of the pathogen to cause disease, and the ability of certain barley cultivars to recognize the fungus and actively restrict infection. Pathogenicity and host specificity appear in this case to be two sides of the same molecular coin!

Bacterial *avr* genes

The first *avr* genes to be successfully isolated were all from bacterial pathogens. These genes were identified

by mutations which altered host range, followed by testing of random clones from genomic libraries to find those which restored the original host specificity. Alternatively, one can change the host range of a particular pathogen race by introducing DNA clones from another, different race. The basic idea behind this procedure is shown in Fig. 10.4, in which a pathogen race (1) initially able to infect cultivar B is converted to an incompatible pathogen by introduction of DNA from an incompatible race (2). According to the gene-for-gene model the piece of DNA responsible for this alteration in host specificity should contain an *avr* gene recognized by cultivar B. These types of experiments quickly led to the cloning and characterization of a whole series of *avr* genes from different pathovars of the major bacterial pathogens *Pseudomonas syringae* and *Xanthomonas campestris*. One expectation of this work was that comparison of the sequences of these genes, and their

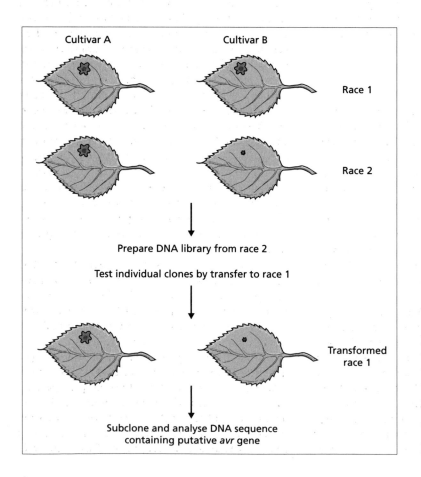

Fig. 10.4 Experimental scheme used to isolate pathogen genes determining host range. Race 2 is incompatible with cultivar B and induces a hypersensitive response. Conversion of the normally compatible race 1 into this phenotype is due to the introduction of a specific avirulence gene.

predicted products, would reveal conserved features suggesting some common mode of action in eliciting host defence. In fact, most bacterial *avr* gene products do not show significant homology, and also lack the characteristics one would expect of a signal molecule exported out of the cell to act as an elicitor. Instead they appear to be cytoplasmic proteins located inside the bacterium (but see below). Critically, the gene products themselves did not prove to be active in inducing a hypersensitive response in hosts containing the corresponding *R* gene; they therefore differ from elicitors such as the fungal peptides described above. One possibility is that these genes are indirectly involved in the production of a specific elicitor. Support for this idea comes from analysis of the *avrD* gene derived from *P. syringae* pathovars infecting tomato or soybean. The avrD protein itself lacks elicitor activity, but bacteria producing this protein secrete a low-molecular-weight compound, syringolide (Fig. 10.2), which induces HR on soybeans containing the corresponding *R* gene. The *avrD* gene product might therefore be an enzyme involved in the synthesis of the active elicitor molecule.

Other functions for avr genes?

Why, if *avr* genes encode products directly or indirectly recognized by the plant, and hence restrict the host range of the pathogen, are these genes not strongly selected against and eliminated from the pathogen population? Surely it would be an advantage to the pathogen to infect as many different host types as possible? The answer to this question is not clear, but one hypothesis is that *avr* genes must be performing some other useful, perhaps vital, function in the life of the pathogen. Loss of the gene might therefore compromise the fitness of the organism. According to this view, the products of *avr* genes are proteins or other specific molecules which the host plant has 'learned' to recognize. Their role in specific recognition is not therefore their primary function. The NIP proteins from *Rhynchosporium secalis* discussed above appear to be examples of this. There is now increasing genetic evidence to support this idea. Unlike their plant counterparts, the *R* genes, different *avr* genes show little homology and appear to be randomly distributed in the pathogen genome. This suggests that they encode diverse products with different functions. The widespread occurrence of non-

functional versions of *avr* genes in pathogen races virulent on plants with corresponding *R* genes indicates that some important property may be conserved, but that the products themselves may have been modified to evade recognition. Comparison of different alleles of the same *avr* gene, for instance *avrD* from *Pseudomonas syringae*, has shown that while the encoded proteins are very similar, differing by only a few amino acids, the position of the gene often varies from race to race, usually occurring on a plasmid but sometimes on the chromosome. Hence there may be small changes in the *avr* gene product itself, combined with locational changes which are likely to influence gene expression. Eventually, when a sufficient number and variety of *avr* genes have been cloned and analysed, it should be possible to deduce their true biological functions.

hrp genes

The starting point for many of the recent molecular studies on host–pathogen interaction is the generation of mutants altered in pathogenicity or host specificity. One interesting group of bacterial mutants found in such studies are those in which pathogenicity to the host plant, and the ability to induce a hypersensitive response (HR) in resistant hosts, or in non-hosts, are both simultaneously lost. These are known as *hrp* (*h*ypersensitive *r*eaction and *p*athogenicity) mutants. Although such mutants have been recognized for several years in most of the important genera of bacterial pathogens, including *Erwinia*, *Pseudomonas* and *Xanthomonas*, the nature of the genes involved was initially obscure. More recent analyses have, however, started to unravel the mystery, with some surprising results.

Xanthomonas campestris pv. *vesicatoria* (Xcv) causes bacterial spot disease of pepper and tomato. Non-pathogenic mutants which had also lost the ability to incite HR on resistant hosts were obtained by chemical mutagenesis of the pathogenic wild-type bacterium. Complementation analysis, in which DNA clones from a genomic library of the wild type (see Fig. 8.1) are tested for their ability to restore function in the mutant, showed that the genes responsible are clustered in the bacterial chromosome spanning a region of about 25 kilobases (kb). Transposon mutagenesis confirmed that there are at least six *hrp* genes, coded A to F, in this region. This is shown in

Fig. 10.5. Inactivation of each *hrp* gene by insertion of a transposon leads to reduced multiplication of the mutant in the host plant, pepper; insertions outside the *hrp* cluster do not affect pathogenicity. Large clusters of *hrp* genes have also been discovered in other plant-pathogenic bacteria, for instance *Ralstonia solanacearum*.

Like *avr* genes, *hrp* genes are not expressed when the bacteria are grown in nutrient-rich culture media, but are switched on when the pathogen grows under low nutrient conditions typical of the environment encountered within the host plant. When *Pseudomonas syringae* is inoculated into tobacco leaves, enhanced expression of *hrp* genes can be detected within 1 h. Filtrates from plant cells are effective inducers of the Xcv *hrp* cluster, suggesting that these genes may in fact be regulated by plant signals. Synthesis of *hrp* gene products therefore rep-

resents an early event in the molecular 'cross-talk' between host and pathogen.

But what are the actual products of these genes? The fireblight bacterium, *Erwinia amylovora*, has a 40 kb chromosomal region containing *hrp* genes; these have been cloned and expressed in the closely related non-pathogen *Escherichia coli*, aiding analysis of their function. One gene in particular (*hrpN*) encodes a 385 amino-acid protein which is associated with the cell envelope of the bacterium (Table 10.3). This protein, named **harpin**, is a potent elicitor of HR, causing plant cells to leak electrolytes and collapse. Mutations in the *hrpN* gene preventing production of harpin abolish both pathogenicity to the normal host plant, pear, and the induction of HR in non-hosts such as tobacco. Harpin therefore plays an essential role in disease development, and may also activate plant defence.

Other harpin proteins have now been isolated and purified, notably from *Pseudomonas syringae*. Sequence analysis shows that these elicitors have little homology with previously described proteins, and they also lack features characteristic of membrane-associated molecules. Furthermore they do not possess typical signal sequences that target proteins for export across the bacterial membrane by the usual secretory pathway. Nevertheless these molecules, to act as elicitors, must be exported out of the cell.

It is at this point that one of the more surprising aspects of the *hrp* gene story emerges. Sequencing of other *hrp* genes present in these large clusters reveals that the gene products are strikingly similar to certain proteins implicated in the pathogenicity of bacteria to animal hosts, including well-known human pathogens such as *Yersinia*, *Shigella* and *Salmonella*. The precise functions of these proteins vary, but most seem to play a part in a secretion pathway required for pathogenicity (Table 10.3). Hence elicitors and other biologically active molecules involved in pathogenicity may be exported by a specialized transport system. This observation also begins to make sense of the apparent conundrum that most *avr* gene products are proteins without signal sequences or obvious transmembrane domains, and therefore appear to be cytoplasmic. How then can they function as signal molecules interacting with host-cell receptors? Export by a protein secretion pathway would solve this dilemma. In fact, recent work with *Xanthomonas* has indicated that Avr proteins may be translocated

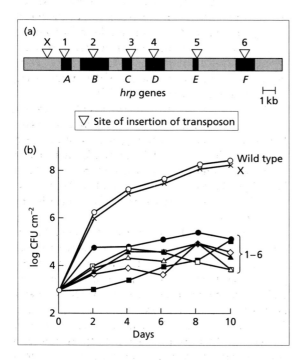

Fig. 10.5 (a) A cluster of *hrp* genes (*A–F*) from *Xanthomonas campestris*. Insertion of a transposon in any one of these genes, to give mutants 1–6, abolishes the ability to induce a hypersensitive response, and reduces multiplication in host leaves. Insertions outside the cluster (e.g. mutant X) have no effect on activity or pathogenicity. (b) Growth curves in leaves for mutants 1–6 compared with wild type and mutant X. (After Bonas *et al.* 1991.)

Table 10.3 Some *hrp* genes and their products.

Gene	Source bacterium	Product	Suggested function
hrPS	*Pseudomonas syringae* pv. *phaseolicola*	Regulatory protein	Activates *hrp* operon in response to plant signals
hrpH	*P. syringae* pv. *syringae*	Membrane protein	Component of protein secretory pathway
hrpC	*Xanthomonas campestris* pv. *vesicatoria*	Membrane protein	Component of protein secretory pathway
HrpZ	*P. syringae* pv. *syringae*	Extracellular protein elicitor	Induces hypersensitive response
HrpN	*Erwinia amylovora*	Extracellular protein elicitor	Induces hypersensitive response

directly into host cells, where they may be targeted to the plant nucleus.

It might be useful to conclude this quite complex story by summarizing the known properties of *hrp* genes. They are linked in large clusters and are transcribed in low nutrient environments or in response to plant signals. While all play a role in pathogenicity, different *hrp* genes perform distinct functions (Table 10.3). Some appear to be regulatory genes involved in induction or repression of the *hrp* operon. Others encode potent protein elicitors, the harpins, inducing HR. Most encode membrane proteins with homology to secretion pathway proteins found in several animal-pathogenic bacteria. This observation is of particular interest as it indicates that at least some of the key determinants of pathogenicity may be conserved among bacterial pathogens of plants and animals. To reinforce this point, recent studies have shown that some strains of the animal pathogen *Pseudomonas aeruginosa* are also capable of causing disease in a plant, *Arabidopsis*, and mutations in genes reducing the ability of this bacterium to infect mice also reduced pathogenicity to the plant host.

Cloning plant resistance genes

While the identification and isolation of pathogen genes controlling host range is an important advance in understanding the molecular basis of specificity, an even greater prize is to clone the corresponding host resistance genes. This is because *R* genes are the raw material selected by the plant breeder, and isolation of such genes will create novel possibilities for their use in practice. Furthermore, understanding how such genes work should suggest alternative ways of engineering crop resistance, and hopefully of improving the durability of resistance in the field.

The existence of dominant genes conferring resistance to pathogens has been recognized by plant breeders for almost a century; indeed, many such genes have been utilized in conventional breeding programmes aimed at crop improvement (see p. 217). Techniques for locating and physically isolating these genes have, however, only become available in the past few years.

The technical problems involved in cloning *R* genes are formidable. The main difficulty is the sheer size of the higher plant genome. With bacteria, assuming the genome is similar in size to that of *E. coli*, less than 1000 random clones are needed to ensure that every gene is represented in a genomic library (Table 10.4). With fungi this figure is closer to 10000, and with plants it is 10 times greater again. As the only way to confirm that a particular clone contains an *R* gene is to transform a susceptible plant and test for conversion to a resistant reaction type, finding the right piece of DNA in a plant gene library by the random process of shotgun cloning is a monumental task. Furthermore, a high proportion of plant DNA is repetitive or non-coding, and valuable time would be wasted in cloning such sequences. What is required is

Organism	Genome size (bp)	Number of clones*
Escherichia coli	4×10^6	700
Aspergillus	4×10^7	7 000
Arabidopsis	2×10^8	35 000
Tomato	7×10^8	125 000
Tobacco	4×10^9	710 000

Table 10.4 Relative genome sizes of organisms and number of clones required to make a genomic library.

* Assuming a phage (lambda) vector which can accommodate 17 kb fragments.

a means of homing in on regions of the plant genome where *R* genes are located, to increase the probability of isolating such genes.

Two main approaches are being used in this search. The first, **map-based cloning**, relies on identifying markers closely linked to a resistance gene on a map of the plant genome. Such markers may be phenotypic, in other words are expressed as a plant character such as flower colour or leaf shape, or more usually are molecular markers, identified as small stretches of DNA which hybridize to random probes, or can be amplified by the polymerase chain reaction (PCR) with particular primers. Examples include restriction fragment length polymorphisms (RFLPs) or randomly amplified polymorphic DNA (RAPD). These provide recognizable anchors or landmarks on the genetic map. If, in a sexual cross, such a marker co-segregates with an *R* gene (i.e. always occurs in progeny containing the *R* gene), it is likely that the landmark is very close to the gene; 'walking' from the marker along the chromosome using overlapping DNA sequences should then lead to the gene itself. As this process may be unfamiliar to some readers, a simple analogy might be helpful. Imagine the plant genome is a railway system consisting of many miles of track. Our *R* gene is one small stretch of track, and we have no idea where it is. However, at intervals along the track there are stations; some of these are large stations at well-known cities and towns. In our analogy these can be regarded as familiar phenotypic markers. At much more frequent intervals there are small stations, which represent molecular markers such as RFLPs or RAPDs. Our gene will lie somewhere between two stations, and it is most likely that these will be of the small, molecular type. Genetic analysis of the whole system is then analogous to positioning all the large and small stations on the map. It will also tell us which stations are nearest to our particular piece of track; by

walking along the track from such a station we can arrive at our 'gene' quite quickly. This is obviously preferable to walking along the whole system in the hope that we might chance upon the stretch we are interested in!

Alternatively, the gene may be 'tagged' (i.e. labelled) by a transposon. **Transposon-tagging**, already discussed in the context of identifying pathogenicity genes, is based on mobile genetic elements which move around the plant genome inserting at random sites. If such an element inserts into an *R* gene, then its function may be inactivated, and the previously resistant plant becomes susceptible to infection. By screening large numbers of plants in which transposons are active one may find rare individuals which possess this altered disease reaction type. By using an appropriate probe which hybridizes with the transposon one can then locate the stretch of DNA containing the *R* gene itself. Once isolated, the gene can be sequenced and the functional properties of the gene product predicted from the encoded amino acid sequence.

Both approaches have now been successfully used to locate and isolate plant resistance genes. At the time of writing, more than 10 such genes have been described. Table 10.5 shows the first eight reported; four of these confer resistance to fungi, three to bacteria, and one to a virus. This is significant as it is now possible to compare the sequences of plant genes effective in the recognition of contrasting types of pathogens, and to ask whether these genes have features in common.

The first *R* gene to be cloned was *Hm1* from maize, conferring resistance to *Cochliobolus (Helminthosporium) carbonum*. *Hm1* is distinct from all the other *R* genes isolated to date as it does not function in pathogen recognition. Instead this gene encodes an enzyme which degrades a low-molecular-weight toxin produced by the pathogen; activity of the

Table 10.5 Some cloned plant
disease resistance genes.

Gene	Host	Pathogen	Location	Function
Pto	Tomato	*Pseudomonas*	Cytoplasmic	Kinase
RPS2	*Arabidopsis*	*Pseudomonas*	Membrane	Receptor
Xa21	Rice	*Xanthomonas*	Membrane	Receptor
N	Tobacco	TMV	Cytoplasmic	Receptor
Cf9	Tomato	*Cladosporium*	Membrane	Receptor
RPP5	*Arabidopsis*	*Peronospora*	Membrane	Receptor
L6	Flax	*Melampsora*	Cytoplasmic	Receptor
Hm1	Maize	*Helminthosporium*	Cytoplasmic	Toxin reductase

TMV, tobacco mosaic virus.

enzyme, a toxin reductase, can only be detected in extracts from maize plants containing the gene. As in this case the pathogenicity of the fungus is completely correlated with production of the host-specific toxin, then plants able to inactivate the toxin will be unaffected by the fungus. Isolation and analysis of *Hm1* adds convincing genetic and biochemical proof to the argument that host-specific toxins are the sole determinants of disease with these types of fungal pathogens (see p. 131).

The next *R* gene to be isolated was *Pto* from tomato, which confers resistance to strains of the bacterial pathogen *Pseudomonas syringae* possessing the avirulence gene *avrPto*. This therefore is the first example of an *R* gene conforming to the gene-for-gene model. *Pto* was located by map-based cloning procedures, using molecular landmarks in the tomato genome to eventually isolate DNA clones containing the gene. Analysis of the sequence indicates that the predicted gene product is a kinase enzyme, rather than a membrane-located receptor protein. Kinases play vital roles in signal transduction pathways, such as those participating in plant defence. The discovery of a kinase function involved in resistance is therefore not surprising, but at the same time it was not clear whether this enzyme interacts directly with an *avr* gene product to switch on the process. Recently it has been shown that expression of the corresponding *avr* gene (*avrPto*) in transgenic plant cells containing *Pto* induces a hypersensitive response. Furthermore, there is now evidence for a direct physical interaction between the host kinase and the *avr* gene protein. Analysis of mutants suggests, however, that there are at least two other plant genes involved in *Pto* function, and current predictions are that these encode partner proteins necessary for signal transduction.

The remaining *R* genes listed in Table 10.5 all have certain properties in common, notably the presence in the predicted protein products of repeated sequences rich in the amino acid leucine. Such leucine-rich repeats (LRRs) are characteristic of several classes of proteins with known receptor functions, for instance hormone receptors in animal cells. It seems likely therefore that these proteins may indeed be receptors which interact with elicitor molecules from the pathogen, thereby triggering the signal transduction pathway leading to active plant defence. A simple model of this process, based on the predicted protein product of the *Xa21* gene from rice, is shown in Fig. 10.6.

Both *RPS2* and *RPP5* were cloned from *Arabidopsis*, taking advantage of the relatively small genome (Table 10.4) and very detailed genetic map now available for this plant. *Xa21* was also isolated by map-based cloning. The other *R* genes listed in Table 10.5 were located by transposon-tagging. The principle of this procedure is simple. First a suitable transposon, usually derived from maize, is introduced into the host plant. After crossing and genetic analysis, plants are selected in which a transposon is known to be present close to the *R* gene. This step is important as transposons usually 'jump' over only short distances. Next the selected plants containing a transposon close to the *R* gene are crossed to others lacking the *R* gene. Usually in such a cross, all the F$_1$ progeny would be expected to be resistant, as the *R* gene is dominant. However, if the transposon has jumped into the *R* gene, thereby inactivating it, such plants can be identified

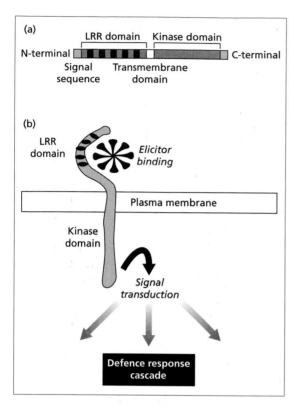

Fig. 10.6 (a) Basic structure of the predicted product of a plant recognition gene (*Xa21* from rice) showing the leucine-rich repeat (LRR), transmembrane and kinase domains. (b) A model of how the product might act as a membrane-located receptor which switches on host defence following binding of a specific pathogen elicitor molecule. (Redrawn with permission from Song *et al.* 1995. Copyright 1995 American Association for the Advancement of Science.)

among the progeny as giving a susceptible disease reaction.

While this approach works fine in principle, in practice there can be problems with it, the most serious of which is the very low frequency at which insertion of the transposon into the gene of interest occurs. It is necessary therefore to screen very large numbers of progeny to find the susceptible plants, and there is also the risk that some of these rare individuals might have an inactive form of the *R* gene caused by a mutation or other type of genetic instability not associated with insertion of a transposon. Exactly this problem occurred in the search for *L*[6] in flax, where out of more than 30 000 plants screened, 29 susceptible mutants were identified, but

of these only one had the transposon inserted in the gene. The remainder of the susceptible plants appeared to have lost the gene altogether by deletion of the piece of DNA concerned. In the case of *Cf9* from tomato and *N* from tobacco, the use of a transposon to inactivate the gene was combined with an ingenious selection procedure to improve the chances of recovering progeny with an altered disease reaction type.

With tobacco, the temperature-sensitivity of *N* resistance was exploited. This gene only operates within a certain temperature range. If plants containing *N* are inoculated with tobacco mosaic virus (TMV) and kept at a higher temperature than usual, for instance 33°C, the gene is ineffective and the virus multiplies and spreads systemically in the host. If these plants are then transferred to a lower temperature, such as 25°C, the gene becomes effective and a hypersensitive response occurs. As the virus is by now present throughout the plant, this necrotic reaction is lethal. In the transposon experiment, all the progeny were inoculated with the virus and kept first at the higher temperature. When the seedlings were then transferred to a lower temperature, almost all of them became necrotic and died. The few plants which survived the shift to the lower temperature were predicted to contain an inactive *N* gene. These rare survivors were then analysed for the presence of the transposon in the resistance gene.

The tomato gene *Cf9* was found by an analogous procedure, taking advantage of the previously recognized interaction between the plant resistance gene and the pathogen elicitor produced by the corresponding avirulence gene *avr9* (see p. 171). The first step was once again to introduce a transposable element into a tomato line containing the *R* gene (Fig. 10.7). At the same time a tomato line lacking the *R* gene was engineered to contain the fungal gene producing the *avr9* peptide elicitor. These plants therefore produced the specific fungal signal molecule which is recognized by the product of the host resistance gene. When a tomato plant containing the *Cf9* gene is crossed with another containing the fungal *avr9* gene the predicted outcome is induction of the hypersensitive defence response in most cells of the resulting hybrid. This is exactly what was observed. Hybrids containing both a functional plant resistance gene and the matching fungal avirulence gene germinated, but became necrotic and died within 2 weeks. However, if the transposon had inserted into and

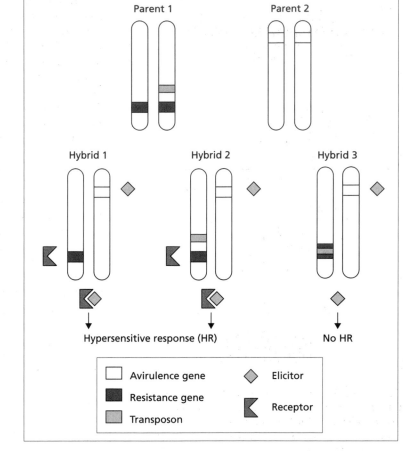

Fig. 10.7 Simplified version of the selection scheme used to isolate the tomato resistance gene *Cf9* by transposon tagging. Parent 1 is homozygous for *Cf9* and parent 2 is homozygous for the corresponding fungal avirulence gene *avr9*. The progeny from this cross (hybrid 1) carry both genes, and as the interaction between the *avr9* elicitor and the *Cf9* gene product leads to hypersensitive necrosis, the seedlings die. When, however, a transposon is close to one copy of the *Cf9* gene, some progeny will be like hybrid 2, where the transposon has not moved into the target gene, but a few will be like hybrid 3, in which the transposon has inserted into *Cf9* and inactivated it. These progeny, which contain a tagged *R* gene, survive. (Redrawn with permission from Jones *et al.* 1994. Copyright 1994 American Association for the Advancement of science.)

inactivated the *Cf9* gene, such hybrids would survive, due to lack of a specific interaction between the plant and pathogen gene products (Fig. 10.7). When seedling survivors from this experiment were analysed, it was found that most had indeed lost the ability to recognize the fungal elicitor due to insertion of the transposon into the host resistance gene. By using molecular probes which hybridized with the transposon it was then possible to locate the *R* gene itself.

The success of these ingenious experiments in isolating several specific recognition genes lays the foundation for understanding how plants perceive the presence of a pathogen, and switch on their active defence systems. The presence of conserved sequences within most of the *R* genes isolated to date also means that it should be possible to design oligonucleotide primers which will hybridize with related sequences and hence can be used in the polymerase chain reaction to locate other similar genes. This will greatly

accelerate the hunt for putative pathogen recognition genes in other plants. However, it is still not known how exactly the protein products of such genes convert perception of the pathogen elicitor molecule into a signal which activates plant resistance. Current work is dissecting the signal transduction pathway with the aim of answering this question.

Organization of *R* genes

A further revelation emerging from work on the mapping and cloning of *R* genes is that they often occur in clusters grouped at particular points in the plant genome. In *Arabidopsis*, specific genes for downy mildew resistance (*RPP* — recognition of *Peronospora parasitica* — genes) are present on all five chromosomes, but map to loci close to other *RPP* genes. A similar clustering together of *R* genes with different specificities for the same pathogen is seen in the interaction of lettuce with downy mildew, *Bremia lactucae*, and of flax with rust, *Melampsora lini*. Furthermore, such clusters can contain not only several *R* genes recognizing different races of the same pathogen, but also genes specific for quite different pathogens, such as fungi, bacteria and nematodes. In maize, for example, there is a cluster of tightly linked genes on chromosome 3 which confer resistance to rust and two different viruses, as well as other genes which influence reaction type to fungal stalk rot and an insect pest, the European corn borer. This suggests that there are particular regions of the plant genome which encode multiple recognition functions. Such regions also appear to be rich in genes for kinase proteins which might function in signal pathways.

What is the significance of this genetic organization? One idea is that different *R* genes might belong to multigene families which have evolved from a common ancestral gene. Duplication of such genes, coupled with mutation, or recombination between genes, could generate new variants capable of recognizing different pathogen signal molecules. According to this theory, a prototype receptor could diversify in time to bind a range of different pathogen elicitors. Conversely, the pathogen is no doubt able to modify the corresponding signal to evade recognition. The extraordinary diversity of host–pathogen specificities might therefore reflect a series of variations based on a rather similar molecular lock and key.

Specificity of virus pathogens

Plant viruses also exhibit specificity in the range of host plants which they can successfully infect. The problem in this case is unique, as successful reproduction of these subcellular parasites depends upon a precise sequence of molecular events (Fig. 10.8). Thus the virus must gain entry to the host cell, the virus genome must be released from its protective protein coat, then be translated and replicated, and finally new particles must be assembled. In addition, continuous spread of the virus to other cells and tissues is necessary if systemic colonization of the host is to occur. In theory, specificity might be controlled at any of these stages in the infection cycle. For instance, the lack of host subunits necessary to form a functional replicase enzyme might prevent more copies of the viral nucleic acid from being produced. This notion envisages a highly specific 'molecular fit' between host and virus.

In practice, however, factors other than an intracellular molecular match may be of greater significance as determinants of virus host range. The need for an appropriate vector to transmit infection adds a further dimension to the equation of virus specificity. The relationship of the virus with the vector, and of the vector with the host population, may limit the host range of a virus to a greater extent than the ability of the pathogen to replicate in host cells. Mechanical inoculation of viruses that are normally transmitted by a vector has been shown to extend their natural host range. Similarly, inoculation of isolated protoplasts with virus can give rise to infection in cells of host species not known to be susceptible in nature. These findings suggest that the limiting factor with many plant viruses is their transmission and introduction into plant cells, rather than their capacity to exploit the biosynthetic machinery of the host.

A further difficulty in evaluating the specificity of plant viruses is the diversity of host reaction types that may occur. For instance, a virus may replicate successfully in one host species without causing any symptoms, while in another severe disease results. Both plant species are hosts, but the consequences of infection are quite different. In other cases, lack of visible symptoms may be due to plant resistance mechanisms preventing the virus from replicating or moving between cells. It has also become apparent that even in a susceptible host, not all cells are infected to the same extent. In tobacco with mosaic caused by TMV, the dark-green areas of infected leaves contain little or no virus, while the chlorotic patches contain high levels. Plants regenerated from protoplasts isolated from the dark-green tissues are often completely free from virus. One explanation for this differential distribution is that some cells in the developing leaf primordium are actually resistant to virus infection.

Some viruses have extremely wide host ranges; for instance, cucumber mosaic virus (CMV) is known to infect nearly 800 plant species in 85 plant families, including both monocots and dicots. Even more remarkably, some plant viruses are able to replicate both in their host plant and within the insect vector. Their host range therefore embraces both the plant and animal kingdoms. This poses fascinating ques-

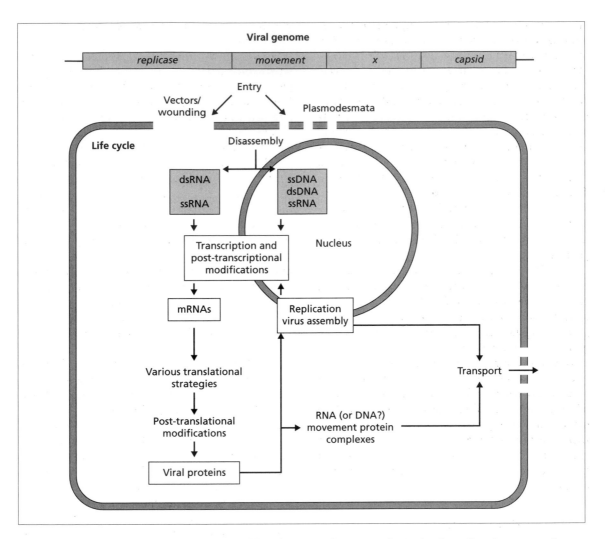

Fig. 10.8 Diagrammatic version of the plant virus life cycle, showing molecular processes involved. The virus genome (top) contains genes for replication, movement and coat (capsid) protein, and various other functions (e.g. gene *x*), such as vector transmission. Upon entry into the cell, disassembly of the virus particle takes place, and the replication cycle occurs, depending on the nature of the viral genome, either in the plant cell nucleus or cytoplasm. Regulation of virus gene expression can occur at the transcriptional or translational level, with post-translational modification of proteins in some cases. Cell-to-cell movement via plasmodesmata may involve either whole virus particles or complexes of the viral genome and movement protein. (After Scholthof *et al.* 1993.)

tions as to how such relationships might have co-evolved.

While comparatively little is known about the factors controlling virus specificity, new insights into virus–host interactions have been gained from recent molecular studies. The key to these experiments is the ability to manipulate, by mutation, deletion or substitution, specific regions of the virus genome (Fig. 10.8). It is even possible to swap parts between different viruses, for instance to replace the gene coding for the movement protein with a similar gene from another virus. When this was done with TMV, by substituting a movement protein gene taken from a related virus infecting members of the orchid family, the hybrid virus gained the ability to spread in orchid hosts, but was reduced in its ability to colonize tobacco, its normal host plant. This result suggests that host range may be determined in some cases by specific interactions between virus gene products and particular host components.

A different level of virus specificity occurs in cases where the plant contains a specific resistance gene. One example, the *N* gene from tobacco, has already been described in the previous section (see p. 180). This gene, derived originally from *Nicotiana glutinosa*, when introduced into the susceptible cultivated tobacco, *N. tabacum*, converts the disease reaction type from a systemic mosaic into local lesions. In effect, the plant is now able to recognize the presence of the virus and activate a hypersensitive defence response. A second gene, *N'*, derived from another related species, *N. sylvestris*, also confers hypersensitive resistance to most strains of TMV. The specific interaction of the virus with these host genes has again been tested by producing mutant strains with altered properties. In the case of *N'* in tobacco, changes in the virus coat protein gene can induce or abolish recognition by the plant. Thus the TMV coat protein apparently acts as an elicitor of the *N'* gene. Experiments with recombinant TMV containing the coat protein gene of potato virus X (PVX) have also confirmed that this coat protein acts as a specific elicitor in resistant potato cultivars. However, the critical features of the virus recognized by the host are not the same in all cases, as the coat protein does not appear to be involved in recognition by the tobacco *N* gene. A further TMV resistance gene found in tomato, known as *Tm-2*, appears to interact instead with the virus movement protein. Hence plants appear to have evolved the ability to recognize and respond to different molecular components of the virus.

The collective evidence from these studies shows that comparatively small changes in virus proteins, affecting only one or a few amino acids, may be sufficient to completely alter the behaviour of a virus in a host plant. The corollary to this is that mutations naturally occurring in a virus may generate variant strains able to bypass a previously effective plant resistance mechanism. This has already proved the case in practice, where resistance has often broken down in the field. For instance, strains of TMV capable of overcoming the *Tm-1* resistance gene in tomato appeared within one year of use of this gene in commercial cultivars.

Susceptibility as an induced state

In the examples discussed so far, the specific event in host–pathogen interaction is recognition of the invading pathogen by the host, followed by activation of plant defence pathways. Susceptibility is therefore regarded as a passive function, resulting from the inability of the host to detect or respond to the pathogen. But an alternative argument might be proposed. Since the majority of plants are resistant to the majority of potential pathogens, and since susceptibility is the exception rather than the rule, understanding the special features of the susceptible interaction might provide the key to specificity. In particular, how does the successful pathogen confound the host's recognition system, or, alternatively, suppress the host's resistance response? According to this view susceptibility, rather than resistance, could be an induced state.

This argument has some attractions, but is there any evidence to support it? Several studies have, in fact, suggested that pathogens can avoid or inactivate host defence systems, by a variety of means. One basic strategy is simply to kill host cells so quickly that active defence is neutralized in the early stages of infection. Fungi that produce potent toxins are the most obvious examples. There is also some evidence that pathogens may have more subtle effects on plant cells, for instance by switching off host gene expression. This might reduce the ability of the host plant to mount an effective resistance response.

Suppression of host resistance

Experiments with several biotrophic fungi have shown that pre-inoculation with a compatible pathogen can lower the resistance of the host plant to subsequent inoculation with an incompatible pathogen. For instance, if barley leaves are inoculated with *Erysiphe graminis* f.sp. *hordei*, and 2 days later the developing colonies are peeled off and replaced by spores of the normally incompatible wheat form of the fungus, *E. graminis* f.sp. *tritici*, some growth and development of the latter will occur. Similarly, rust species normally incapable of infecting non-host leaves may penetrate and produce haustoria in tissues adjacent to colonies of a compatible rust species. Observations like these have been interpreted as showing that a virulent pathogen can in some way 'condition' host tissues to accept another pathogen, which in the normal course of events would fail to establish. The obvious conclusion is that the host's recognition process or defence response is inhibited or impaired by the virulent pathogen.

Recently some more direct evidence that susceptibility involves suppression of host cell activities has come from studies of gene expression in tissues infected by quite diverse pathogens, including viruses, nematodes, and parasitic angiosperms. In healthy cells there is a basal level of gene expression associated with metabolic turnover and the biosynthesis required to maintain the essential functions and fabric of the cell. The genes involved are often described as 'housekeeping genes'. Because such genes are continually being transcribed the cells contain a more or less constant level of the complementary mRNA. This can be detected by northern blotting to give a hybridizing signal. It has now been shown that in certain compatible infections by pathogens, for instance in root tissues adjacent to nematode penetration sites, or infected by the root-parasitic angiosperm witchweed (*Striga* spp.), the amount of message produced by such housekeeping genes is reduced, sometimes to undetectable levels. A similar inhibition, or 'down-regulation' of host genes, is seen in some virus infections. Whether these global effects on gene expression prevent the plant from responding in any effective way to the invading pathogen is not yet clear.

Overall it is an appealing idea that certain pathogens, especially biotrophs, might directly manipulate host metabolism and biosynthesis in such a way as to disarm defence. So far, however, there is relatively little specific evidence to confirm the idea, or to reveal the molecular mechanisms whereby such a state of susceptibility might be induced. In barley leaves infected by powdery mildew, the amount of host mRNA is reduced, suggesting a shut-down of host biosynthesis. This might affect the ability of the plant to produce defence compounds. Some work on the fungus *Phytophthora infestans* has shown that water-soluble glucans from compatible races are able to suppress the hypersensitive response of potato protoplasts exposed to hyphal wall components of the pathogen. In other words, these glucans somehow prevented the elicitation of an active defence response. More evidence is required, however, to confirm whether such an interplay between the elicitation and suppression of host resistance mechanisms can explain specificity.

Destruction of host resistance

Necrotrophic pathogens typically secrete hydrolytic enzymes or toxic compounds during invasion of the host (see p. 125). These may aid penetration and softening of host tissues, inactivate preformed inhibitors, disrupt host metabolism, and ultimately kill host cells. If the latter occurs quickly enough it might effectively pre-empt any defence mechanism dependent upon active metabolism.

When broad beans are inoculated with *Botrytis cinerea,* the fungus is usually restricted to small necrotic lesions; with the closely related species *B. fabae* the fungus continues to colonize leaf tissues, giving rise to much larger, spreading lesions (Fig. 10.9a). If one measures the level of the predominant broad-bean phytoalexin, wyerone acid (see Fig. 9.8), accumulating in these different types of lesion, an interesting contrast emerges. In the case of the weakly pathogenic *B. cinerea* the amount of phytoalexin continues to increase over the first 3 days to reach a level sufficient to account for inhibition of the pathogen. With the aggressive *B. fabae*, wyerone acid starts to accumulate but subsequently declines to a level where it can no longer be detected (Fig. 10.9b). Two factors seem to determine this difference. First, *B. fabae* is less sensitive to the phytoalexin than *B. cinerea*. This is critical in the early stages of infection when the pathogen must establish itself in the face of a mounting host response. Secondly, *B. fabae* kills host cells much more rapidly than *B. cinerea*. The lethal factor(s) have not yet been identified, but they might be pectolytic enzymes or toxins. Once it has gained a foothold, the aggressive pathogen kills cells around the penetration site sufficiently quickly to limit phytoalexin production (Fig. 10.9c). In fact, as the lesions expand, the fungus is able to metabolize the relatively small amounts of phytoalexin to which it is then exposed.

Other host–pathogen interactions in which the death of host cells seems to be a prerequisite for susceptibility includes those involving host-specific toxins. Virulence in these systems is correlated with production of a toxin which specifically kills cells in the susceptible host (see Chapter 8). Interestingly,

(a)

(b)

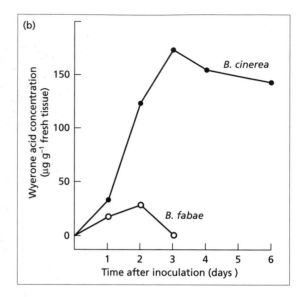

(c)

Fig. 10.9 Chocolate spot of broad bean, caused by *Botrytis fabae*. (a) Lesions on bean leaves. (b) Accumulation of the phytoalexin wyerone acid in leaf tissues and droplets containing spores of either *B. cinerea* or *B. fabae* placed on leaves. (After Hargreaves *et al.* 1977.) (c) Scanning electron micrograph of penetration site showing single spore of *B. fabae* and collapse of epidermal cells around site. Scale bar = 100 μm. (From Daniels & Lucas 1995. Reproduced with permission.)

some of these toxins may have other, more subtle, effects. Recent work on HC-toxin from *Cochliobolus carbonum* suggests that the biochemical mode of action of the toxin is inhibition of an enzyme, histone deacetylase, which interacts with histones, the basic proteins associated with DNA in chromatin. Reversible acetylation of these histone proteins is believed to play a role in modulating gene expression. One theory therefore is that the pathogen toxin directly targets a host enzyme involved in switching on genes. As the majority of induced resistance responses depend on *de novo* gene expression, it appears that this pathogen has evolved an elegant biochemical mechanism for blocking active plant defence.

A model of plant–pathogen interaction

To conclude this section on host–pathogen interaction it might be helpful to summarize some of the ideas discussed with a simple model (Fig. 10.10). This shows that there are several types of genes, and hence gene products, involved in the encounter between a plant and a microbial pathogen. On the pathogen side there are specific functions, such as infection structures, enzymes or toxins, enabling entry and colonization of a living host plant. These are encoded by pathogenicity genes and constitute the overall armoury of weapons involved in microbial attack. On the plant side are constitutive resistance features (not shown in Fig. 10.10) such as physical barriers or

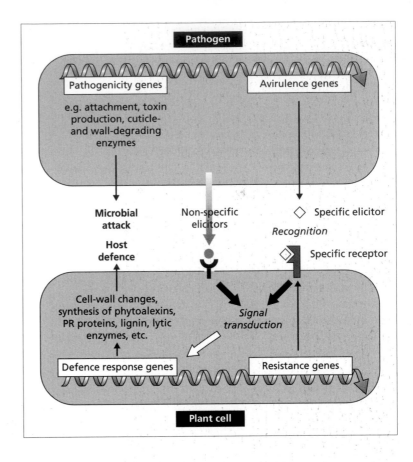

Fig. 10.10 Model of molecular interactions.

preformed inhibitors, and inducible defence pathways switched on once the plant detects the presence of an invader. The recognition system may be non-specific, as in the detection of elicitor molecules such as pathogen cell-wall components, or highly specific, as in the interaction between the products of avirulence genes and host resistance genes. The model predicts the presence of different types of receptor molecules at the host cell surface which bind these different types of pathogen elicitors. The discovery of a bacterial secretion pathway capable of translocating pathogen molecules into plant cells means that some protein interactions involved in specific recognition may also occur within the cell, such as the Avrpto–Pto interaction described above. A signal transduction pathway links such interactions to activation of defence response genes in the plant nucleus. Coordinated induction of these genes then leads to the complex cascade of molecular and biochemical changes underlying active defence.

This model is undoubtedly an oversimplification, omitting, for instance, cytoplasmic receptors detecting virus elicitors, or pathogen products specifically suppressing defence functions. It is also clear that other types of genes, such as those encoding regulatory proteins, may influence the expression of any of the functions shown in Fig. 10.10. We have already seen that pathogen genes encoding products playing key roles in pathogenicity or host specificity may be switched on by plant signals. It should also be noted that the distinctions between these different classes of interacting molecules may not be as clear as the diagram suggests. There are already examples, discussed above, of pathogen molecules involved in microbial attack (i.e. virulence) which can also serve as specific elicitors of plant defence.

This apparent complexity should not be surprising, as one key feature of host–pathogen interaction is its dynamic nature. Both host and pathogen can evolve in time, and the genetic changes underlying all biological variation will also influence the processes determining pathogenicity, host resistance and susceptibility. It has already been suggested that these interactions are the result of an evolutionary 'arms race' between plants and their pests and parasites, and represent one part of a wider biological struggle for survival. This might be a melodramatic view, but there is some evidence for its validity. In one example already discussed, maize infected by the leaf spot pathogen *Cochliobolus carbonum*, we have seen that the pathogen produces a highly active toxin which appears to disarm the host by blocking the defence response genes. Certain maize plants are, however, able to resist the pathogen by producing an enzyme which inactivates the toxin. Will this defence strategy prove effective indefinitely, or can pathogen strains evolve which produce chemical variants of the toxin, or other toxins not degraded by the enzyme? Already this resembles a game of chemical 'cat and mouse', and given the quite astonishing range of biological diversity possible, further developments in the plot are to be expected!

Further reading

Reviews and papers

Bendahmane, A., Kohm, B.A., Dedi, C. & Baulcombe, D.C. (1995) The coat protein of potato virus X is a strain-specific elicitor of *Rx*1-mediated virus resistance in potato. *The Plant Journal* **8**, 933–941.

Bonas, U. & Van den Ackerveken, G. (1997) Recognition of bacterial avirulence proteins occurs inside the plant cell: a general phenomenon in resistance to bacterial diseases? *The Plant Journal* **12**, 1–7.

Bonas, U., Schulte, R., Fensalau, S., Minsavage, G.V., Staskawicz, B.J. & Stahl, R.E. (1991) Isolation of a gene cluster from *Xanthomonas campestris* pv. *vesicatoria* that determines pathogenicity and the hypersensitive response on pepper and tomato. *Molecular Plant–Microbe Interactions* **4**, 81–88.

Brosch, B., Ransom, R., Lechner, T., Walton, J.D. & Loidl, P. (1995) Inhibition of maize histone deacetylases by HC toxin, the host-selective toxin of *Cochliobolus carbonum*. *The Plant Cell* **7**, 1941–1950.

Büschges, R., Hollricher, K., Panstruga, R. *et al.* (1997) The barley *Mlo* gene: a novel control element of plant pathogen resistance. *Cell* **88**, 695–705.

Dangl, J.L. (1995) Piece de resistance: Novel classes of plant disease resistance genes. *Cell* **80**, 363–366.

de Wit, P.J.G.M. (1995) Fungal avirulence genes and plant resistance genes: unraveling the molecular basis of gene-for-gene interactions. *Advances in Botanical Research* **21**, 148–178.

Feuillet, C., Schachermayr, G. & Keller, B. (1996) Molecular cloning of a new receptor-like kinase gene encoded at the *Lr10* disease resistance locus of wheat. *The Plant Journal* **11**, 45–52.

Gopalan, S. & Yang He, S. (1996) Bacterial genes involved in the elicitation of hypersensitive response and pathogenesis. *Plant Disease* **80**, 604–609.

Heath, M.C. (1991). Evolution of resistance to fungal parasitism in natural ecosystems. *New Phytologist* **119**, 331–343.

Innes, R.W. (1995) Plant–parasite interactions: has the gene-for-gene model become outdated? *Trends in Microbiology* **3**, 483–485.

Jakobek, J.L., Smith, J.A. & Lindgren, P.B. (1993) Suppression of bean defense responses by *Pseudomonas syringae*. *The Plant Cell* **5**, 57–63.

Jones, D.A. & Jones, D.G. (1997) The role of leucine-rich repeat proteins in plant defences. *Advances in Plant Pathology* **24**, 90–167.

Kang, S., Sweigard, J.A. & Valent, B. (1995) The *PWL* host specificity gene family in the blast fungus *Magnaporthe grisea*. *Molecular Plant–Microbe Interactions* **8**, 939–948.

Keith, L.W., Boyd, C., Keen, N.T. & Partridge, J.E. (1997) Comparison of *avr*D alleles from *Pseudomonas syringae* pv. *glycinea*. *Molecular Plant–Microbe Interactions* **10**, 416–422.

Kooman-Gersmann, M., Honée, G., Bonnema, G. & de Wit, P.J.G.M. (1996) A high-affinity binding site for the AVR9 peptide elicitor of *Cladosporium fulvum* is present on plasma membranes of tomato and other solanaceous plants. *The Plant Cell* **8**, 929–938.

Lamb, C. (1996) A ligand-receptor mechanism in plant–pathogen recognition. *Science* **274**, 2038–2039.

Lawrence, G.J., Finnegan, E.J., Ayliffe, M.A. & Ellis, J.G. (1995) The *L6* gene for flax rust resistance is related to the *Arabidopsis* bacterial gene *RPS2* and the tobacco viral resistance gene *N*. *The Plant Cell* **7**, 1195–1206.

McMullen, M.D. & Simcox, K.D. (1995) Genomic organization of disease and insect resistance genes in maize. *Molecular Plant–Microbe Interactions* **8**, 811–815.

Manners, J.M. & Scott, K.J. (1985) Reduced translatable messenger RNA activities in leaves of barley infected with *Erysiphe graminis* f.sp. *hordei*. *Physiological Plant Pathology* **26**, 297–308.

Meeley, R.B., Johal, G.S., Briggs, S.P. & Walton, J.D. (1992) A biochemical phenotype for a disease resistance gene of maize. *The Plant Cell* **4**, 71–77.

Pryor, T. (1987) The origin and structure of fungal disease resistance genes in plants. *Trends in Genetics* **3**, 157–161.

Reignault, P., Frost, L.N., Richardson, H., Daniels, M.H., Jones, J.D.G. & Parker, J.E. (1996) Four *Arabidopsis* RPP loci controlling resistance to the *Noco2* isolate of

Peronospora parasitica map to regions known to contain other *RPP* recognition specificities. *Molecular Plant–Microbe Interactions* **9**, 464–473.

Song, W.Y., Wang, G.L., Chen, L.L. *et al.* (1995) A receptor kinase-like protein encoded by the rice disease resistance gene, *Xa21*. *Science* **270**, 1804–1806.

Staskawicz, B.J., Ausubel, F.M., Baker, B.J., Ellis, J.G. & Jones, J.D.G. (1995) Molecular genetics of plant disease resistance. *Science* **268**, 661–667.

Tang, X., Frederick, R.D., Zhou, J., Halterman, D.A., Jia, Y. & Martin, G.B. (1996) Initiation of plant disease resistance by physical interaction of AvrPto and Pto kinase. *Science* **274**, 2060–2065.

Van den Ackerveken, G., Marois, E. & Bonas, U. (1996) Recognition of the bacterial avirulence protein AvrBs3 occurs inside the host plant cell. *Cell* **87**, 1307–1316.

Part 3
Disease Management

'Plant pathology has many successes to its credit during its brief history, and methods have been developed to prevent the great losses some diseases used to cause. However, it would be vain to maintain that all is well.' [F.C. Bawden, 1908–1972]

Agricultural systems are in many ways the opposite of natural biological communities. Modern crop husbandry entails growing huge numbers of genetically similar plants crowded together over large areas. Nutrients, water or both are applied in quantities sufficient to ensure vigorous crop growth. Any pest, pathogen or weed able to thrive under these ideal conditions therefore has the potential for explosive multiplication and spread, free from the usual constraints encountered in a diverse natural community. One important aspect of crop production is to prevent such disease outbreaks, or if they occur to restrict populations of pathogens or pests to levels which do not adversely affect the performance of the crop. Strategies for disease management are therefore an essential part of any crop production system.

Options for disease control

A number of different approaches are available to the grower to prevent or limit the damage caused by plant pathogens. These are grouped in Table 1 under three main headings. First it may be possible to avoid infection altogether by ensuring that the pathogen is excluded from the crop. One obvious strategy here is to grow the crop in a region in which the pathogen does not occur; alternatively, vectors required for transmission of a pathogen may be absent. In the UK, for example, seed potatoes are mainly produced in Scotland where the chances of aphids introducing viruses are much less than in warmer southern regions. Exclusion of the pathogen is an important strategy in crops grown in controlled environments such as glasshouses. Maintaining a pathogen-free area usually requires constant vigilance to avoid introduction of disease agents via seed or propagation material. Hence there may be legislative measures to ensure clean seed, or to quarantine any plants being moved between regions or

countries. These aspects have already been discussed in Chapter 5.

A second set of control options concern reducing the amount of pathogen inoculum available to cause disease. Measures such as soil sterilization or disinfection of seeds are aimed at eradicating or reducing the number of pathogen propagules present before the crop is planted. Many cultural practices, such as crop rotation, also reduce the carry-over of spores or other pathogen survival structures between seasons. This decline in pathogen populations occurring between crops depends to a large extent upon natural processes of biological control, such as predation or parasitism of propagules by other organisms. There is considerable interest therefore in devising methods for enhancing this natural antagonism in soil or on crop residues. Such methods of biological control are discussed in Chapter 13.

Once a crop is exposed to a pathogen, a further set of control options must be considered (Table 1). These are all aimed, directly or indirectly, at reducing the rate of disease development in the crop. Genetic resistance to the pathogen is one of the most important and effective ways of achieving this (see Chapter 12). In cases where resistance is not fully effective against all components of the pathogen population, benefits may be obtained by growing mixtures of different cultivars, or diversifying crops between fields, farms or regions (see p. 226). Another highly effective strategy to restrict epidemic development is to treat the crop with a chemical active against the pathogen or its vector. Traditionally such treatments were best applied to protect the crop before significant exposure to a pathogen, but the newer generation fungicides have curative properties and thus may be worth applying even after infection is established in a crop (see Chapter 11). Concerns over the possible environmental effects of pesticides have recently placed increased emphasis on biological approaches to

Table 1 Options for control of plant disease.

---

1 Exclusion of the pathogen
(a) Pathogen-free area
(b) Clean seed or propagation material (certification)
(c) Quarantine

2 Reduction of inoculum
(a) Soil treatment
(b) Treated seed or propagation material
(c) Husbandry practices (e.g. rotation)
(d) Biological control

3 Reduce rate of pathogen multiplication
(a) Rate-reducing plant resistance
(b) Cultivar diversification/mixtures
(c) Chemical control
(d) Biological control

4 Integrated control programmes
Combinations of several approaches

---

disease control, although developing biological agents with an efficacy comparable to the best available chemicals remains a difficult task.

Finally, the most cost-effective, durable and environmentally acceptable strategies for disease management usually entail the use of several, complementary approaches. Such integrated control programmes require a sound understanding of the diverse factors underlying disease development, including pathogen ecology, crop growth and physiology, soil status and climatic conditions. For most plant diseases, the ideal scenario of a fully integrated control programme remains elusive, but progress towards this goal is discussed in Chapter 14.

The economics of control

The options available to the grower for controlling plant disease depend not only upon the type of crop and the nature of the pathogen concerned, but also upon the economic context. Almost all control measures require significant expenditure in terms of manpower and materials, and these costs need to be set against the benefits gained by applying such measures. There is no point recommending a control strategy, however effective, if the resources required cost more than any likely return in crop yield or quality. The only exception to this rule is in cases

where some wider strategic significance applies, for instance where eradication of a pathogen removes a long-term threat to crop production in a particular region.

It is also important to appreciate that agricultural systems differ widely, and disease control options appropriate for one system may not be economic in another. Figure 1 shows a simple classification of agricultural systems based upon the level of the inputs into the system, and the market value of the product. In subsistence farming, which is still practised over large areas of the developing world, the value of the crop is insufficient to sustain expensive inputs of fertilizers or pesticides. Control measures must be cheap and easy to apply. In contrast, intensive production of

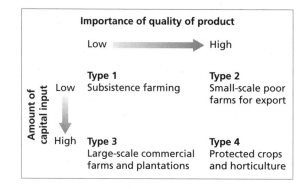
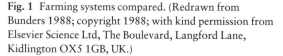

Fig. 1 Farming systems compared. (Redrawn from Bunders 1988; copyright 1988; with kind permission from Elsevier Science Ltd, The Boulevard, Langford Lane, Kidlington OX5 1GB, UK.)

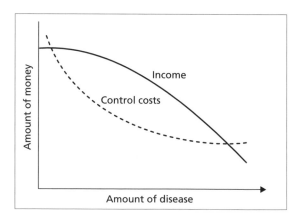

Fig. 2 Crop income vs. control costs.

high-value horticultural crops in glasshouses involves sophisticated systems for regulation of the crop environment, including the use of biological and chemical disease- and pest-control agents. Most agricultural systems fall between these two extremes, and the level of inputs varies depending upon market conditions. In some regions, such as within the European Community, certain crops attract subsidies which maintain the market value of the product above its normal baseline. In such instances the grower may be able to justify the use of extra inputs to control disease. But market conditions change from year to year, and options for disease control have to be adjusted accordingly. Recently grain prices have fallen both in Europe and North America, and hence there is increasing pressure to reduce the costs of inputs by, for instance, improving the efficiency of use of fungicides and fertilizers.

A generalized relationship between control costs and crop income is plotted in Fig. 2. This is a simplified model, but shows that increasing expenditure is required to improve the degree of control achieved. While it may be possible to reduce the amount of disease, or populations of pests or weeds, to insignificant levels, this may entail repeat treatments, or much larger doses of pesticides. The economic threshold—the point at which the cost of control is balanced by the income from the extra crop produced—is usually at an intermediate point, where

some disease is still present. Practical disease management in most crops therefore involves an element of compromise, in which some damage by pests or pathogens is tolerated.

The model above is based solely on direct costs, such as the manpower and materials used to apply a treatment. Increasingly, however, these analyses are being required to include calculations of indirect costs, for instance of chemical residues in food or groundwater, or effects on non-target species in agricultural ecosystems. Obtaining such estimates is, of course, a complex and often controversial business, but it does highlight the need for a sophisticated approach to disease control, taking account of environmental as well as economic considerations.

Further reading

Books

Fry, W.E. (1982) *Principles of Plant Disease Management.* Academic Press, New York.

Maloy, O.C. (1993) *Plant Disease Control. Principles and Practice.* John Wiley & Sons, New York.

Paper

Bunders, J. (1988) Appropriate biotechnology for sustainable agriculture in developing countries. *Trends in Biotechnology* 6, 173–180.

11   Disease Management by Chemicals

'It was, finally, a matter of trying a large number of remedies, assuring myself of their merits, making them easy of execution, of taking into special account their economy of application, and of winning the confidence of the farmer.' [M. Tillet, 1714–1791]

In an age of increasing environmental awareness, the use of chemicals to control pests, pathogens and weeds is now questioned. This is part of the wider debate about intensive agriculture and its effects on the environment, but the issue of chemicals has become particularly emotive. There is nowadays no shortage of critics eager to discredit the manufacturers and users of pesticides. But this state of affairs is a recent development, and should not obscure the relief and excitement which greeted the discovery of the first effective pesticides, which provided growers with a quick, economic means to control previously destructive infestations and diseases. The stability and security of food supplies in the developed world is due in part to the success of this strategy. While assessing the current status of the chemical control of plant diseases it is important to maintain this historical perspective, and to consider the achievements as well as the problems posed by the use of chemicals in crop management.

The following account will focus on fungicides, as these are the chemicals most frequently used to control plant diseases. Many of the basic principles discussed, however, apply equally well to other important types of crop protection chemicals, such as insecticides and herbicides.

Fungicides

The evolution of fungicides

The fungicidal properties of certain chemicals have been known for many years. The first fungicides, based on sulphur and copper, were discovered in 1846 and 1882, respectively. The discovery of Bordeaux mixture, based on copper sulphate and lime, by Pierre Millardet in France, is one of the most familiar stories in plant pathology, starting with the chance observation that copper salts applied to grapevines to deter thieves also controlled infection by the downy mildew pathogen, *Plasmopara viticola*. Millardet's achievement was to translate this observation into practical use by developing formulations of copper for effective commercial application on crops. During the century since this discovery, fungicides have diversified and changed dramatically (Fig. 11.1; Tables 11.1 & 11.2). The early inorganic compounds have now been superseded by organic chemicals which are active at very low doses, are effective against a wide range of fungal pathogens, and can be applied with precision by machinery appropriate to a small plot or a 1000-ha plantation. However, if pioneers such as Millardet were still alive today, two features of the current fungicide market (Fig. 11.2) would surprise them. Firstly, many of the old, original compounds are still widely used. Secondly, almost all the modern generation of fungicides were discovered by a process not dissimilar to Millardet's initial observation, with the most effective compounds being selected by random screening for activity against a few fungi chosen to represent the most important target pathogens. Only in the past few years, with advances in molecular biology, structural chemistry, and computer technology, has the prospect of designing molecules to perform specific tasks become a reality.

The perfect fungicide

It is relatively easy to compile a list of the desirable properties one would like any new fungicide to

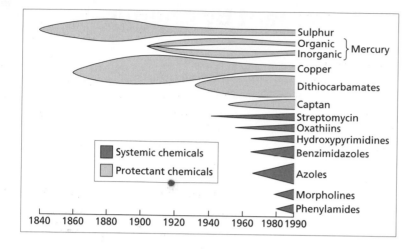

Sulphur
Organic
Inorganic } Mercury
Copper
Dithiocarbamates
Captan
Streptomycin
Oxathiins
Hydroxypyrimidines
Benzimidazoles
Azoles
Morpholines
Phenylamides

Systemic chemicals
Protectant chemicals

1840 1860 1880 1900 1920 1940 1960 1980 1990

Fig. 11.1 Evolution of fungicides, types available, origins and relative importance.

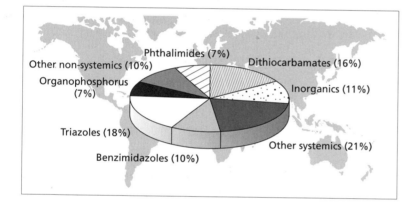

Phthalimides (7%)
Other non-systemics (10%)
Organophosphorus (7%)
Triazoles (18%)
Benzimidazoles (10%)
Dithiocarbamates (16%)
Inorganics (11%)
Other systemics (21%)

Fig. 11.2 Fungicides world market 1991. (After Lyr 1995.)

possess (Table 11.3). In practice, it is difficult to satisfy all these requirements. The biological considerations listed are attainable for many fungicides applied to aerial tissues or seeds, but targeting pathogens found in the soil is more problematical. Soil sterilants such as chloropicrin and methyl bromide are broad-spectrum biocides which have drastic effects on the soil microflora and fauna, and have now been withdrawn in many countries due to concerns about their safety. Similarly, the use of certain systemic compounds which are taken up by plants raises questions over the persistence of residues in crop products. Such residues may be beneficial in terms of reducing post-harvest diseases, such as fruit rots, but persistence into the food chain is regarded as less acceptable. All new agrochemicals are subjected to rigorous toxicology testing (see below), and the most common reason for a compound failing to make

it to the market is some question mark, however, small, about safety. Alternatively the cost of production, or the economics of use by comparison with existing products, may lead to the demise of otherwise promising novel compounds.

The discovery process

The starting point in the search for compounds with potentially useful biological activity is synthetic chemistry. Most agrochemical companies employ teams of chemists who continuously make new compounds. Each compound is screened for activity against a range of target organisms, including plant pathogens, pests and weeds. A selected few are then chosen for more intensive evaluation as candidate compounds for possible commercial development.

Table 11.1 Major groups of protectant fungicides, with examples.

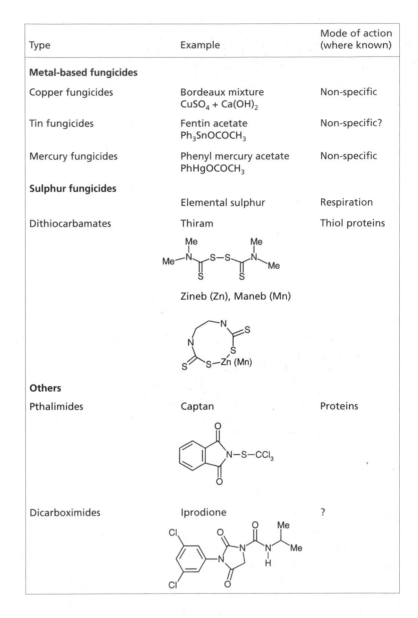

Type	Example	Mode of action (where known)
Metal-based fungicides		
Copper fungicides	Bordeaux mixture $CuSO_4 + Ca(OH)_2$	Non-specific
Tin fungicides	Fentin acetate $Ph_3SnOCOCH_3$	Non-specific?
Mercury fungicides	Phenyl mercury acetate $PhHgOCOCH_3$	Non-specific
Sulphur fungicides		
	Elemental sulphur	Respiration
Dithiocarbamates	Thiram	Thiol proteins
	Zineb (Zn), Maneb (Mn)	
Others		
Pthalimides	Captan	Proteins
Dicarboximides	Iprodione	?

In practice this apparently random process, known as **empirical screening**, is carefully designed and controlled (Fig. 11.3). Because developing agrochemicals is a commercial enterprise, it is vital not to waste time and resources on compounds which, for one reason or another, are unlikely to make it to the market. The initial biological screens used are designed to identify chemicals with the most promising activity. Many highly active compounds are discarded at this stage if their mode of action is shown to be identical to existing products already on the market. Rigorous toxicology tests reject those which might fail on safety grounds. Only a tiny proportion of chemicals screened eventually become commercial products. Original estimates of one in 10 000 chemicals tested have been revised downwards to less than one in 20 000. This no doubt is a reflection of the difficulty of finding new compounds with a significant advantage over existing products, along with the increased stringency of registration requirements, particularly

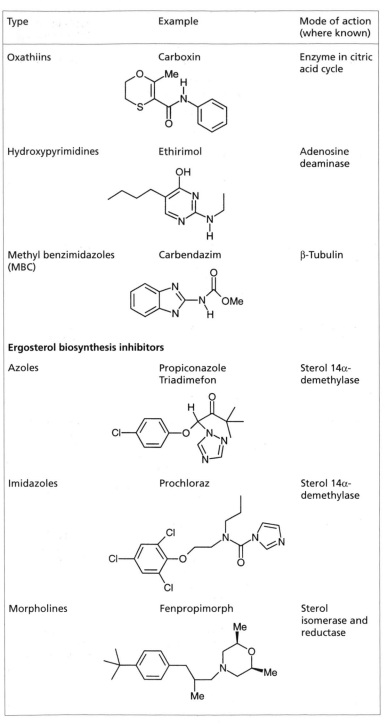

Type	Example	Mode of action (where known)
Oxathiins	Carboxin	Enzyme in citric acid cycle
Hydroxypyrimidines	Ethirimol	Adenosine deaminase
Methyl benzimidazoles (MBC)	Carbendazim	β-Tubulin
Ergosterol biosynthesis inhibitors		
Azoles	Propiconazole Triadimefon	Sterol 14α-demethylase
Imidazoles	Prochloraz	Sterol 14α-demethylase
Morpholines	Fenpropimorph	Sterol isomerase and reductase

Table 11.2 Major groups of systemic fungicides, with examples.

Continued

Table 11.2 *Continued*

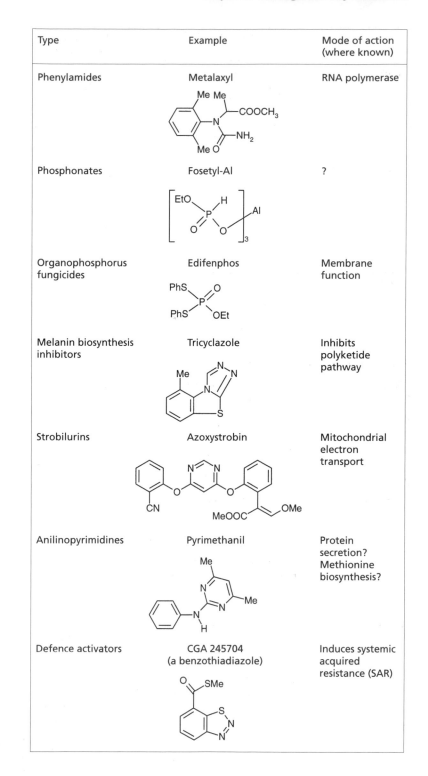

Type	Example	Mode of action (where known)
Phenylamides	Metalaxyl	RNA polymerase
Phosphonates	Fosetyl-Al	?
Organophosphorus fungicides	Edifenphos	Membrane function
Melanin biosynthesis inhibitors	Tricyclazole	Inhibits polyketide pathway
Strobilurins	Azoxystrobin	Mitochondrial electron transport
Anilinopyrimidines	Pyrimethanil	Protein secretion? Methionine biosynthesis?
Defence activators	CGA 245704 (a benzothiadiazole)	Induces systemic acquired resistance (SAR)

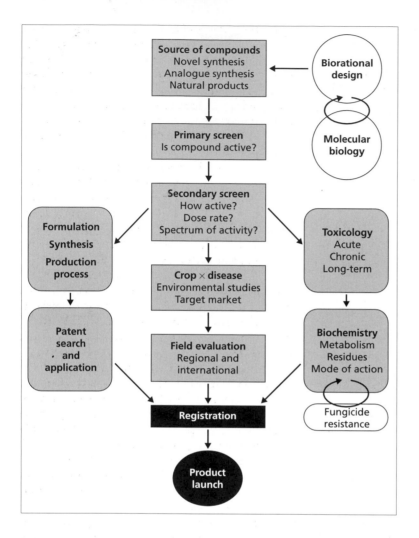

Fig. 11.3 Activities involved in the discovery and development of new agrochemicals.

concerning product safety. As a consequence, agrochemical development is a multi-million dollar exercise which can only be tackled by large, international companies. The notion that agrochemical companies are all involved in some fiercely competitive discovery race is, however, misleading. In reality companies cooperate as well as compete, through licensing agreements to permit manufacture or marketing of each other's compounds, often in mixtures. The virtue of such collaboration is not only commercial, as it can encourage concerted action to combat problems such as fungicide resistance.

The main steps in fungicide development are shown in Fig. 11.3. There are in fact several possible sources of the new molecules required in phase 1. Novel, or 'blue sky' synthesis, is only one approach.

Another is analogue synthesis, in which chemists produce a series of modified molecules related to compounds with known activity. The best recent examples of these are the azoles (Table 11.2), a family of fungicides with the same mode of action (inhibition of sterol biosynthesis), which account for a significant proportion of the world fungicide market (Fig. 11.2). Alongside synthesis of analogues is the more contentious business of 'patent-busting', in which rival companies try to identify chemical loopholes in existing patents which might allow production of a similar, but legally distinct, compound. An alternative, and increasingly popular, approach to chemical synthesis is to screen natural products, usually obtained from microbial cultures. This has of course been particularly fruitful in the discovery

Table 11.3 Specifications of a perfect fungicide.

Biological considerations
It must offer effective and consistent disease control
It should not be phytotoxic at the recommended dose
  level
It should not adversely affect other parts of the crop
  ecosystem

Toxicological considerations
It must not constitute a hazard during application
Residues in the crop should not pose a problem for
  the consumer

Formulation factors
It should be safe to store and transport
It should be simple to apply at a precise dosage level
The method of formulation should increase its
  efficiency as a fungicide

Economic considerations
The financial return should exceed the cost of the
  fungicide and its application

---

of antibiotics for clinical use, but some recently developed fungicides, such as the strobilurins (Table 11.2), also derive from natural chemicals, in this case metabolites found in certain species of mushrooms. The biochemical diversity of living organisms is now seen as a potentially limitless source of novel activity.

Whatever the origin of the new molecule, identification of useful biological activity depends upon a phased selection process (Fig. 11.3). The aim of the primary screen is to rapidly identify compounds with promising properties. Typically this will include a range of fungal pathogens representing on the one hand biological diversity, and on the other, known economic importance. There is little commercial purpose in selecting compounds with excellent activity against minor pathogens, for which there is no significant market. Hence most screens will include a downy mildew such as *Plasmopara*, a *Phytophthora* species, a powdery mildew, a rust, and several other fungi known to be significant on a world scale, such as rice blast (*Magnaporthe grisea*), *Botrytis*, apple scab (*Venturia*), and *Rhizoctonia*.

There are of course many variations on this theme, and no two company screens will be identical. There may also be subtleties in the type of activity searched for. Originally, many companies screened compounds *in vitro*, against cultures of pathogenic fungi. Quite apart from the problem of testing non-culturable biotrophs, this approach potentially overlooked useful interactions on or in the plant. For instance, a compound might be converted into a more active form through plant metabolism, or might activate endogenous plant defences. For these reasons almost all screens nowadays use more natural systems which assess the effects of the test chemical on the particular plant–pathogen interaction.

Once a promising compound has been selected in the primary screen, more intensive evaluation begins (Fig. 11.3). The secondary screen seeks to estimate the potency of the chemical and more clearly define its spectrum of activity. Additional target pathogens are often included at this stage. Further properties of the compound, such as its uptake, movement and persistence in the plant, will be investigated. Work may also start on structure–activity relationships, with chemical modification of the molecule to optimize its activity. Next it is important to assess the field performance of the new compound by comparison with existing standards, and to initiate toxicology tests to evaluate safety. Environmental studies are included to monitor the fate of the chemical in the ecosystem, and to detect any adverse effects on non-target species. Due to the increasing awareness of the potential problem of fungicide resistance (see p. 208), there will most likely be some work aimed at assessing the risk of resistance developing, for instance genetic studies using fungal mutants. It is also essential for the company to protect its discovery as early as possible by filing an appropriate patent.

Further development aims to progress the compound from a promising candidate molecule to a commercial product in practical use. This involves several different activities, from development of a chemical production process, through formulation and recommendations for use, to the final stages of product registration and marketing. Each country has its own regulatory requirements and in practice this can prove one of the most time-consuming and difficult steps in launching a new agrochemical.

In spite of the complexity and expense of this long process, the agrochemical industry has been remarkably successful in finding and developing a succession of new types of molecules to aid the fight against plant disease. Predictions of the rapid demise of chemical control through a combination of economic, environmental and biological changes, including pathogen evolution, have proved premature. But a number of problems remain. To date there are no commercially effective compounds active against plant viruses, and few effective bacteriocides. Several fungal targets remain elusive, including some vascular wilts, root pathogens, and other soil-borne diseases. Only one class of compounds, the phosphonates, has significant shoot to root mobility, providing control of root infection following aerial application. And added to the commercial and legal pressures of patent life and product registration are the increasing problems of environmental acceptability and pathogen resistance.

Rational design of fungicides

A quite different, and potentially more efficient means of inventing pesticides than the hit-and-miss business of empirical screening, is to specifically design molecules with a particular target in mind. This is really an extension of structure–activity studies into the realm of molecular modelling and the prediction of chemical configurations with optimum biological activity. Given the power of modern computers this is, in theory at least, now feasible. This approach, however, depends upon a fundamental understanding of the specific target, and the likely effects of interference with the target on the pathogen.

The basic idea is shown in Fig. 11.4. First a biochemical process essential for the growth, development or pathogenicity of the fungus is identified. This might be a particular step in a metabolic pathway leading to production of a vital molecule such as a component of a membrane or a cell wall, or a pathogenicity factor such as a toxin. It should be obvious that this biochemical step should not be present, or at least not be essential, in the host plant, otherwise any intervention will damage the crop as well as the pathogen. The next step is purification of the protein, usually an enzyme, responsible for carrying out the process. Three-dimensional modelling of the protein then identifies the active site, and permits the design of molecules which should fit into the site and thereby disrupt normal function. The ultimate test is to synthesize the predicted molecule and assess its activity against the pathogen.

This approach may be refined by using the tools of molecular genetics. If one can identify the gene encoding the target protein (Fig. 11.4) then there are a number of possibilities. Sequencing the gene will assist prediction of the encoded protein structure and conformation. The gene may be expressed in another organism, usually a bacterium, to produce large quantities of the protein. Such recombinant DNA technology might also permit the development of alternative, or 'smart' screens, in which a microorganism expressing the gene of interest, rather than the pathogen itself, is used to rapidly test compounds for their biological effects.

Molecular biology can also aid target validation. It is vital to have information on the likely biological effects of inhibiting the target process. Replacing the gene encoding the target protein with a defective copy (known as gene disruption) will abolish function altogether. If the biochemical step concerned is vital to the fungus, such genetic intervention will be lethal. But while this may confirm the importance of the target, it is not a definitive test. The argument is as follows. It is unlikely, in practical terms, that any fungicide will completely abolish the activity of a target enzyme, due mainly to the difficulty of getting a sufficient dose to the specific cellular site. A more likely scenario is that the compound will reduce rather than abolish activity. What is required is a more precise test of what happens if target protein X is reduced in activity by 10%, or 50%, etc. One way of doing this is to reduce the amount of protein X by attenuating expression of the gene encoding the protein. It may be possible to modulate gene expression by means of antisense RNA, which effectively ties up a proportion of the message for the target protein. This approach is now routinely used to attenuate gene expression in plants, but it should be noted that to date its application in fungi remains largely unproven.

Types of fungicides

There are several different ways of classifying the diverse range of chemical compounds currently used as commercial fungicides. One is by chemical class: for instance, inorganic vs. organic compounds. Another is by mode of action: for example, compounds which have toxic effects on a variety of cellular processes vs. those which interfere with a specific process (see below). A further classification is based on the behaviour of the compound on or in the plant.

Prior to 1960, nearly all the fungicides discovered came under the general description of **protectant**

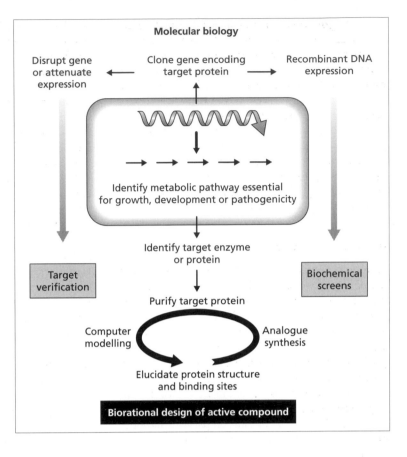

Fig. 11.4 Rational approach to fungicide design.

compounds (Table 11.1). These materials supplement the defences of the plant by forming a superficial chemical barrier to prevent, or protect against, infection. While protectant compounds are effective against a wide range of fungi, they have limitations in practical use. By the very nature of their mode of action they must be applied before the pathogen attempts to penetrate the host. Hence there is a need for reliable, early warning of an infection risk (see p. 64) if protectant compounds are to be used effectively and economically. Because they form surface coatings, such fungicides are subject to degradation and erosion by light, rain and other environmental factors. Last, but not least, applications to growing plants rapidly become ineffective as new leaves, flowers and fruits continue to develop. For this reason protectant fungicides may need to be applied to a crop at regular intervals throughout the season.

Due to these limitations there was a sustained hunt for compounds with a different type of activity, which are actually taken up by the plant and poison the pathogen from within. The advantages of such systemic chemicals should be obvious (Table 11.2). These compounds offer opportunities for therapy, i.e. they may eradicate an established infection. In this way the use of chemicals in plant pathology would resemble human medicine, where the emphasis has always been on developing cures as well as preventing disease.

Some of the first compounds to be used as systemic pesticides were in fact antibiotics, such as streptomycin, but the real breakthrough came in the 1960s with the advent of the benzimidazoles, such as benomyl and carbendazim. These compounds are active against many different plant-pathogenic fungi, and also move through tissues so that accurate and widespread distribution of the chemical over the host is not necessary (Table 11.4). Better still, it is possible to apply such fungicides in the form of seed-dressings, where continual uptake by the growing seedling can protect the plant for many weeks after germination.

Table 11.4 Effect of distribution on the efficiency of the protectant fungicide dinocap and the systemic fungicide benomyl on control of cucumber powdery mildew, *Sphaerotheca fuliginea*. For each fungicide the same amount of material was applied per leaf in either one or a number of drops. After a period of incubation in an atmosphere laden with pathogen spores, the effects of the fungicides were assessed in terms of the area of leaf affected by the pathogen. The results are expressed as percentage reductions from control values. (After Evans 1977.)

No. drops per leaf	Fungicide concentration per drop	Disease control (% reduction)	
		Dinocap	Benomyl
1	0.0250	5	20
2	0.0125	10	40
4	0.0067	20	100
8	0.0033	55	100
16	0.0017	100	100

The large majority of systemic compounds, however, move only in the apoplast, and hence tend to travel from the base to the top of the plant, accumulating in leaves and shoot apices. For this reason they are usually ineffective against soil-borne pathogens infecting roots or other subterranean organs. The notable exceptions are the phosphonates, such as fosetyl-Al (Table 11.2), which are phloem-mobile and can therefore travel from the shoot to the root, providing effective treatment for diseases such as root rot caused by *Phytophthora* species (see p. 208).

The discovery of systemic fungicides provided an enormous step forward in the chemical control of plant disease. Previously intractable problems such as deep-seated infections of fruits or seeds could now be effectively tackled. But one of the apparent strengths of these new compounds, their highly specific mode of action, soon proved to be a weakness. Within a short period of time after their introduction a significant number of target pathogens began to develop resistance to these chemicals.

Mode of action of fungicides

Most first-generation, protectant fungicides (Table 11.1) are known to be **multisite** inhibitors, which interfere with the central metabolic processes of the target fungus. Indeed the majority of these fungicides appear to affect the production of energy or ATP, either by inhibiting respiration or by uncoupling oxidative phosphorylation. Metal-based fungicides such as copper or mercury inhibit a wide range of enzymes involved in various metabolic pathways. Similarly, dithiocarbamates complex with thiol groups on proteins, thereby inactivating enzymes and ultimately causing cell death. This fatal disruption of core processes probably explains why few fungi have evolved systems able to circumvent the toxic effects of these fungicides. Thus for decades the copper, sulphur and dithiocarbamate fungicides have remained as effective as when they were first discovered. One remarkable exception to this rule is the case of *Pyrenophora avenae*, which managed to overcome the toxic effects of mercury applied as a seed-dressing to oats.

By contrast, most of the systemic compounds discovered to date act at a **single site** in the cell, inhibiting a specific enzyme or process. For example, early work on the mode of action of benzimidazole fungicides showed that these compounds inhibited cell division. It was later shown that the specific site of action is β-tubulin, a polymeric protein found in microtubules — essential components of the cytoskeleton. Binding of the fungicide to the tubulin molecule prevents polymerization, and hence disrupts the normal activities of the cytoskeleton, including spindle formation during cell division. The widely used azole fungicides interfere with the biosynthesis of sterols, which are molecules found in fungal cell membranes. These 'sterol biosynthesis inhibitors' (SBIs) affect membrane structure and function, with widespread consequences for the cell. The specific interaction is with an enzyme protein catalysing a single demethylation step in the sterol biosynthesis pathway. Such azole fungicides are therefore referred to more precisely as 'demethylation inhibitors', or DMIs. A second class of SBIs are the morpholines (Table 11.2), which act on the same pathway but at different steps affecting sterol isomerization and reduction. This property is useful as fungi which have become insensitive to DMIs are often still sensitive to morpholine fungicides. Other examples of single-site inhibitors listed in Table 11.2 include the phenylamides, which affect nucleic acid synthesis, and the recently introduced strobilurin fungicides, which block mitochondrial electron transport.

Table 11.5 Comparison of protectant and systemic fungicides.

	Protectant (multisite inhibitors)	Systemic (single-site inhibitors)
Action	Prophylactic	Therapeutic
Basis of toxicity	Many metabolic systems affected	Few metabolic systems affected
Phytotoxicity	Common, especially if applied to wrong tissue or an inappropriate host	Rare
Pathogens affected	Numerous	Variable — some extremely specific, others are effective against a broad spectrum
Pathogen resistance	Rare	Common
Movement	Confined to redistribution on surfaces	Translocated, usually in apoplast (xylem, cell walls)

The highly specific mode of action of single-site inhibitors means, perhaps inevitably, that small changes in the fungus, for instance in the target protein, may alter the efficacy of the compound. In many cases only a single mutation in the fungus is sufficient to abolish activity, and hence lead to resistance. The implications of this are discussed in more detail below.

The contrasting properties of protectant and systemic fungicides are summarized in Table 11.5.

Formulation and application

Discovering chemicals with useful biological activity is only one part of fungicide development. It is also necessary to produce the compound in a form suitable for storage and subsequent application to the crop. Therefore, alongside the constant hunt for better compounds, efforts are continually made to optimize activity in the field by improving formulation and devising better ways of delivering the chemical to the target fungus. The goal is not only to ensure effective control of the disease, but also to reduce the amount of fungicide applied to the crop. This lowers costs and also minimizes any risk to non-target species in the crop ecosystem.

Much of the mystique of the agrochemical industry concerns formulation. The tricks of how to get insoluble compounds into a form suitable for application, and then distribute them over a crop in such as way that they stick to water-repellent plant surfaces and remain active for days or even weeks, are closely guarded secrets. With protectant compounds the main problem is to ensure an even coverage of the plant, and to prevent loss of the active ingredient through weathering or degradation. The biologically active chemical is therefore mixed with carrier compounds which aid dispersion and adhesion to the crop. Such ingredients are often described as 'stickers' and 'spreaders'. They include surface-active detergents and polymers such as carboxymethylcellulose and alginates. With systemic compounds, uptake by the plant is important, and ingredients may therefore be added to aid penetration across the external cuticle of the plant.

The problems of formulation are closely allied to the method of application. Clearly this depends in large part on the type of pathogen to be controlled. Formulations designed to deliver an active ingredient to the leaf surface are unlikely to be fully effective as seed-dressings or treatments for soil. There is also the question of scale. Applying fungicides in a controlled environment such as a glasshouse is a very different proposition to treating a whole field or plantation. In the former case it may be possible to add a chemical to the irrigation system, or to fumigate the crop atmosphere, while the latter usually requires spray application from a tractor or the air. Figure 11.5 illustrates some of the different methods used to apply agrochemicals. The most common approach is to spray a diluted solution or suspension of the active ingredient through a hydraulic nozzle. Such conventional, high-volume sprays are relatively inefficient because a wide range of droplet sizes are generated and much of the active ingredient misses the target. The larger droplets tend to run off the plant, while the smaller droplets may be carried away in turbulent air, a phenomenon known as 'drift'. Less than 1% of the

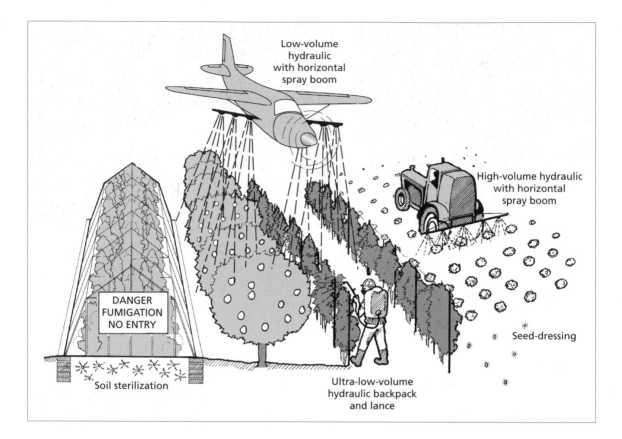

Fig. 11.5 Some methods used to apply pesticides to various crops.

chemical may reach the correct place. A cynic once observed that this method of delivery is analogous to treating the common cold by dropping aspirin tablets from an aeroplane and hoping to find a proportion of the human population looking skywards with their mouths open! While this is obviously an exaggeration it does illustrate the difficulty of reaching disease agents which may be present on the underside of leaves or, as in the case of the eyespot pathogen of wheat, *Pseudocercosporella herpotrichoides*, infecting the base of the stem. Interestingly, control of this disease has been shown to be more effective in instances where rainfall follows application of a fungicide spray, presumably due to the compound being washed down the leaf sheaths to the infection site. Due to the inefficiency of conventional hydraulic sprays, much effort has been expended on developing techniques for controlled droplet application, using

improved nozzle designs or rotating discs to generate sprays of defined droplet size. Such improvements permit a lower volume of pesticide to be applied to the crop. Recently, ultra-low-volume methods have been developed, such as electrodynamic spray techniques (Fig. 11.6a). This technology generates electrically charged droplets which are attracted to plant surfaces, giving improved coverage from a smaller volume of liquid. The active chemical can be formulated in oil rather than water, which prevents evaporation and is also an advantage in regions where rainfall and hence the water supply is limiting. Electrostatic spraying has been successfully used to apply pesticides in tropical crops such as cotton and cowpea.

While spraying is the most widely used way of applying pesticides, other methods may be more suitable depending on the compound or the target pathogen. Sulphur, for instance, is easily applied as a dust rather than in suspension, while fumigation can be highly effective in protected cultivation such as glasshouses. Seed-dressing is a very efficient way of

(a)

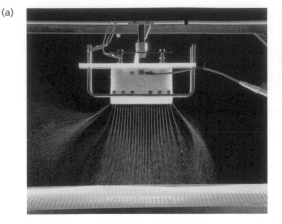

(b)

Fig. 11.6 Some novel ways of applying fungicides. (a) An electrostatic sprayer generating charged droplets of controlled size. (Courtesy of Eric Hislop.) (b) Trunk injection of an avocado tree using a plastic syringe containing an aqueous solution of fosetyl-Al to control

Phytophthora root rot. (Reprinted from Coffey 1987; copyright 1987, with kind permission from Elsevier Science Ltd, The Boulevard, Langford Lane, Kidlington OX5 1GB, UK.)

applying compounds to prevent seedling diseases such as damping-off, and, provided the fungicide is sufficiently mobile in the growing plant, can also protect against infection by airborne pathogens. For instance, systemic compounds such as ethirimol and some azoles, applied to cereal seeds, provide good control of powdery mildew, *Erysiphe graminis*. The advantage of this approach is that the compound can be applied in an appropriate formulation to batches of seed by the supplier rather than the grower, which saves time and money. There can, however, be drawbacks, as some fungicides are phytotoxic when applied to seed, and it has been argued that the more prolonged exposure of the pathogen to a fungicide originating from seed can increase the risk of resistance (see below).

Alternative methods are often necessary when the pathogen occurs in soil, or is difficult to target due to its growth habit in the host plant. Drenching soil with a pesticide, or treating by fumigation, may be costly and also can affect non-target species. In this situation, granular formulations of the chemical may be more appropriate, designed to gradually release the active ingredient into soil over a period of time. Pathogens which grow in inaccessible parts of the plant, such as deep within seeds, or in vascular tissues, may also present problems. Earlier this century smut fungi such as species of *Ustilago* and

Tilletia were among the most damaging disease agents of cereal crops, due to transmission from season to season through seed. Techniques were devised to soak seeds in solutions of fungicide, but the real breakthrough in control of these diseases came with the introduction of organomercury treatments, which effectively eradicated most seed-borne fungi. Some smut fungi, however, such as loose smut (*Ustilago nuda*), infect the embryo rather than the seed coat, and are therefore not affected by treatments which are restricted to surface tissues. Control of loose smut was not achieved until the advent of truly systemic compounds such as carboxin which could penetrate and move within host tissues. Ironically, some smut diseases are now staging a comeback due to the withdrawal of organomercury fungicides on environmental grounds. Effective alternatives are available but are more expensive, and hence the use of seed treatments is declining in some crops.

Vascular wilt fungi are difficult to control with chemicals, especially those infecting woody perennial hosts. A good example is Dutch elm disease, caused by *Ophiostoma novo-ulmi*, which has decimated elm populations across much of the northern hemisphere. Attempts were made to protect specimen elms, e.g. those in city parks, by injecting fungicides into the sapwood, where the chemicals are carried upwards in the transpiration stream. Some success was achieved using benzimidazole fungicides such as benomyl, but the treatment proved ineffective on large trees, mainly due to the relative insolubility of these compounds, which prevented them moving in sufficient concentrations to prevent infection of the upper branches. However, in some instances injection can be an extremely effective way of applying a fungicide to a woody plant. The most destructive disease of avocado groves is root rot, caused by *Phytophthora cinnamomi*. This soil-borne pathogen attacks the small, feeder roots of the tree, with often fatal consequences (see p. 251). *Phytophthora* root rot was not amenable to chemical control until the introduction of systemic fungicides active against oomycete fungi such as metalaxyl. This compound, applied as a root drench, controlled disease in young trees, but the prognosis for more mature trees remained gloomy. The advent of phosphonates such as fosetyl-Al, which can move from shoot to root, radically changed the situation, with foliar sprays at monthly intervals giving good control. Even better results were achieved by applying the compound by injection with a syringe which is left inserted in the trunk of the tree (Fig. 11.6b). The fungicide solution is gradually taken up by the tree, and only two injections a year may be sufficient to ensure control of this previously lethal disease.

It should be obvious that labour-intensive application methods, such as injection, are only feasible with high-value crops where protection of individual plants can be justified on economic or amenity grounds.

Fungicide resistance

Prior to the discovery of the first systemic, selective fungicides, there were very few instances when application of a protectant compound, at the correct time and dose rate, failed to control a pathogen. Thus for decades the copper, sulphur and dithiocarbamate fungicides have remained as effective as when they were first discovered. The few exceptions to this rule include the development of resistance to mercury-based seed-dressings in *Pyrenophora*, dodine in apple scab (*Venturia inaequalis*), and problems with the use of diphenyl compounds to control post-harvest rots of citrus fruits caused by *Penicillium* species (Plate 8, facing p. 12). But in general, the protectant, multisite inhibitors, have given long-term, durable control of many crop diseases.

Practical experience with the newer, systemic compounds has been different. There have been numerous cases in which an initially highly effective fungicide has subsequently failed to control a

Table 11.6 Some examples of fungicide resistance in practice.

Fungicide group	Pathogen	Crop	Date first reported
Organomercury	*Pyrenophora avenae*	Oats	1970
Dodine	*Venturia inaequalis*	Apple	1969
Pyrimidine	*Sphaerotheca fuliginea*	Cucumber	1970
Benzimidazoles	*Botrytis cinerea*	Various	1971–1973
	Cercospora spp.	Peanut, sugarbeet	1974
	Pseudocercosporella herpotrichoides	Wheat	1981
Dicarboximides	*Botrytis cinerea*	Grapevine	1978
	Monilinia fructicola	Stone fruit	1986
Phenylamides	*Phytophthora infestans*	Potato	1980
	Peronospora tabacina	Tobacco	1981
	Peronospora parasitica	Brassicas	1983
	Bremia lactucae	Lettuce	1983
Demethylation inhibitors (DMIs)	*Erysiphe graminis*	Cereals	1982–1986
	Pyrenophora teres	Barley	1985
	Penicillium digitatum	Citrus	1987

pathogen in a crop (Table 11.6). Sometimes such failures have occurred within a short time of first use of the compound. For example, the pyrimidine fungicide dimethirimol, introduced in 1968, showed outstanding activity against powdery mildews, and was recommended to control the cucumber powdery mildew pathogen *Sphaerotheca fuliginea* (Fig. 11.7) in glasshouses. Intensive use quickly led to the emergence of highly resistant strains of the pathogen, and by 1971 the compound was withdrawn in The Netherlands. Similarly, the phenylamide fungicide metalaxyl (Table 11.2) was launched in 1977 and recommended for control of many important oomycete pathogens such as the downy mildews, *Phytophthora* and *Pythium*. By 1979, isolates of cucumber downy mildew, *Pseudoperonospora cubensis*, able to tolerate 20 times the initially effective dose of the fungicide had been recorded, and, more dramatically, in the following season failures to control potato blight occurred in Ireland and The Netherlands. This was shown to be due to the incidence of metalaxyl-resistant strains of *Phytophthora infestans* in the field, and shortly after, formulations of fungicide containing metalaxyl alone were withdrawn. Similar experiences occurred with blue mould of tobacco, *Peronospora tabacina*, and other downy mildews such as *Peronospora parasitica* and *Bremia lactucae*. Table 11.6. lists some further examples where resistance to systemic fungicides has occurred.

Why did this happen, and what lessons can be learned from these setbacks? More importantly, can anything be done to prevent such 'boom-and-bust' episodes in the future?

Some definitions

Before attempting to answer these questions, some terms used to describe the problem should be defined. **Insensitivity** and **tolerance** have both been used to indicate changes in the response of fungi to fungicides, but **resistance** is now the preferred term. It is essential, however, to distinguish between instances where the sensitivity of a target fungus to a particular chemical has changed, and the actual loss of efficacy of a fungicide in practical use. There are numerous examples of resistance being detected in some strains of fungi, yet the fungicide still gives effective control of the disease in the field. The phrase **resistance in practice** has therefore been recommended to describe situations where reduced sensitivity of a fungal pathogen to a fungicide results in poor disease control in the field. **Cross-resistance** is the phenomenon whereby development of resistance to one chemical in a particular class also confers resistance to other, related chemicals. For example, strains of *Botrytis cinerea* (Plate 7, facing p. 12) resistant to benomyl are also resistant to other benzimidazole fungicides such as carbendazim and thiabendazole. This has important practical implications, as strains altered in sensitivity to one fungicide in a particular group may simultaneously become resistant to all other compounds in that group.

The risk of resistance

Three factors determine the risk of resistance arising and the extent to which it will spread in the pathogen population and hence become a practical problem:
1 the nature of the fungicide;
2 the way in which the fungicide is used;
3 the biology of the pathogen concerned.
Single-site fungicides are much more likely to encounter problems of resistance since only a single mutation in the pathogen may be sufficient to counter the action of the compound. With multisite fungicides numerous mutations may be required, which greatly reduce the probability of resistance developing to such compounds.

Important aspects of fungicide use are the frequency of application, and whether a compound is

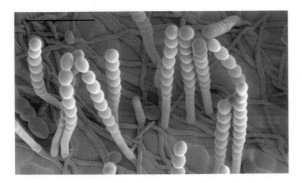

Fig. 11.7 Scanning electron micrograph of cucumber powdery mildew, *Sphaerotheca fuliginea*, showing surface mycelium, conidiophores and chains of spores. Scale bar = 100 μm. (Courtesy of Alison Daniels, AgrEvo UK.)

applied on its own rather than in combination with other chemicals. Ultimately, this is about selection pressure—the degree to which the presence of a fungicide in the crop ecosystem will favour strains of a pathogen less sensitive to the chemical concerned. If, for example, a pathogen population comprises a mixture of strains which differ in sensitivity to a chemical, exposure to that chemical will select those individuals able to withstand the treatment. It should be obvious that the more frequently a fungicide is used, and the more often a pathogen is exposed to it, the greater the likelihood that such resistant individuals will survive rather than the originally predominant sensitive strains.

Biological factors influencing risk are to do with the genetic system of the fungus, and its life cycle. The first consideration is the frequency of mutation to a resistant form; this provides the genes, or alleles, conferring resistance. However, for resistance in practice to occur, such genes must persist and spread in the pathogen population. If, for example, mutation to resistance to a fungicide has other effects on the fungus, such as reducing its growth rate or reproductive capacity, then resistant individuals may not survive in competition with sensitive strains. Genetic recombination may, of course, mix up genes and provide opportunities for fitter, resistant strains to arise. Then there is the extent to which such resistant strains are likely to spread. The types of spores produced and their mode of dispersal are therefore important.

Given these various considerations, it should be possible, at least in theory, to distinguish between 'high-risk' and 'low-risk' pathogens. Fungi with rapid reproductive cycles, producing large numbers of wind-dispersed spores, are more likely to pose problems of resistance than slowly reproducing fungi which are dispersed over only short distances. Practical experience to some extent supports such predictions, with resistance often developing rapidly in polycyclic airborne pathogens such as downy and powdery mildews. But not all incidences of resistance conform to this model. Eyespot disease of cereals (Plate 6, facing p. 12), caused by *Pseudocercosporella herpotrichoides*, developed resistance to methyl benzimidazole (MBC) fungicides during the 1980s, with field populations shifting to a predominance of resistant individuals within only a few seasons. Yet this pathogen was believed to be asexual and splash-dispersed, and the speed with which resistance built

up took many people by surprise. More recently a sexual stage has been discovered in this fungus, but it is not known whether sexual inoculum played any part in the emergence of MBC resistance. It might simply be that mutations to MBC resistance can occur at a fairly high frequency, that such mutations have no discernible effect on fitness, and that MBC fungicides were widely used on a high proportion of cereal crops. The moral of this story is that risk assessment is far from simple, and with highly variable and adaptable organisms such as fungi a useful guiding principle is to expect the unexpected!

The evolution of resistance

The development of resistance to fungicides is an example of an evolutionary change in a fungal population caused by a human activity, the application of a chemical to a crop. The raw material for this evolution is genetic variation in the fungus, but the driving force, the selection pressure, is provided by crop management practices. Understanding the nature of this change in the pathogen, and how it came about, is essential if we are to counter the problem of fungicide resistance.

Monitoring changes in the response of fungal populations to fungicides, season by season, has become an important activity, both in defining the problem, and predicting future events. This relies on surveys of fungal isolates taken from the field to determine their dose–response to the chemical concerned. The usual way to measure fungicide sensitivity is to determine the dose which inhibits 50% of a particular physiological parameter. For example, if growth or spore germination is reduced by half, then this is described as the ED_{50} (effective dose) or EC_{50} (effective concentration) which gives 50% inhibition. Alternatively the minimum inhibitory concentration (MIC), the lowest concentration completely preventing growth of the pathogen, may be determined. Different isolates of a fungus can therefore be compared in a standardized dose–response test.

An important conclusion arising from this type of research is that not all pathogens or fungicides behave in the same way. For instance, in the case of MBC fungicides, the development of resistance was associated with the emergence of fungal strains which could withstand huge doses of the compound, 1000-fold more than sensitive strains. But with some other fungicides, such as the SBIs, resistance has developed as a gradual shift in sensitivity, rather than as a dramatic change (Fig. 11.8). It is clear that more than one phenomenon is likely to be involved. These contrasting patterns in the development of resistance have been described in population terms as either **disruptive selection,** in which the population diverges into sensitive and resistant classes, or **directional selection,** in which the population overall shifts towards reduced sensitivity. This is shown in diagrammatic form in Fig. 11.9. Application of a fungicide either selects a subset of the population which is highly resistant to the compound, or alternatively eliminates

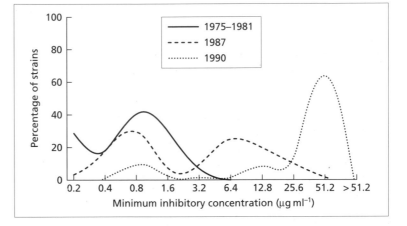

Fig. 11.8 Changes in the sensitivity of the barley pathogen *Rhynchosporium secalis* to the sterol biosynthesis inhibitor (SBI) fungicide triadimenol in populations of the fungus sampled between 1975 and 1990. (Redrawn from Kendall *et al.* 1993; copyright 1993, with kind permission from Elsevier Science Ltd, The Boulevard, Langford Lane, Kidlington OX5 1GB, UK.)

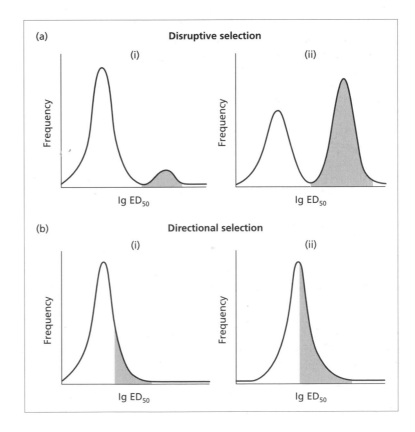

Fig. 11.9 Selection of resistance to fungicides in pathogen populations. Shading represents the more resistant individuals in the population. (a) Disruptive selection: (i) distribution before application of the fungicide; (ii) distribution after prolonged selection. (b) Directional selection: (i) distribution before application of the fungicide; (ii) 'shifting' of the distribution of ED_{50} values after fungicide application. (After Berg *et al.* 1990.)

the least resistant individuals. The former change causes a quantum leap in resistance (as in the case of MBC and phenylamide fungicides), while the latter causes a gradual erosion in the efficacy of the compound over a more extended time scale (e.g. Fig. 11.8).

These different scenarios are ultimately due to differences in the mechanism of resistance, and in its underlying genetic basis. Figure 11.10 shows some of the possible mechanisms involved in the resistance of a fungus to a particular chemical. Comparison of these mechanisms should clarify why the degree of resistance varies. Changes in the target site can, for instance, completely abolish the activity of the fungicide. This is essentially what happened with MBC fungicides, once the tubulin protein had mutated to prevent binding of the chemical. Some of the other mechanisms are likely to have less effect on the sensitivity of the pathogen, for instance changes in the rate of uptake, efflux or detoxification of the compound. Such alterations will reduce, rather than completely abolish, activity. This scheme also shows that more than one gene may affect the response of a fungus to a fungicide. Resistance due to changes in a target protein is usually encoded by a single gene, while other types of resistance may involve several processes encoded by multiple genes. This may explain why some forms of resistance develop gradually with incremental shifts in sensitivity over a period of time.

Combating fungicide resistance

Resistance to pesticides is now an established fact of life in the agrochemical industry, so the question is not so much will resistance occur, as when will it occur, and can it be managed? In fact, even where major reductions in the efficacy of a fungicide have taken place, experience has shown that the compound may still have a useful role to play in disease management, provided the rules of the game are understood.

The early, dramatic failures in fungicide use were due almost entirely to a 'quick fix' mentality in which continuous application of a single compound on its own led, perhaps inevitably, to the selection of resistant forms of the pathogen. To a large extent the first systemic, single-site compounds were victims of their own success. These fungicides were so active and effective that they were quickly adopted by growers and used sometimes indiscriminately in crops throughout the season. The selection pressure on pathogens was therefore considerable. With the benefit of hindsight manufacturers, advisors and growers are now aware of the potential scale of the problem, and measures have been put in place to minimize the risk of resistance occurring. The agrochemical industry has coordinated its response by establishing the Fungicide Resistance Action Committee (FRAC), which is a forum aiming to prolong

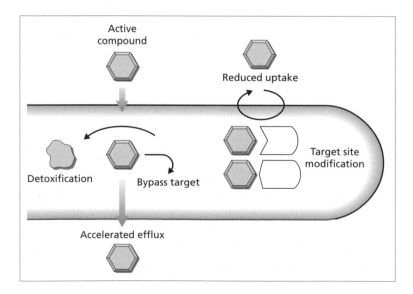

Fig. 11.10 Some mechanisms of fungicide resistance.

the effectiveness of fungicides liable to encounter resistance problems, and to limit crop losses during the emergence of resistance. Bodies such as FRAC identify existing and potential resistance problems, but their most important role is to develop guidelines for fungicide use which minimize the risk of resistance in practice.

Several commonsense strategies to combat fungicide resistance are summarized in Table 11.7. These aim to reduce selection pressure either by reducing the frequency of use of a fungicide, or by diversifying the chemicals the pathogen population is exposed to during the season. Thus, fungicides are used as mixtures with different modes of action, or treatments are alternated between compounds with different modes of action. One perhaps ironic aspect of this is that some of the older, multisite fungicides now play key roles as partners in mixtures, or as treatments in 'fungicide rotation'.

These ideas can be illustrated with examples from practical experience with different pathogens and contrasting compounds. The grey mould fungus, *Botrytis cinerea*, is a major threat to grape production in many parts of the world, and fungicides are routinely used to control this pathogen in vines. When MBC fungicides were first introduced they gave excellent results with *Botrytis*, but quickly lost their efficacy due to the emergence of highly resistant strains. Such resistant strains are as fit as sensitive strains, so they are able to persist in the pathogen population; reducing or abolishing use of MBC fungicides does not therefore significantly reduce the threat of resistance to these compounds. However, MBC resistance is often correlated with increased sensitivity to another group of chemicals known as phenylcarbamates. Hence for a while, mixtures of MBC and phenylcarbamate fungicides were recommended for control of *Botrytis*. This strategy has now been undermined by the emergence of strains resistant to both groups of chemicals. However, alternative fungicides with contrasting modes of action are available. These include the dicarboximides, such as iprodione (Table 11.1) and vinclozolin, and the recently introduced anilinopyrimidines (Table 11.2). *Botrytis* has also adapted to dicarboximides, but in this case resistant strains appear to be less competitive than sensitive strains, and hence the level of resistance in the pathogen population can be reduced, or at least stabilized, by rotation to other compounds. The currently highly effective anilinopyrimidines are from the outset being recommended as one treatment in an alternated spray programme, or formulated as mixtures with other compounds. Thus, even in a very difficult case, such as *Botrytis* in vineyards, effective options still exist for chemical management of the disease.

Another highly adaptable pathogen capable of exerting high disease pressure on a crop is *Phytophthora infestans* (Plates 1 & 2, facing p. 12). The rapid demise of the phenylamide fungicide metalaxyl in the potato crop in Europe in 1979–1980 has already been described above. However, metalaxyl is still used for the control of potato blight, formulated in mixtures with dithiocarbamates such as mancozeb, since such mixtures give control superior to the protectant compound alone. Figure 11.11 shows the incidence of metalaxyl-resistant strains of *P. infestans* in The Netherlands during the 1980s. Following the initial problem, use of the fungicide was suspended from 1981 to 1984, and the proportion of resistant strains declined. Metalaxyl was then reintroduced for use in mixtures co-formulated with multisite compounds, and the proportion of resistant strains rose again, albeit to a level lower than at the outset (Fig.

Table 11.7 Strategies to minimize the risk of fungicide resistance in practice.

Reduce fungicide use

Apply fungicides only when and where necessary, ideally based on a disease risk prediction

Use fungicides as part of an integrated control programme, e.g. combined with disease-resistant cultivars and cultural measures to reduce inoculum

Diversify fungicide treatments

Avoid repeated use of fungicides with the same mode of action

Use mixtures of fungicides with different modes of action

In a spray programme, alternate fungicides with different modes of action

Include multisite fungicides in mixtures or alternations

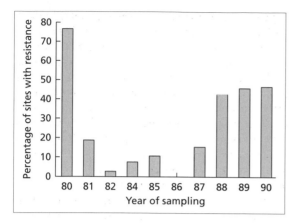

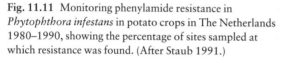

Fig. 11.11 Monitoring phenylamide resistance in *Phytophthora infestans* in potato crops in The Netherlands 1980–1990, showing the percentage of sites sampled at which resistance was found. (After Staub 1991.)

11.11). A rather similar pattern has also occurred in Ireland. The increases in resistance recorded in the late 1980s have been linked to several seasons when disease pressure was high, and the use of metalaxyl on seed crops, with the possibility of resistance being carried over to the next season. Current recommendations are to use alternative multisite fungicides on seed potato crops. This example illustrates the importance of understanding pathogen epidemiology in the management of fungicide resistance.

The final case histories concern experiences with SBIs. Unlike MBC and phenylamide fungicides, resistance to SBIs has developed more slowly, conforming to the directional selection model (Figs 11.8 & 11.9). In many cases such shifts in sensitivity have not led to any actual reduction of control in the field. Nevertheless, strategies to prevent any further erosion in the efficacy of SBI compounds are being pursued. One key component in these strategies is the morpholine fungicides, which act at different sites to DMIs in the sterol biosynthetic pathway. Thus, any reduction in sensitivity to DMIs does not affect sensitivity to morpholines. This provides a useful option for mixing or alternating SBI fungicides in the control of important diseases such as powdery mildew of cereals, *Erysiphe graminis*, or Sigatoka of bananas, *Mycosphaerella fijiensis*. The latter pathogen initially developed resistance to MBC compounds, while declining sensitivity to azoles has now been reported from banana crops in Central and South America. However, morpho-

lines such as tridemorph still give consistent control, and are used in resistance management strategies which include alternation with different fungicides, and reductions in the number of sprays, and area of crop sprayed, whenever possible.

Putting resistance to practical use

Paradoxically, the occurrence of resistance can be an aid to understanding the mode of action of different fungicidal chemicals. Molecular genetic analysis of mutants altered in sensitivity to fungicides can provide detailed information on the target site of the fungicide, as well as clues as to why different classes of fungi differ in their response to particular compounds. For instance, analysis of laboratory mutants of model fungi such as *Aspergillus* confirmed that the site of action of MBC fungicides is β-tubulin. Isolating and sequencing the genes conferring resistance showed that loss of activity is due to changes in one or a few critical amino acids in the β-tubulin protein.

There are important practical spin-offs from such fundamental molecular studies. Firstly, detailed structure–activity interactions can be defined, indicating how particular chemicals bind to a target protein. This may aid rational design of molecules with improved activity. Secondly, molecular probes based on DNA sequences from specific resistance genes can be used to analyse resistance in natural fungal populations. Such probes can, for instance, detect the presence of individual resistance genes in field isolates of pathogens, thereby providing information on the distribution of resistance, and the likelihood of problems occurring with the use of a fungicide to control an epidemic. When combined with PCR it may even be possible to devise diagnostic kits for the rapid detection of particular resistance genes, thereby aiding decisions on practical disease control. Lastly, some genes for fungicide resistance have proved to be valuable tools for the genetic manipulation of commercially important fungi, as they can be used as dominant selectable markers for transformation. In other words, resistance to the fungicide confirms that the gene of interest has been successfully introduced into the fungus.

The future of fungicides

An appropriate comment to conclude this chapter is

that rumours of the complete demise of fungicides have been exaggerated! Gloomy predictions that the rapid adaptation of fungi to chemical selection would undermine the whole basis of control by fungicides have proved to be unduly pessimistic. Other worrying scenarios, such as exhausting the supply of new, safer chemicals to the extent that no new fungicides would ever be launched, have also been discredited. Following the comparative lull in discovery during the 1980s, the current decade has seen the introduction of several new chemical classes with high activity, low environmental impact, and novel modes of action. Compounds such as the strobilurins are likely to provide improved control of important diseases, such as *Septoria* on cereals, at lower dose rates and with greater flexibility in time of application, than previously available chemicals. In most major crops, growers now have a greater choice of effective fungicides than ever before. The status of chemical control of plant disease is therefore more encouraging than the current situation in clinical microbiology, where antibiotic resistance now threatens effective treatment of several previously tractable bacterial diseases.

There is, however, no room for complacency, both in terms of the ability of pathogens to adapt to new circumstances, and the continuing concerns about the environmental acceptability of chemicals. The age of blanket treatment of a crop with a single compound to combat one disease is to a large extent over. Chemicals are increasingly viewed as only one element in an integrated crop management system aimed at maximizing output while reducing inputs. It is of interest that many agrochemical companies have now merged with, or bought stakes in, seed companies marketing particular crop cultivars. This is part of the trend towards offering the grower an integrated control package, including host genetic resistance. At the same time several of the larger companies are investing in plant biotechnology, in the expectation that novel approaches to control, for instance using transgenic crops (see p. 228), will provide additional options alongside their portfolio of chemicals.

But what of the chemicals themselves? Already there is a greater emphasis on 'natural' approaches to pesticide discovery, not only through screening natural products for biological activity, but also in the search for chemicals which act via the endogenous defence systems of the plant. One of the most interesting recent developments has been the launch of CGA 245704, a benzothiadiazole compound (Table 11.2) which has no innate fungistatic or fungitoxic properties, but which nevertheless is active against a range of diseases. This and related chemicals, described as plant activators, switch on systemic acquired resistance in the host plant. There are hopes that because this is a completely different mode of action to conventional fungicides, such compounds will be invaluable in combating fungicide resistance. There may also be exciting possibilities in looking for chemical analogues of other, natural signal molecules, for instance hormones regulating fungal growth, so that processes such as sporulation or sexual reproduction might be inhibited. This would not prevent infections but would instead limit the rate of epidemic development by reducing inoculum. Such behaviour-modifying chemicals already form the basis for effective strategies for insect control, using analogues of volatile signal molecules known as pheromones. Another fruitful area may be to exploit natural chemistry which already serves a defence function. Recently, many small antifungal peptides have been isolated from plant seeds (see p. 143), and some of these show promise as natural fungicides, or even as defence proteins for engineering into transgenic plants. Thus the fields of fungicide discovery, biotechnology and biological control are now converging. Add to this the possibility of rational design of active molecules, and the prospects for further improvements in chemical control look bright.

Further reading

Books

Dixon, G.K., Copping, L.G. & Hollomon, D.W. (eds) (1995) *Antifungal Agents: Discovery and Mode of Action*. Bios Scientific Publishers, Oxford.

Heaney, S., Slawson, D., Hollomon, D.W., Smith, M., Russell, P.E. & Parry, D.W. (eds) (1994) *Fungicide Resistance*. BCPC Monograph No. 60. British Crop Protection Council, Farnham.

Lyr, H. (ed.) (1995) *Modern Selective Fungicides*, 2nd edn. Gustav Fischer Verlag, Jena.

Lyr, H., Russell, P.E. & Sisler, H.D. (eds) (1996) *Modern Fungicides and Antifungal Compounds*. Intercept, Andover.

Matthews, G.A. & Hislop, E.C. (eds) (1993) *Application Technology for Crop Protection*. CAB International, Wallingford.

Reviews and papers

Brown, J.K.M., Jessop, A.C., Thomas, S. & Rezanoor, H.N. (1992) Genetic control of the response of *Erysiphe graminis* f.sp. *hordei* to ethirimol and triadimenol. *Plant Pathology* **41**, 126–135.

DeWaard, M.A., Georgopoulos, S.G., Hollomon, D.W., Ishii, H., Leroux, P. & Schwinn, F.J. (1993) Chemical control of plant diseases—problems and prospects. *Annual Review of Phytopathology* **31**, 403–421.

Elad, Y., Yunis, H. & Katan, T. (1992) Multiple fungicide resistance to benzimidazoles, dicarboximides and diethofencarb in field isolates of *Botrytis cinerea* from Israel. *Plant Pathology* **41**, 41–46.

Georgopoulos, S.G. & Skylakakis, G. (1986) Genetic variability in the fungi and the problem of fungicide resistance. *Crop Protection* **5**, 299–305.

Görlach, J., Volrath, S., Knauf-Beiter, G. *et al.* (1996) Benzothiadiazole, a novel class of inducers of systemic acquired resistance, activates gene expression and disease resistance in wheat. *The Plant Cell* **8**, 629–643.

Kendall, S., Hollomon, D.W., Cooke, L.R. & Jones, D.R. (1993) Changes in sensitivity to DMI fungicides in *Rhynchosporium secalis*. *Crop Protection* **12**, 357–362.

Schwinn, F.J. & Morton, H.V. (1990) Antiresistance strategies. Design and implementation in practice. In: *Managing Resistance to Agrochemicals. From Fundamental Research to Practical Strategies* (eds M.B. Green, H.M. LeBaron & W.K. Moberg). ACS Symposium Series No. 421, pp. 170–183. American Chemical Society, Washington, D.C.

Staub, T. (1991) Fungicide resistance: practical experience with antiresistance strategies and the role of integrated use. *Annual Review of Phytopathology* **29**, 421–442.

Steffens, J.J., Pell, E.J. & Tien, M. (1996) Mechanisms of fungicide resistance in phytopathogenic fungi. *Current Opinion in Biotechnology* **7**, 348–355.

12 Disease Management by Host Resistance

'Those who believed only in genes postulated the existence of an eternal quality, R, which they could take from wild plants, build into the genetical constitution of cultivated ones, so as to make them disease-resistant forever. Those who thought ... of the green flux of ever-changing nature, saw little hope of such permanency, and no end to man's labours in defending the crops' [E.C. Large, 1902–1976]

The natural genetic resistance of plants to pests and pathogens has no doubt played a key role in crop protection since the dawn of agriculture. Indeed, it is doubtful if systematic cultivation of crops would have been possible were it not for the ability of most plants to withstand attack by pathogens. Up until the end of the 19th century, plant breeding was an empirical process. Nevertheless it must have been important in restricting disease losses. In years when disease epidemics wiped out most of the crop, only the most resistant plants survived to provide seed for the next season. This sequence was merely an extension of the process of natural selection, which maintains an equilibrium between a pathogen and its host in natural communities. At the same time the gradual development of the intuitive art of plant breeding, in which higher-yielding and more resilient individuals were progressively selected, undoubtedly included an element of selection for resistance to disease (Fig. 12.1).

The foundations of the scientific process of breeding for disease resistance were laid by Rowland Biffen early in the 20th century. By 1912 he had shown that resistance to yellow stripe rust, *Puccinia striiformis*, in the wheat cultivar Rivet was determined by a single recessive gene which was inherited according to Mendelian laws. The realization that disease resistance was a genetic trait which could be manipulated in a breeding programme triggered a sustained effort to introduce similar resistance into other major crops.

The introduction of crop cultivars containing new genes has since become a normal part of modern agriculture, and, apart from concerns about the conservation of genetic diversity, and more recently recombinant DNA, the process does not have serious environmental implications. Indeed, at one time breeding for resistance seemed to promise an ideal and permanent solution to the problem of plant disease. What had not been anticipated, however, was the ability of pathogens themselves to adapt and evolve in response to genetic changes in the crop. The permanent solution proved to be elusive, and instead resistance breeding began to resemble a race between plant breeders and the pathogens.

Breeding for disease resistance

We saw in Chapter 9 that there are several different ways in which plants defend themselves against pathogens. For the plant breeder, resistance can be defined as any inherited characteristic of the host plant which lessens the effect of disease. From a practical viewpoint, any character which improves the performance of the plant in the presence of a pathogen may be useful, provided that it is heritable and stable. It is also worth pointing out that resistance to disease is only one objective of the breeder, and for many crops is of lower priority than other characters such as increased yield or improved quality.

The basic requirements to produce a novel, disease-resistant crop cultivar are:
1 a source of genetic resistance;
2 a method for identifying and selecting this resistance;
3 a method for combining this resistance with other, agronomically desirable characters to produce a commercially acceptable crop genotype.

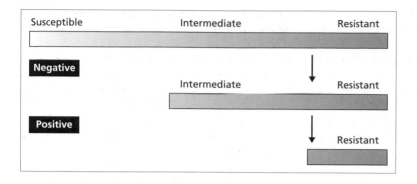

Fig. 12.1 Selection for disease resistance in a population of plants. The disease reaction type of individuals in the initial population varies from susceptible to resistant. Negative selection, as in natural disease epidemics, removes the most susceptible individuals, while positive selection, as practised by the breeder, chooses only the most resistant plants.

Sources of resistance

The raw material for any breeding programme is genetic variation. In the case of disease resistance there may already be significant variation within the crop itself. For instance, *Verticillium* wilt is an important disease of the forage legume alfalfa (lucerne), but because this crop is outbreeding and genetically variable, individual plants differ in their susceptibility to infection. In this case, recurrent selection, season by season, of the most resistant plants, can lead to significant improvements in the overall wilt resistance of the crop.

But what if the crop itself lacks any useful variation in disease reaction type? In this case it is often necessary to look for alternative sources, such as in wild relatives of the crop species. This approach has been widely used in many important crops, including cereals, potatoes and tomatoes, and tropical plantation trees such as coffee and avocado. When potato blight, *Phytophthora infestans*, swept Europe and North America in the 19th century, contemporary potato varieties were almost universally vulnerable to disease. Later attempts to introduce resistance therefore focused on wild relatives in Central and South America. One species in particular, *Solanum demissum*, found in Mexico, proved to be a useful source of resistance (*R*) genes, which were introduced into the cultivated potato, *S. tuberosum*, by hybridization. The genes were dominant, gave a high level of resistance, and were therefore easily selected in progeny plants from a cross. Unfortunately the resistance proved to be specific to particular races of the blight pathogen, and broke down once new races of the fungus appeared (see below). Other wild *Solanum* species are a useful source of genes for insect resistance. For instance, *S. berthaultii* possesses surface hairs (trichomes) which impair the activity of aphids. This not only reduces damage due to the aphid itself, but also restricts transmission of important virus pathogens, such as potato leaf roll virus (PLRV).

The most fruitful source of novel resistance to a disease is usually assumed to be the geographic centre of origin of the pathogen, where co-evolution with the host has taken place over a long period. Thus considerable diversity for blight resistance has been found in wild potatoes in Central America, while for rust and powdery mildew diseases of wheat and barley, useful sources of resistance have been discovered in wild grasses in the Middle East. Provided the wild species will hybridize with the crop, such genes can be introduced into commercial genotypes. The timespan concerned is, however, a drawback of this process, as several rounds of back-crossing and selection are usually necessary to dilute out other, unwanted genes from the wild relative, to end up with a disease-resistant cultivar suitable for commercial use. It may take more than 10 years to complete this process, depending on the life cycle of the crop concerned.

The important issue of the conservation of plant genetic resources is of course directly relevant to resistance breeding. Germplasm collections of both cultivated species and their wild relatives should, ideally, embrace all of the natural variation for resistance occurring in such species. One concern is that with most of the current emphasis being placed on elite, high-yielding genotypes of a crop species, much of the useful variation for other characters, including resistance to pests and pathogens, may be discarded. Furthermore the continuing destruction of natural plant communities, and consequent loss of biodiversity, might also limit future options for introducing novel variation into crops.

Two further sources of novel resistance genes should be considered. The first is to try to actually induce or create a new resistance within the crop using techniques designed to cause genetic changes. For instance, **mutation breeding** is based on the use of mutagenic treatments, such as irradiation or chemical mutagens, which induce changes in the DNA. This approach was in vogue for a period following World War II, but overall has produced relatively little variation of potential value in breeding programmes. One difficulty is that most mutations cause a loss of function, rather than a gain of a new property such as increased resistance. It is also necessary to screen very large numbers of plants to detect potentially useful mutations. More recently, techniques of plant tissue culture have been used to extend the range of variation found in crops. If one regenerates plants from protoplasts, or callus cultures maintained *in vitro*, the regenerants often show much greater variation than equivalent populations grown from seed. Such **somaclonal variation** includes alterations in the reaction of plants to pathogens. This approach is of value in extending the range of variation available to the breeder within a commercially acceptable genetic background, but to date there are few examples of useful resistance being obtained by this means.

Another recent development is the introduction of techniques to transfer foreign genes into plants via genetic transformation. This innovation opens up a completely new set of possibilities for engineering disease resistance, and will be discussed in more detail later (see p. 228).

Selection of disease resistance

Traditionally, resistance breeding has relied on selection of genotypes with improved resistance, either in the field or in mass-inoculation experiments where plants are exposed to the pathogen concerned. These methods have often succeeded in identifying the most promising individuals, but they are time-consuming and relatively inefficient. It would be an advantage if one could select resistance to pathogens in some other way.

One possibility which has been explored is to select cell cultures rather than intact plants. In theory this should permit screening of much larger numbers of individuals, and should also accelerate the selection process. In practice there are difficulties, mainly because the correlation between the resistance of individual cells and whole plants is often poor. The only convincing exceptions are diseases in which symptoms are mainly or exclusively due to production of a toxin (see p. 129). In these cases insensitivity to the toxin in culture is usually correlated with insensitivity in the plant, and hence the toxin can actually be used as a chemical selection agent to identify resistant cells which are then regenerated to obtain resistant plants.

A more generally applicable procedure, increasingly used by breeders, is **marker-assisted selection**. This is based on the increasingly powerful techniques now available for detecting variation at the genomic level. Rather than looking for the phenotype — i.e. individuals with less disease in an infection trial—one searches for a genetic marker which segregates with this phenotype. The basic idea is simple. Let us assume that after crossing two parent plants, one resistant and the other susceptible to a particular pathogen, the progeny derived are segregating for resistance. If the resistant progeny are then pooled in one group, and the susceptible in another, and the DNA from these two pools is then compared, any differences should be linked to the presence or absence of resistance. This method relies on identifying a DNA polymorphism (i.e. a sequence difference) which is correlated with resistance or lack of resistance (i.e. susceptibility). If such a molecular marker can be found, then selection for the marker can be used instead of selection for resistance in an infection trial. The main advantage here is the speed and convenience of selection, which can be applied to large numbers of seedling plants or even seeds. An example is shown in Fig. 12.2, in which a DNA fragment amplified by the polymerase chain reaction (PCR) identifies rice plants containing a gene for resistance to blast disease, caused by the fungus *Magnaporthe grisea*. This DNA sequence is closely linked to the resistance gene and can be used with a high degree of accuracy (>95%) to select resistant individuals.

Resistance in the field

Plant breeders are engaged in a constant drive to produce better crops showing improvements in yield, quality or other agronomic characters including disease or stress resistance. In most respects their efforts have been remarkably successful, leading to

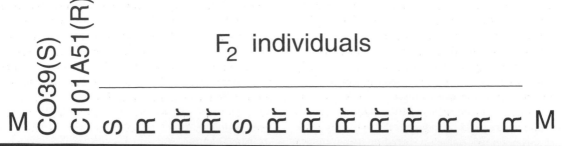

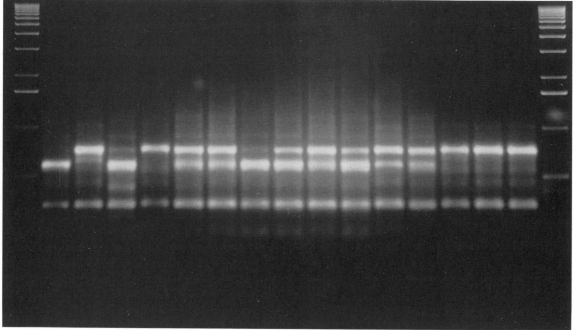

Fig. 12.2 Identification of a gene for resistance to rice blast, *Magnaporthe grisea*, in a population of plants by means of a DNA fragment amplified by the polymerase chain reaction (PCR). Tracks 1 and 2 in the gel show the pattern of the susceptible (S) and resistant (R) parent plants. The remainder are F_2 progeny from a cross between these parents. Segregation of the marker bands identifies susceptible (S) individuals, and those both heterozygous (Rr) and homozygous (R) for the resistance gene. (From Hittalmani *et al.* 1995.)

the so-called 'Green Revolution' in which the major food crops such as rice, wheat and maize have been progressively bred to give higher and higher yields. New, dwarf cultivars, that respond well to fertilizers and irrigation, mature early, and can be cropped at high densities, have replaced the older varieties. This intensification of agriculture, especially in developing countries, has raised some controversial issues, such as genetic uniformity and the loss of locally adapted crop types, but illustrates nevertheless the power of a sustained breeding programme. In the UK, for instance, wheat yields have risen as much as 5% each year since 1950; about half of this increase can be attributed to improved cultivars, and the remainder to improvements in the management of the crop, including disease-control measures.

The 'boom-and-bust' cycle

Efforts to boost the resistance of crop species to their major pests and pathogens through plant breeding have a more chequered history. Breeders have succeeded in identifying and introducing novel sources of resistance to many diseases, but all too often the protection provided by such genetic resistance has proved short-lived in the field. The introduction of a new host gene for resistance was regularly followed by the appearance of a new pathotype (race) of the pathogen capable of overcoming the effects of the gene. This in turn led to a renewed search for alternative resistance genes which were again frequently countered by changes in the pathogen. An example of this type of sequential introduction of host genes and subsequent pathogen evolution is seen in the development of barley cultivars resistant to powdery mildew (Fig. 12.3); resistance genes from a variety of sources were countered with monotonous regularity.

The rate at which the pathogen responds to newly created selection pressures depends upon several factors. Foremost amongst these is the proportion of total crop area occupied by the novel cultivar (Fig. 12.4). The widespread planting of a particular cultivar containing one or a few resistance genes will inevitably favour genotypes of the pathogen possessing the matching virulence genes, a process known as directional selection. The recent history of yellow stripe rust, *Puccinia striiformis*, on wheat in the UK illustrates this point. During the 1980s the winter

wheat cultivar Slejpner was widely grown due to its high yield. In 1988–1989 there was a major yellow rust epidemic which was particularly severe on Slejpner; crops of this cultivar not treated with fungicide suffered yield losses as high as 75%. Slejpner contains the resistance gene *Yr9*, and the epidemic coincided with the emergence and spread of a new race of the rust pathogen possessing the matching virulence gene, *Yv9* (Fig. 12.5). Annual surveys showed that by 1986 over 90% of the rust population sampled contained this virulence. A second race with combined virulence to *Yr9* and a second resistance gene, *Yr6*, found in another popular wheat cultivar, Hornet, subsequently emerged, so that by 1989 both Slejpner and Hornet were vulnerable to rust infection. This change was reflected in the relative disease ratings of the two cultivars, which slumped from a score of 9 (highly resistant) to 2 (highly susceptible) within two to three seasons (Fig. 12.5).

The capacity of pathogens such as *P. striiformis* to evolve new virulent forms has been graphically demonstrated by the recent history of yellow stripe rust in Australia. This disease was unknown in the country until 1979, when a single race of the pathogen was detected, presumably introduced from overseas. During the following 10 years, regular surveys of pathogen isolates in Australia and New

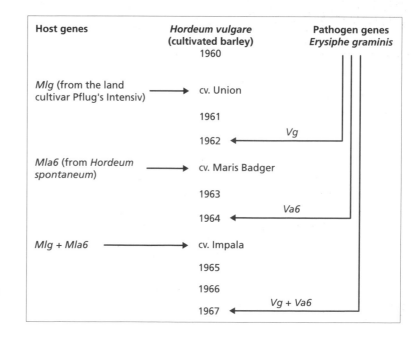

Fig. 12.3 Successive introductions into the UK barley crop of cultivars containing genes for resistance to powdery mildew and the subsequent selection for virulence genes in the *Erysiphe* population.

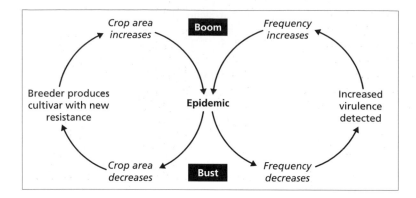

Fig. 12.4 The boom-and-bust cycle. (After Priestley 1978.)

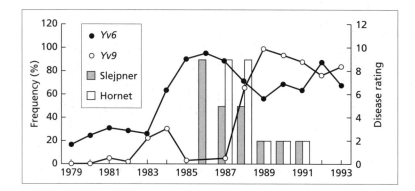

Fig. 12.5 Annual incidence of virulence (*Yv6* and *Yv9*) matching rust resistance genes *Yr6* and *Yr9* in the yellow rust (*Puccinia striiformis*) population in the UK. Sharp increases in the incidence of virulence between 1983 and 1989 led to a revision of the disease ratings of wheat cultivars Slejpner and Hornet given by the National Institute of Agricultural Botany.

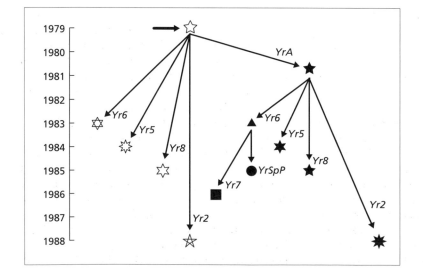

Fig. 12.6 Evolution of virulence in the yellow rust (*Puccinia striiformis*) population in Australia following introduction of the pathogen in 1979. The proposed phylogenetic tree shows the year of detection of virulence to particular rust resistance (*Yr*) genes, and apparent relationships between the different pathotypes. (After Wellings & McIntosh 1990.)

Zealand (to where the fungus is believed to have travelled by wind transport of spores) revealed the emergence of at least 15 pathogenic variants with different virulences. In most cases each variant seemed to arise from a previous form by a single mutational event, so that a family tree of pathotypes could be constructed (Fig. 12.6). Such rapid evolution of novel virulence, in this case in the absence of any sexual process in the fungus, highlights the difficulties facing the plant breeder attempting to introduce lasting resistance into a crop.

Types of plant resistance

The phenomenon of boom-and-bust, typified by the rapid loss of effectiveness of a resistance gene in the field, may be to a large extent a legacy of breeding for particular types of plant resistance. For obvious reasons, breeders have usually selected single genes which have a major effect on disease development. Such genes are expressed in seedlings as well as adult plants, and can therefore be easily detected in progeny segregating in a breeding programme. Usually, these *R* genes conform to the gene-for-gene model of host–pathogen interaction (see p. 27), and most likely encode a pathogen recognition function. Failure of recognition, i.e. susceptibility, can occur readily through loss of the corresponding avirulence (*avr*) gene in the pathogen.

The demise of many major resistance genes in practice has prompted attempts to find alternative types of plant resistance. The goal is to develop forms of resistance which will be effective against all the different genetic variants of the pathogen. Such resistance is often described as **race non-specific** as opposed to **race-specific**. An alternative, and useful way of thinking about this is the concept of **horizontal** vs. **vertical** resistance. Figure 12.7 illustrates diagrammatically the interaction of a host plant containing a single *R* gene with five races of a pathogen. The gene confers a high degree of resistance to some races, but very little to others (Fig. 12.7a). Such vertical resistance can be contrasted with the situation in Fig. 12.7b where a lower degree of protection is expressed more or less equally against all five races. Figure 12.8 shows segregation for disease reaction type for leaf rust infection (*Hemileia vastatrix*) in coffee plants from two different crosses. In the first, the progeny are grouped into two separate populations, suggesting segregation of a single gene of major effect; while in the second, segre-

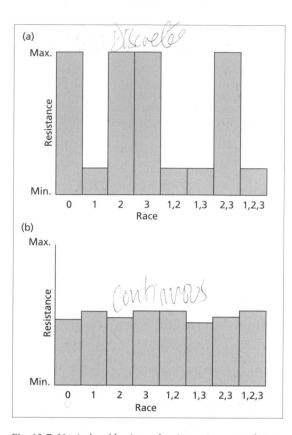

Fig. 12.7 Vertical and horizontal resistance compared. (a) A cultivar with a single gene for vertical resistance shows a high level of resistance to some pathogen races, but little to others. (b) A cultivar with horizontal resistance shows an intermediate level of resistance to all pathogen races. (After Vanderplank 1963.)

gation is more or less continuous from complete susceptibility to resistance. This more complex pattern suggests the interaction of several genes, and is typical of horizontal resistance. Nevertheless, it is still possible to select the most resistant individuals as parents for use in a breeding programme.

The terminology used to describe these different types of resistance is potentially confusing, as it includes both the genetic control of resistance, and observations on the expression or performance of resistance in the field. It is generally assumed that race-specific resistance is encoded by one or a few genes, whereas race non-specific resistance is a more complex genetic trait. This is the basis of the classification into **monogenic** and **polygenic** types of resistance shown in Table 12.1. It should be stressed,

Table 12.1 Main points of contrast between monogenic and polygenic resistance. (From Hayes & Johnston 1971.)

	Monogenic	Polygenic
Genetic control	One or few genes involved	Usually many genes involved
Description	Generally clear-cut; functions throughout life of plant, or may be expressed in mature plants only	Variable; seedlings generally less resistant, but resistance increases as plants mature
Mechanism	Generally a hypersensitive host reaction	Reduced rate and degree of infection, development and/or reproduction of pathogen
Efficiency	Highly efficient, but often extremely low resistance to other races	Variable, effective against all races of pathogen
Stability	Liable to suddenly break down due to the development of new physiological races of the pathogen	Not affected by changes in virulence genes of the pathogen
Commonly used approximate synonyms	Vertical Major gene Seedling Race-specific	Horizontal Minor gene Adult plant Race non-specific Field Rate-reducing

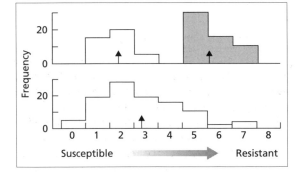

Fig. 12.8 Segregation for resistance to leaf rust (*Hemileia vastatrix*) in coffee shown as frequency distributions of different reaction types from susceptible (0) to resistant (8) in plants from two crosses. In the upper cross two discrete progeny populations are observed, while in the lower cross segregation for resistance is more continuous. Arrows show mean disease reaction score. (After Eskes 1983; with kind permission from Kluwer Academic Publishers.)

however, that in many cases the genetic basis of a particular phenotype, for instance adult plant or rate-reducing resistance (see p. 225), is not clearly defined, and hence these various types of resistance may not be directly comparable.

Durable resistance

The ephemeral nature of resistance provided by certain genes or even combinations of genes once deployed in the field has focused attention on alternative, more lasting forms of resistance. In practical terms, what matters is not so much the nature or even the genetic basis of resistance, but whether it will remain effective over a long period. Hence the concept of **durable resistance**, defined simply as resistance which continues to provide control even after an extended period of exposure to the pathogen.

It is often assumed that resistance based on single genes of major effect is doomed to break down if deployed on a commercial scale over several seasons. In fact there are examples of single genes which have remained effective year after year even when widely used. The *mlo* gene for resistance to barley powdery mildew, *Erysiphe graminis* f.sp. *hordei,* was introduced in 1979 but still gives satisfactory control of the disease throughout Europe. Interestingly, this gene was first discovered in mutant barley lines produced by exposure to X-rays, but was later detected occurring naturally in a few barley samples from Ethiopia. The gene *Sr26* conferring resistance to stem rust (*Puccinia graminis*) in wheat has been used in Australia since 1969, but to date no pathogen isolates with matching virulence have been detected in the field. Similarly, the *Lr34* gene in wheat for resistance to leaf rust, *Puccinia recondita*, first identified in South American cultivars, has remained effective for many years in several different countries.

Why certain genes prove to be durable in practice is not fully understood. One idea is that such genes encode resistance mechanisms which cannot easily be overcome by changes in the pathogen. The *mlo* gene, for example, triggers formation of a cell-wall barrier, or papilla (see p. 145) at sites of attempted penetration by the fungus. This mechanism appears to be effective against all genetic variants of the pathogen. The *Lr34* gene for rust resistance seems to reduce infection and delay development of the pathogen without any associated hypersensitive necrosis of host cells. There is also evidence that this gene interacts with other plant genes to confer effective resistance.

Ultimately the durability or otherwise of single resistance genes will depend upon the frequency of mutations to virulence in the pathogen, and whether such mutations can survive and spread in the pathogen population. It is possible that mutations countering host resistance genes might in certain cases incur a fitness penalty in the pathogen. Such mutants would therefore fail to survive in competition with the remainder of the population. Evidence supporting this idea has come from recent studies on bacterial pathogens. Resistance genes such as *Xa21* from rice and *Bs2* from pepper are effective against all known races of the corresponding pathogens *Xanthomonas oryzae* and *X. campestris* pv. *vesicatoria* (Xcv); every race apparently contains a copy of the matching *avr* gene. In the case of Xcv, loss of the gene concerned, *avrBs2*, leads to a severe reduction in the vigour of the bacterium. It would be extremely useful if the durability of particular resistance genes could be predicted prior to commercial release. One hope is that detailed analysis of the resistance mechanisms encoded by genes known to be durable might aid identification of further similar genes. To date, however, the only reliable test of durability is field exposure over an extended period. Even then, there are no guarantees that pathogen evolution will not eventually generate new genotypes capable of overcoming such resistance.

Due to the recurrent problems with vertical (usually monogenic) resistance, the main emphasis in resistance breeding in many crops is now on horizontal resistance. In potatoes, for example, breeders have abandoned attempts to introduce single genes for blight resistance, such as those obtained from wild *Solanum* species, and are instead evaluating potato lines which, while not fully resistant to infection, nevertheless withstand severe blight epidemics. Trials in Mexico, where the greatest genetic diversity exists in the pathogen population, have shown that some potato genotypes perform better than others, year after year. Such lines can therefore be selected and used in breeding programmes. The genetic control of this resistance is not fully understood, but probably involves several genes. One key feature of such resistance is that it delays development of the disease. The pathogen is able to infect the plant, and even to cause disease lesions, but the time from initial infection to the production of further spores is extended. Such resistance is often described as **rate-reducing** (Table 12.1). The longer disease cycle reduces the risk of serious damage to the crop, and, when combined with other measures (see p. 250), may be sufficient to provide effective and durable control of the disease.

Breaking the 'boom-and-bust' cycle

The lack of durability of genetic resistance in the field is not simply due to the use of single, vertical forms of resistance. Such resistance may be intrinsically unstable but, nevertheless, if used with care it can be highly effective. The crucial issue is to understand the selection pressures leading to the 'boom-and-bust' scenario, and to find ways of reducing them. In many respects modern intensive agriculture has made life easy for pathogens. A few, elite cultivars of the crop are planted in monoculture over a large area, favouring any genetic variants of a pathogen able to attack such cultivars. Directional selection for virulence, akin to the changes in fungicide sensitivity described in Chapter 11, is likely to occur.

There are a number of ways in which the boom-and-bust cycle might be averted or broken. One is to try to find durable sources of resistance, as described above. The others are all, to a greater or lesser extent, based on the concept of **genetic diversification** (Table 12.2). Horizontal resistance under the control of several genes is one option. Another is to breed for combinations of vertical, race-specific genes, in a

Table 12.2 Strategies for avoiding directional selection for virulence.

> Breed for horizontal resistance under polygenic control
> Combine different *R* genes in one cultivar
> Diversify deployment of *R* genes
>   Spatial diversity between regions
>   Spatial diversity between fields
>   Multiline cultivars
>   Cultivar mixtures
>   Intercropping

single cultivar. This process, known as **pyramiding** *R* genes, is controversial, as it may select for complex races of the pathogen possessing several matching virulences. There is already evidence from annual surveys of some pathogen populations, such as *Phytophthora infestans*, powdery and downy mildews, and rust fungi, that the number of virulence genes present in isolates can increase over time, in response to selection. Surveys of stem rust (*Puccinia graminis tritici*) populations in Australia over 30 years have shown that while most isolates in the 1950s possessed on average one matching virulence, by 1968 this had risen to two to three, and by 1980 the predominant race in south Australia had virulence matching five host genes. In view of this trend, it may be better to confront the pathogen with a constantly changing host population. Put simply, the idea is to 'confuse' the pathogen, so that any change in virulence on one host does not automatically lead to a selective advantage. Rather than the pathogen dictating events, this strategy aims to stabilize selection, so that the effectiveness of certain *R* genes is preserved.

Genetic diversity for resistance can be achieved in several ways (Table 12.2). One is to deploy different *R* genes in an organized manner across regions or between farms or fields. Needless to say this requires a high degree of coordination between farmers. In the USA, for instance, the normal annual pattern of rust epidemics on cereals is for the disease to spread from south to north during the season. Early outbreaks arise from inoculum overwintering in the warmer, southern states or Mexico, and spores are then carried northwards to infect later-maturing crops. In this case it is possible to plant different cultivars, with different *R* genes, along the disease pathway. Selection for virulence on a particular host genotype does

not, then, confer an advantage once the pathogen spreads to the next geographic zone. A similar strategy can be applied on a smaller scale to districts, farms or individual fields. Experiments have shown that growing cultivars with different resistance genes in a patchwork of small plots can reduce disease levels by comparison with monocultures. The ultimate version of this diversification is intercropping, in which different host crops are mixed together within a small area. This approach aims to recreate some of the genetic diversity found in natural plant communities, which are less prone to explosive disease outbreaks (see p. 13). While such genetic mosaics may be desirable for limiting disease, they are not compatible with most of the labour-saving techniques of modern, mechanized agriculture, and require more intensive management.

An alternative possibility is to mix *R* genes together in multiline cultivars or cultivar mixtures. A multiline is a series of near-isogenic (i.e. genetically identical) plant lines which differ in a single character, such as disease resistance. These lines can be grown together like a conventional crop, and retain the agronomic advantages of a normal cultivar, such as uniformity, while presenting the pathogen with a genetic puzzle. In theory, the number of *R* genes which might be present in a multiline is almost limitless. In practice, it takes considerable breeding effort to develop such cultivars, and they may be based on relatively few *R* genes. However, the specific components of the multiline can be changed year by year, so that vulnerable genes are replaced by resistances which are currently effective.

The concept of cultivar mixtures is similar to multilines, but easier to put into practice. In this case one simply mixes together seed of several genetically distinct cultivars and grows them as a single crop. Each cultivar in the mixture contains one or more different *R* genes. Trials have shown that such mixtures often develop less disease than would be expected from levels of infection observed on the component cultivars grown alone (Fig. 12.9). Several mechanisms might account for this reduction in disease in the mixture, including less efficient spread from plant to plant, and possibly induction of host resistance by incompatible pathotypes attempting to infect the different hosts. Once again one can vary the mixture from season to season, and hence keep one step ahead of the pathogen. Figure 12.10 shows such a scheme, which also includes a strategic fungicide treatment

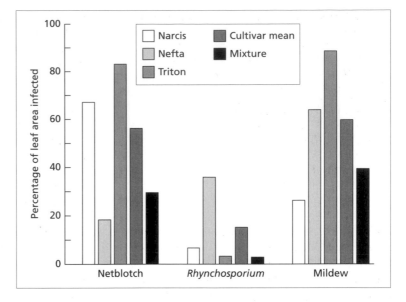

Fig. 12.9 Amounts of foliar infection by three fungal pathogens of barley in three cultivars grown separately, or in a mixture. In each case the level of infection in the mixture is lower than the expected mean based on scores for individual cultivars. (After Wolfe 1993; with kind permission from Kluwer Academic Publishers.)

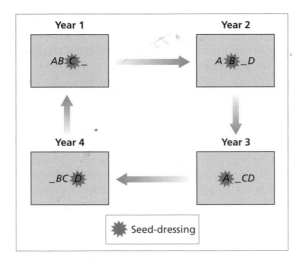

Fig. 12.10 Cultivar diversification scheme based on four host lines with different *R* genes (*ABCD*), planted as a three-component mixture. One line is changed each year in the cycle, and the line in the third year of use is seed-dressed with a systemic fungicide to safeguard against infection.

alternated between the host components as an insurance policy.

Cultivar mixtures have been used in cereal crops, for instance barley, in parts of Europe, but there is still some reluctance on the part of growers to adopt these policies. The root of the problem is that markets demand a particular product, and variation in the quality of that product is seen as a weakness. Barley, for instance, is used in malting for beer production, and very stringent standards are applied to the quality of the grain to be processed. It should be obvious that grain from a mixture of genotypes is more likely to vary than that from a single pure-bred cultivar. It is difficult, for instance, to devise mixtures in which all the components mature at exactly the same rate. Until the philosophy of modern mass production and quality control itself changes, concepts of genetic diversification may remain ideas rather than practical realities.

There are other, biological concerns about diversification. These all revolve around the spectre of the 'super-race', a form of a pathogen which contains a combination of numerous virulence genes, and hence is able to indiscriminately attack all of the different host lines, irrespective of their *R* gene composition. Could such a super-race evolve? There are some biological precedents for such a development, for instance in medical microbiology where drug-resistant strains of bacteria able to withstand cocktails of antibiotics have become a reality. The gradual shift to more complex virulence combinations, mentioned above, is also worrying. How complex can a pathogen race become? As in any evolutionary debate, it is difficult to predict all eventualities, but the current consensus is that the ultimate super-race is unlikely to emerge. Much of this debate is based on quite complex population genetic models, but the crucial issue concerns the fitness of pathogen races with multiple virulences. The experimental evidence is still far from conclusive, but several studies have shown that fungal isolates with complex virulence characteristics are often less competitive than comparable simple races, and, in the absence of strong selection, disappear from the population. There may be simple physiological reasons why this happens. In the gene-for-gene model, resistance depends on recognition of an avirulence gene product of the pathogen (see p. 168). Mutations to virulence involve loss of that product, and hence evasion of specific recognition by the host. While the real biological functions of avirulence genes are mainly unknown, it is highly unlikely that they play a role only in specific recognition. More probably, such genes perform useful physiological tasks in the growth and development of the pathogen. As discussed earlier, loss of these genes may therefore reduce the overall fitness of the organism. But the ultimate test will again be practical experience with crop mixtures, especially if adopted on any large scale.

Engineering disease resistance in crops

The advent of techniques for the stable transformation of plants has opened up apparently limitless opportunities for engineering novel pest- and disease-resistant crops. It is now possible to introduce genes from unrelated plants, or more exotic sources such as animals or even microorganisms. What difference is this likely to make to resistance breeding? The technology is still in its infancy, and for some crops, such as the cereals, is not yet routine, but the first generation of transgenic crops are now being tested in the field and, apart from the legislative questions concerning release of recombinant organisms, there is no reason why this approach should not become a normal part of the breeding process.

The techniques involved in genetic engineering are beyond the scope of this book, but the basic principle is straightforward. A piece of foreign DNA, usually described as a construct, is introduced into the plant cell and becomes integrated into the plant genome. Hopefully this new DNA sequence will be expressed and remain a stable part of the genome, so that the new gene(s) will be passed on in the usual manner to progeny. In this way the transformed plants will have a new phenotype and also the novel trait can be manipulated in a conventional programme of cross-ing and selection. The real tricks of genetic engineering concern firstly, how to get the new DNA into the cell and stably integrated, and secondly, designing constructs which will ensure good levels of expression of the foreign gene, and correct processing and targeting of the gene product. The latter problem is particularly pertinent to engineering resistance to pathogens, as most colonize specific areas of host tissues or cells (see p. 104), and it is important that any defence product is present in the optimum place to inhibit pathogen growth. It is no good, for instance, dumping an antimicrobial protein in the cell vacuole when the microbe concerned is multiplying in intercellular spaces.

There are several ways of getting foreign DNA into plant cells, including uptake into isolated protoplasts, and biolistic methods where the DNA is coated on small particles which are then shot into a suitable tissue. The preferred method for the majority of crop species is, however, to utilize the natural capacity of *Agrobacterium* to introduce the Ti plasmid into plant cells (see p. 136). The novel gene is engineered into a modified plasmid which is then transferred by infection with the bacterium.

First-generation transgenic crops

The first transgenic crops to be tested in the field were engineered for resistance to insects or herbicides. The insecticidal protein toxin produced by the bacterium *Bacillus thuringiensis* (Bt) was introduced into crops such as tobacco, tomato and cotton, to protect against attack by lepidopteran pests whose larvae feed on the plant. Caterpillars grazing on leaf tissues expressing the foreign protein are rapidly poisoned. The strategy has proved highly effective, although there are concerns about the possible development of resistance to Bt toxin in some target pests.

There are now numerous examples of plants engineered for improved resistance to pathogens (Table 12.3). They include genes conferring resistance to fungi, bacteria or viruses. The large majority are still at an experimental stage, and are being tested in model crops such as tobacco and potato. Nevertheless, in the next phase commercial crops and species less amenable to transformation, such as the major cereals, will become available.

The examples shown in Table 12.3 are mainly based on introduction and constitutive expression of a single gene whose product interferes with growth or

Table 12.3 Some examples of genetically engineered crops with enhanced resistance to pathogens.

Introduced gene(s)	Plant	Target pathogen(s)
Fungi		
Chitinase	Tobacco	*Rhizoctonia solani*
Chitinase + glucanase	Tobacco	*Cercospora nicotianae*
PR-1	Tobacco	*Peronospora tabacina*
Ribosome-inactivating protein	Tobacco	*Rhizoctonia solani*
Stilbene synthase	Tobacco	*Botrytis cinerea*
Bacteria		
Hordothionin	Tobacco	*Pseudomonas tabaci*
Lysozyme	Potato	*Erwinia carotovora*
Viruses		
Virus coat protein	Tobacco	Tobacco mosaic virus
	Potato	Potato virus X and Y
Virus replicase	Tobacco	Tobacco mosaic virus
	Potato	Potato virus X and Y
		Potato leaf roll virus
Antisense RNA	Potato	Potato virus Y
		Potato leaf roll virus
Antiviral protein	Tobacco	Cucumber mosaic virus
	Potato	Potato virus X and Y

replication of the pathogen. For instance, chitinase is an enzyme active against the cell walls of many fungi (see p. 144), while lysozyme is similarly active towards bacteria. Hordothionin is a small antimicrobial protein derived from barley seeds. PR-1 is one of the pathogenesis-related proteins induced in response to infection (see p. 154), but when expressed constitutively can enhance the level of resistance to oomycete pathogens such as downy mildews or *Phytophthora*. Improved resistance to viruses has also been obtained by expressing inhibitory proteins derived from plants, but the preferred strategy is to use a gene taken from the virus itself (see below). Another interesting exception to introduction of a gene with a directly inhibitory product is the case of stilbene synthase (Table 12.3). This enzyme catalyses a key step in the synthesis of the phytoalexin resveratrol (see Fig. 9.13), which is found in plants such as grapevine, but not in tobacco. However, the chemical precursors of resveratrol are present in tobacco, so if the gene coding for stilbene synthase is engineered into, and constitutively expressed in tobacco, the phytoalexin is produced. When this was done, the transgenic plants were found to be more resistant to colonization by the fungus *Botrytis cinerea*. This is an example of metabolic engineering, in which the biosynthetic capacities

of the plant are altered by diverting a pathway or otherwise changing its regulation.

The notion that producing a single antimicrobial protein or compound in a plant will render that plant highly resistant to infection is, of course, naive, given that resistance is usually based on many interacting components (see Chapter 9). However, transgenic plants expressing these new products often show improvements in resistance, such as a reduced rate of disease development. Such resistance might therefore be useful in an integrated control programme. Alternatively, more than one novel gene may be introduced. An example is shown in Fig. 12.11. Initially tobacco was transformed with separate constructs for the enzymes chitinase and glucanase; transformants expressing one or other enzyme were then crossed to give hybrids which contained both novel genes. These hybrids expressing both enzymes were significantly more resistant to the leaf spot pathogen, *Cercospora nicotianae*, than plants with only one transgene.

Engineering resistance to viruses

The most promising results to date using transgenes to control disease have been obtained with

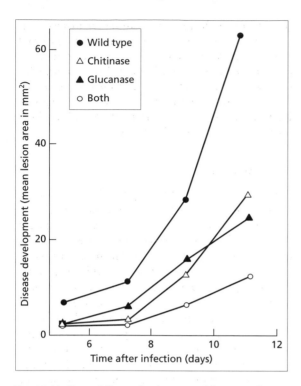

Fig. 12.11 Rate of disease development of frog-eye of tobacco (a leaf spot disease caused by *Cercospora nicotiana*) in transgenic tobacco plants. The data show the mean lesion area in wild-type plants, compared with plants expressing either chitinase or glucanase, or both enzymes. (After Zhu *et al.* 1994.)

engineered resistance to viruses. It has been known for a long time that infection of a plant with one virus can interfere with infection by a second, related virus, a phenomenon known as 'cross-protection' (see p. 162). The presence of the first virus somehow inhibits the replication cycle of the second. Starting from this observation it was reasoned that expression of foreign viral genes in plant cells might have the same effect. The first practical demonstration of the feasibility of this approach was given in 1986, when it was shown that transgenic tobacco plants expressing the coat protein gene from tobacco mosaic virus (TMV) were resistant to infection by the virus. Since then similar results have been obtained with a whole series of other plant viruses, including alfalfa mosaic, cucumber mosaic and potato viruses X and Y.

The term **pathogen-derived resistance** (PDR) has been proposed to describe resistance based on genes taken from the virus. While most cases have utilized coat protein genes, it has now been shown that other parts of the virus genome may also give effective protection, for instance the genes encoding the replicase function or movement protein. In principle it seems that any part of a plant viral genome can potentially give rise to PDR.

Some important features of PDR should be noted. First, the level of resistance is often variable, depending on the viral sequence used, and also between individual transformant plants. In some cases there is only slight attenuation of symptoms, while in others almost complete resistance to infection has been achieved. Secondly, PDR is often quite specific for the virus from which the transgene was derived. Little protection may be observed against unrelated viruses.

The exact mechanism of PDR is still debated. In some cases, it seems that expression of the protein is necessary to confer effective resistance, whereas in others the presence of the viral RNA alone may be sufficient. There has also been some limited success expressing an antisense version of virus RNA, which binds to the complementary sense strand, and interferes with replication, although this type of resistance appears to break down at high inoculum concentrations, most likely because there is insufficient antisense RNA present to tie up all the viral message. However, the various types of PDR together offer a promising new way to control virus infections. The full potential of this resistance, and its long-term stability in the field, have yet to be evaluated, but the early results are encouraging.

Future prospects for transgenic resistance

The molecular genetic dissection of the recognition and response systems involved in plant defence, discussed in Chapter 10, is already suggesting new ways in which resistance might be manipulated. Most of these ideas are speculative, but given the very rapid progress currently being made in many areas of plant molecular biology, at least some may be feasible within the next 10 years. Plant resistance (*R*) genes involved in pathogen recognition are being mapped, cloned and analysed at an increasing rate. The interesting homologies between many of these genes will, on the one hand, aid the detection and isolation of families of similar genes, and on the other, provide clues as to how the genes' products function in specific recognition. It is already possible to transfer cloned *R* genes into novel genetic backgrounds, thereby extending the options available to the plant breeder. A more interesting possibility, however, is to try to alter the specificity of the system. Once the recognition domains in the *R* gene protein product have been characterized, then it may be possible to engineer new protein conformations able to interact with a different range of pathogen elicitor molecules. Ideally, such a modified receptor would bind a widely conserved molecular 'signature', such as a common component of the fungal or bacterial cell wall, thereby switching on defence on contact with a whole range of pathogens. An alternative idea is to use the existing interaction between the *R* gene product and the specific pathogen elicitor, but to trigger it less specifically. This concept arose from work on the peptide elicitors encoded by specific *avr* genes in the tomato pathogen *Cladosporium fulvum*. As discussed in Chapter 10 (see p. 171), the interaction of the elicitor with the particular *R* gene induces hypersensitive cell death. The proposed strategy is to engineer the *R* gene, and corresponding *avr* gene, into the same plant, but to regulate expression of the latter so that the elicitor is only produced in certain circumstances. The key to this two-component system is to link the *avr* gene to a promoter which is only switched on in response to a particular event, such as penetration by a pathogen. This challenge would induce expression of the peptide in the plant cell, leading to a rapid hypersensitive response (HR).

Even more radical possibilities concern the introduction of genes from unrelated sources, such as components of the animal immune system. Recently it has been demonstrated that functional antibody molecules, retaining their ability to recognize a specific antigen, can be produced in transgenic plants. Such 'plantibodies' have many potential uses, for instance as vaccines manufactured by crops. It may also be possible to express antibodies recognizing pathogen products essential for pathogenicity, thereby inhibiting disease development. Such an approach is already being tested using plantibodies specific for the digestive enzymes secreted by nematodes during colonization of plant roots.

One of the technical problems limiting the type of transgenic crops produced is the relatively small amount of genetic information that can be introduced by transformation. This is because the constructs used cannot accommodate more than a few genes at a time. A potentially important advance in this technology is the development of vectors based on bacterial artificial chromosomes (BACs), which can carry much larger chromosomal inserts. Experiments with these vectors have already broken the size barrier for plant DNA transfer, so that in future it may be possible to introduce a whole group of genes, including, for instance, an *R* gene cluster (see p. 181).

Conclusions

The recent history of breeding for disease resistance has interesting parallels with developments in chemical control (see p. 208). 'Boom-and-bust' applies equally well to resistance genes and selective fungicides. In both cases the 'quick fix' mentality, based on simple solutions, has been superseded by a better understanding of the nature of the enemy. The variability and adaptability of pathogens demands a more subtle approach to disease control, in which more than one strategy is deployed. Major questions remain over the extent to which novel approaches to engineering resistance, based on transgenic plants, will provide lasting solutions to the problem of disease. Such resistance will undoubtedly be of value in extending the genetic options for disease control, but has yet to be exposed on a large scale in the field, the only laboratory which can provide the ultimate test of pathogen evolution! The smart advice now is to consider host resistance as only one component of an integrated disease management strategy. Alternative cultural and biological approaches and their combined use are discussed in the remaining two chapters.

Further reading

Books

Birch, A.N.E., Isaac, A.M., Marshall, E.J.P., Thomas, W.T.B. & Thompson, A.K. (eds) (1994) *The Impact of Genetic Variation on Sustainable Agriculture*. Aspects of Applied Biology 39. Association of Applied Biologists, Horticulture Research International, Wellesbourne.

Gatehouse, A.M.R., Hilder, V.A. & Boulter, D. (eds) (1992) *Plant Genetic Manipulation for Crop Protection*. CAB International, Wallingford.

Jacobs, Th. & Parlevliet, J.E. (eds) (1993) *Durability of Disease Resistance*. Kluwer, Dordrecht.

Johnson, R. & Jellis, G.J. (eds) (1993) *Breeding for Disease Resistance*. Kluwer, Dordrecht.

Reviews and papers

Baulcombe, D. (1994) Novel strategies for engineering virus resistance in plants. *Current Opinion in Biotechnology* 5, 117–124.

Beachy, R.N., Loesch-Fries, S. & Turner, N.E. (1990) Coat protein-mediated resistance against virus infection. *Annual Review of Phytopathology* 28, 451–474.

Cornelissen, B.J.C. & Melchers, L.S. (1993) Strategies for control of fungal diseases with transgenic plants. *Plant Physiology* 101, 709–712.

Hain, R., Reif, H.J., Krause, E. *et al.* (1993) Disease resistance results from foreign phytoalexin expression in a novel plant. *Nature* 361, 153–156.

Hamilton, C.M., Frary, A., Lewis, C. & Tanksley, S.D. (1996) Stable transfer of intact high molecular weight DNA into plant chromosomes. *Proceedings of the National Academy of Sciences of the USA* 93, 9975–9979.

Hammond-Kosack, K.E., Jones, D.A. & Jones, J.D.G. (1996) Ensnaring microbes: the components of plant disease resistance. *New Phytologist* 133, 11–24.

Hittalmani, S., Foolad, M.R., Mew, T., Rodriguez, R.L. & Huang, N. (1995) Development of a PCR-based marker to identify rice blast resistance gene, *Pi-2(t)*, in a segregating population. *Theoretical and Applied Genetics* 91, 9–14.

Jones, A.T. (1987) Control of virus infection in crop plants through vector resistance: a review of achievements, prospects and problems. *Annals of Applied Biology* 111, 745–772.

Kawchuk, L.M., Martin, R.R. & McPherson, J. (1991) Sense and antisense RNA-mediated resistance to potato leafroll virus in Russet Burbank potato plants. *Molecular Plant–Microbe Interactions* 4, 247–253.

Lamb, C.J., Ryals, J.A., Ward, E.R. & Dixon, R.A. (1992) Emerging strategies for enhancing crop resistance to microbial pathogens. *BioTechnology* 10, 1436–1445.

Lomonossoff, G.P. (1995) Pathogen-derived resistance to plant viruses. *Annual Review of Phytopathology* 33, 323–343.

Michelmore, R. (1995) Molecular approaches to manipulation of disease resistance genes. *Annual Review of Phytopathology* 33, 393–427.

Panopoulos, N.J., Hatziloukas, E. & Afendra, A.S. (1996) Transgenic crop resistance to bacteria. *Field Crops Research* 45, 85–97.

Simmonds, N.W. (1991) Genetics of horizontal resistance to diseases of crops. *Biological Reviews* 66, 189–241.

Wellings, C.R. & McIntosh, R.A. (1990) *Puccinia striiformis* f.sp. *tritici* in Australasia: pathogenic changes during the first 10 years. *Plant Pathology* 39, 316–325.

Wilson, T.M.A. (1993) Strategies to protect crop plants against viruses: Pathogen-derived resistance blossoms. *Proceedings of the National Academy of Sciences of the USA* 90, 3134–3141.

Wolfe, M.S. & Barrett, J.A. (1980) Can we lead the pathogen astray? *Plant Disease* 64, 148–155.

Young, N.D. (1996) QTL mapping and quantitative disease resistance in plants. *Annual Review of Phytopathology* 34, 479–501.

Zhu, Q., Maher, E.A., Masoud, S., Dixon, R.A. & Lamb, C.J. (1994) Enhanced protection against fungal attack by constitutive co-expression of chitinase and glucanase genes in transgenic tobacco. *BioTechnology* 12, 807–812.

13   Biological Control of Plant Disease

'The future of biological control will be limited only by our imagination.' [R.J. Cook, 1984]

Throughout most of the long history of agriculture, prior to the discovery of chemical pesticides and understanding of the genetic basis of heredity, crops were grown without any specific strategy for disease control. Practices evolved, however, which tended to restrict the losses caused by disease; these included time-honoured rotations using different sequences of crops. Also, the crops themselves were often genetically diverse, and consisted of a mixture of genotypes with different properties, including differences in susceptibility to disease. This type of small-scale mixed cropping is still practised in many developing countries, for instance with legume crops in many parts of Africa.

The demands of an ever-increasing world population have inevitably led to larger and more intensive agricultural regimes. The development of highly effective pesticides, which seemed to offer instant answers to the threat of disease, shifted attention to non-biological methods of control. The application of genetic principles to plant breeding further transformed agricultural practice. Selection for increased yield and responsiveness to fertilizers led to greater uniformity and the development of agricultural systems relying on routine inputs of chemicals. These approaches have been highly successful in boosting crop production and maintaining a reliable supply of the major commodities, but there is now concern over the costs and environmental side-effects of such intensive regimes. Hence there is renewed interest in developing sustainable systems of production, with reduced inputs, which maintain some of the biodiversity of the agricultural landscape. This shift has refocused attention on the application of ecological principles to disease control.

Cultural practices and disease control

Understanding the effects of cultural practices on the incidence of disease is vital if we are to develop more natural control measures. A whole range of agronomic factors influence disease, including sowing date, cultivations, nutrients and water, soil organic matter, management of crop residues such as straw, and crop rotations. Altering cultural practices can both increase or decrease disease risk. Sowing date provides a good example. In northern Europe there has been a trend towards growing winter cereals, sown in the autumn, as the final yields exceed those of spring-sown crops. Further yield increases may be gained by sowing early in the autumn, for instance in September. However, early-sown crops are more at risk of infection by certain pathogens such as barley yellow dwarf virus (BYDV) (Fig. 13.1). This is because the early crop is emerging during a period when aphid vectors that may be carrying the virus are still active (see Fig. 4.7). Later-sown crops avoid this risk period. The potential advantages of early sowing must therefore be balanced against the increased risk of yield penalties or control costs due to virus disease. In some regions, intensive cereal production has actually created a situation in which late-harvested crops overlap with the next crop emerging, so that host plants are continuously available. This so-called 'green bridge' can encourage transfer of biotrophic pathogens such as powdery mildew (*Erysiphe graminis*) or rust fungi, which require living host tissue to develop.

Crop rotation

Varying the type of crop grown on a particular plot of land has long been known to influence the amount of disease present; indeed, this is arguably the oldest form of disease control. Figure 13.2 shows

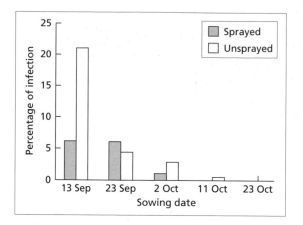

Fig. 13.1 Effect of sowing date on the incidence of barley yellow dwarf virus in an autumn-sown winter cereal crop the following season (May). The sprayed crop was treated with an insecticide at the end of October. (After Plumb 1986.)

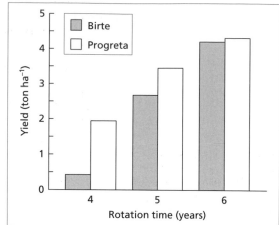

Fig. 13.2 Effect of rotation on the yield of two pea cultivars. (After Rogers-Lewis 1985; © Crown copyright.)

yield data for two varieties of pea grown on the same land with a 4-, 5- or 6-year break between crops. Rotations less than 6 years in duration gave reduced yield, primarily due to infection by a mixture of fungal pathogens attacking the roots and stem base of the plants.

Continuous cropping with a single species of host plant will, almost always, lead to a build-up of pathogens adapted to that crop, especially soil-borne pathogens. Black root rot, caused by the fungus *Thielaviopsis basicola*, is a disease of increasing importance in cotton crops grown in California. The pathogen survives in soil mainly as chlamydospores, and there is a direct relationship between the amount of inoculum present and disease severity (Fig. 13.3). Soil samples from fields cropped continuously with cotton for 3 or 4 years contained on average more than 100 propagules per gram of soil, whereas soils rotated to other crops contained less than 20 propagules per gram. This difference in pathogen inoculum levels is critical in terms of symptom severity in the crop, with severe infection in the former case compared with only trace symptoms in the latter. A similar trend is seen with other soil-borne fungi, such as *Fusarium solani* f.sp. *phaseoli* infecting beans (Fig. 13.4).

Previous crop history also affects the incidence of stem-base and root-infecting pathogens of cereal crops, such as eyespot disease (see Plate 6, facing p.

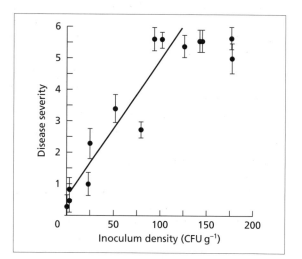

Fig. 13.3 Relationship between inoculum density of *Thielaviopsis basicola* in soil and disease severity in cotton seedlings. Bars indicate standard errors. (After Holtz & Weinhold 1994.)

12), which is generally more severe in continuous wheat cultivation. Populations of the take-all fungus, *Gaeumannomyces graminis*, can be substantially reduced by a 2–3-year rotation to grass leys. The eradication of *Gaeumannomyces* has been shown to result from an increase in the population of a fungal antagonist, *Phialophora radicicola*, which is especially common on the surfaces of grass roots. Many

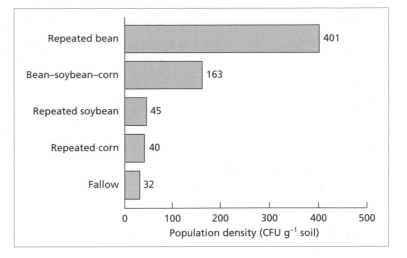

Fig. 13.4 Effect of crop sequence on the population density of *Fusarium solani* f.sp. *phaseoli* in soil after 14 years of continuous cropping with different crops, or a bean–soybean–corn rotation. (After Hall & Phillips 1992.)

of the beneficial effects of crop rotation on disease levels may operate via similar mechanisms of microbial antagonism, which raises the possibility of enhancing such processes by means of appropriate cultural practices (see p. 237).

Clearly rotation is only feasible as a disease management tool where the pathogen concerned is unable to survive for extended periods away from the host. Fungi with very durable survival structures may persist at damaging levels beyond any economically acceptable rotation scheme. The best examples are fungal species forming sclerotia. The vascular wilt pathogen, *Verticillium dahliae*, survives as microsclerotia, and the minimum period required to effectively reduce inoculum between susceptible crops, such as potatoes, is 5–10 years. The shorter 2–3-year rotations favoured by growers are in this case ineffective. Other sclerotial fungi are even better survivors; the white rot fungus of onion, *Sclerotium cepivorum*, can persist in soil for at least 20 years (see p. 52), so that once land becomes infested it is, for all practical purposes, unusable for this crop.

Soil solarization

Crop rotation as a means of disease control relies on natural factors to gradually reduce the level of inoculum in soil. Can anything be done to accelerate this process, and to reduce the viability of inoculum? Chemical treatment of soil with fumigants such as methyl bromide is one possibility, but in addition to the expense these toxic sterilants kill non-target organisms, and their use is now restricted. A more benign method is needed. One promising approach is to cover the soil with materials which raise the surface temperature above ambient. Traditionally, such **mulches** have used compost, sawdust or other waste products. These organic treatments accelerate crop growth by several means, including increased water retention and nutrient availability as well as effects on soil temperature. More recently, plastic sheeting has replaced such traditional materials, a method described as **soil solarization**. The usual procedure is to overlay the soil for a period of one or two months prior to planting of the crop. Both transparent and black polyethylene sheets can be used, and when laid over soil during periods of high air temperature, these treatments can reduce or eradicate a variety of pathogens and pests, including soil-borne fungi and nematodes, as well as weed seeds. The method has proved most successful in regions with hotter climates, such as southern Europe, Israel and South Africa. This is because the soil temperature needs to rise above a certain critical threshold before pathogen propagules and weed seeds are actually killed. In the central valley of California, for instance, surface soil layers may reach 40°C during summer, and deeper layers 30°C, but solarization can raise these temperatures by around 10°C. Most pathogen propagules are unable to withstand temperatures of 50°C for more than a few hours, and longer periods at lower temperatures may also be lethal. Best results are achieved if the soil is moist, so solarization is often combined with an irrigation treatment.

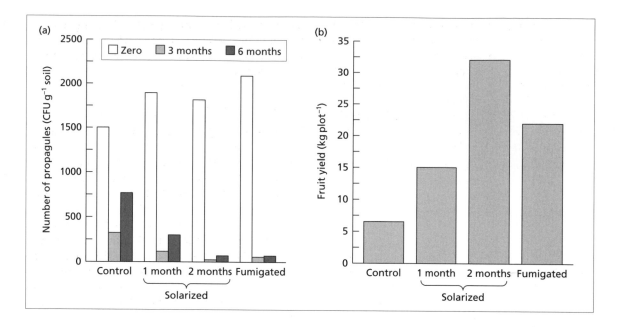

Fig. 13.5 Effects of soil solarization and fumigation treatments on (a) numbers of propagules of *Fusarium oxysporum* in soil before treatment, or 3 or 6 months after treatment, and (b) yield of a subsequent melon crop. (After González-Torres *et al.* 1993.)

Flooding soil can in any case control certain pathogens, but the combination of heat and water is more effective.

Soil solarization compares favourably with chemical fumigation as a means of controlling soil-borne diseases. Figure 13.5 shows results from a trial in Spain in which soil was artificially infested with the vascular wilt fungus *Fusarium oxysporum*, and then fumigated or solarized for 1 or 2 months. Levels of pathogen inoculum declined in all treatments, but the highest yields of a subsequent melon crop were obtained on soil which had been solarized for 2 months. Solarization can also control diseases caused by sclerotial pathogens such as onion white rot, *Sclerotium cepivorum*. Treatment of 2–3 months duration can reduce sclerotial populations in surface soil to undetectable levels, with major effects on epidemic development, even in highly susceptible crops such as garlic (Fig. 13.6).

Thermotherapy

The sensitivity of many pathogens to high tempera-tures has also been put to good use in the treatment of seeds or propagation material to eradicate disease agents prior to planting. The optimum combination of time and temperature depends on the type of plant material and its relative ability to withstand heat without a reduction in viability. The fireblight bacterium, *Erwinia amylovora*, is unable to withstand temperatures of 45–50°C for much more than 1 hour,

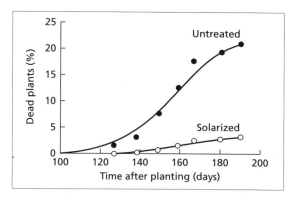

Fig. 13.6 Effect of soil solarization on the disease progress curves of garlic white rot in fields naturally infested by *Sclerotium cepivorum*. (Redrawn from Basallote-Ureba & Melero-Vara 1993; copyright 1993, with kind permission from Elsevier Science Ltd, The Boulevard, Langford Lane, Kidlington OX5 1GB, UK.)

and moist and dry heat treatments may therefore be used to control the disease during propagation of apple and pear shoots. Many other bacterial diseases are seed-borne, and treatment with hot air or water can be a relatively cheap and efficient way to disinfect seed stocks. Thermotherapy can also cure plant tissues of virus infection, and is especially effective when combined with tissue culture, for instance by maintaining excised shoot tips or buds for a period at a temperature around 50°C during micropropagation.

Biological control

The fact that cultural practices such as crop rotation can reduce disease indicates that pathogen populations may be regulated by natural means, rather than by human intervention. During dispersal, survival and the early stages of infection on the surface of the host, pathogens are exposed to other organisms which may affect their growth or viability. The starting point for biological control is to identify these natural constraints on pathogens and exploit them wherever possible to limit disease outbreaks.

The term **biological control** has been applied in a broad sense to cover the use of any organism to control a pathogen. This definition includes host plant resistance as a natural, and highly effective, form of biological control. More usually, biological control is defined as the reduction in attack of a crop species by a pathogen achieved using another living organism, or organisms. This concept itself includes both direct and indirect effects, due either to introduced antagonists, or manipulation of existing populations to reduce disease (Table 13.1). Agronomic practices such as soil amendment with organic matter to boost microbial populations are therefore covered by this definition.

Classical biocontrol

The earliest examples of successful biological control almost all involved pests or weeds. The first breakthrough in the field concerned control of the cotton cushion-scale insect in the USA. This pest was first recorded in 1868 in California, where it spread rapidly through the newly established citrus groves. Entomologists suspected that the pest had come from Australia on imported nursery stock, and a search was made there for natural predators. Two of these, a lady beetle and a fly, were selected, imported and released in California in 1888, and within a few months the problem of cotton cushion-scale had been solved.

The principle involved in this example is simple yet extremely effective. Pests almost invariably have natural enemies somewhere in their geographic range, and introduction of such enemies can reduce the size of the pest population. There are now numerous successful case histories where natural predators or parasites have been released to control insects or weeds. But there are biological constraints to this classical, inundative approach. Potential natural control agents may fail to establish in the environment, or to multiply to levels which significantly reduce the pest or pathogen population. The dynamics of predator–prey relationships, in which there is often a cyclical equilibrium between the two organisms, means that the degree of control achieved may be less than with a chemical agent. It is worth noting that many of the most successful biological control systems concern protected environments such as glasshouses, where conditions can be regulated to favour the introduced antagonist. Control in field crops is more difficult, and requires a good understanding of the population biology and ecology of both the target pest or pathogen, and the biocontrol agent.

Despite these problems a number of microbial products have been released onto the market for the control of insects and weeds (Table 13.2). Microbial pesticides are based on the toxin-producing bacterium, *Bacillus thuringiensis*, and insect-pathogenic fungi and viruses. All the products for weed control

Table 13.1 Approaches used for the biological control of plant pathogens.

Direct
Introduction of antagonists able to suppress activity of the pathogen

Indirect
Manipulation of the microbial balance by means of crop practices, cultural measures, soil supplements, etc.

Host-mediated
Induction of localized and systemic resistance
Virus cross-protection

Agent	Target pest or weed
Microbial insecticides	
Bacteria	
Bacillus thuringiensis	Lepidoptera (moths/butterflies)
	Diptera (flies)
	Coleoptera (beetles)
Fungi	
Verticillium lecanii	Aphids and whiteflies
Metarhizium anisopliae	Spittlebugs
Metarhizium flavoviride	Locusts and grasshoppers
Beauveria bassiana	Colorado beetle
Virus	
Nuclear polyhedrosis virus	Various
Mycoherbicides	
Colletotrichum gloeosporioides	Northern jointvetch (*Aeschynomene virginiana*)
Colletotrichum orbiculare	Round-leaved mallow (*Malva pusilla*)
Phytophthora palmivora	Milkweed vine (*Morrenia odorata*)
Alternaria cassiae	Sicklepod (*Cassia obtusifolia*)
Puccinia chondrillina	Skeleton weed (*Chondrilla juncea*)

Table 13.2 Some microbial insecticides and herbicides.

are based on plant-pathogenic fungi and are therefore described as **mycoherbicides**.

To date there are far fewer examples of biological agents developed for the control of plant pathogens. There have been numerous demonstrations of promising control in model systems, but scaling this up into crops in the field has often given disappointing results. Why is this? The following discussion will focus on the search for effective biocontrol agents (BCAs) for plant pathogens, and consider both the problems and prospects for commercialization of such agents.

Microbial antagonism

Studies on several soil-borne plant pathogens have shown that other microorganisms resident in the soil can have a large influence on the incidence and severity of disease. It has been known for many years that certain soils can reduce disease caused by vascular wilt fungi, such as *Fusarium*, and that the suppressive properties of such soils can be removed by sterilization. It is also possible to convert a 'conducive' soil into a 'suppressive' one by mixing in a proportion of the suppressive type. This evidence suggests that living organisms present in some soils are capable of reducing the viability or infectivity of plant pathogens. Further indirect evidence for such microbial antagonism comes from surveys of disease in continuous monoculture. For instance, the normal pattern of take-all (*Gauemannomyces graminis*) in successive wheat crops is an increase in severity over the first few seasons, followed by a decline. Logic suggests that in the absence of any crop rotation the disease should become more and more prevalent as inoculum levels continue to build up in the soil. But instead the amount of disease actually falls. Similar patterns of incidence have been reported with other soil-borne fungal pathogens (Fig. 13.7). The simplest explanation for this phenomenon is that as the pathogen population increases, there is a corresponding increase in the amount or activity of antagonistic microorganisms. The exact mechanism of such antagonism is still debated. It is known that certain types of bacteria found in the root zone, or rhizosphere, for instance fluorescent pseudomonads, can interfere with infection of roots by fungi, and over time these types may come to predominate. Alternatively there may be increasing microbial competition for the nutrient substrates on which the pathogen sur-

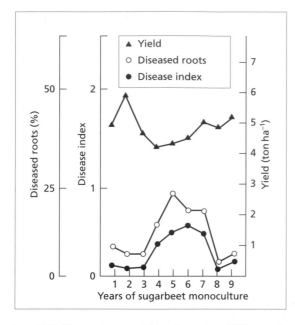

Fig. 13.7 Changes in crop yield and severity of *Rhizoctonia* root rot in sugarbeet during 9 years of monoculture. (After Hyakumachi *et al.* 1990.)

vives between crops, or even an increase in the level of predation or parasitism of the pathogen by other living agents, including viruses. Disease suppression may well be due to a combination of several of these processes. But the important conclusion is that such natural antagonism might be exploited to control disease, either by managing the crop to maximize the impact of resident antagonists, or by isolating and identifying the most effective agents for use in specific biocontrol programmes.

Selection of biocontrol agents for plant disease control

There is little difference, in principle, between screening chemicals or testing biological agents for disease control. Both selection processes require an appropriate assay system to detect activity against the target pathogen, or pathogens. Random screening of large numbers of microorganisms for suppressive effects on plant disease has often been used to select the most promising strains for further development. But the chances of success can be improved by adopting a more rational approach based on ecology. The argu-

ment is as follows. To be effective, a BCA must be able to colonize a particular habitat, or to occupy a specific niche, in sufficient numbers to interfere with the growth or survival of the target pathogen. Rather than introducing a randomly selected microbial antagonist, it would be better to introduce one known to be adapted to the habitat concerned. Hence the best place to look for potential BCAs is in the specific environment in which they are to be used. For instance, if the target is a pathogen which infects plant roots, the logical place to look for an antagonist is in the rhizosphere. Any microorganism isolated from this habitat is likely to have biological and physiological attributes enabling successful multiplication in the root zone, and is therefore 'rhizosphere competent'. Similar concepts apply to leaf surfaces or other substrates such as straw where a particular stage in the pathogen life cycle might be disrupted. Screening for BCAs should therefore focus on specific natural sources chosen to optimize the chances of isolating strains with the correct biological properties.

Production and formulation of biocontrol agents

An important difference between a chemical and a BCA is that the latter is a living organism and usually needs to be metabolically active to be effective. One of the most difficult steps in the commercialization of a BCA is to produce, package and deliver sufficient quantities of the agent in a viable and stable form. Variation is a normal feature of microorganisms, but each batch of a BCA needs to have similar activity. This can be a particular problem in scaling-up production, as microbial populations can change their properties during growth in fermenters. The BCA then has to be harvested and distributed in a formulation which ensures viability. To be of practical use, a BCA needs to have a reasonable shelf-life so that it can be stored for a period without significant loss of activity. It then needs to be applied to the crop, or into soil, in a way which ensures that the antagonist grows and persists in the environment for sufficient time to exert control. Several different approaches have been used to solve these various problems, including freeze-drying microbial cultures, or mixing cells with inert carriers such as clay or talc. Alternatively, biomass may be immobilized or encapsulated in an alginate polymer. A food source such as wheat bran can be added, which not only acts as a carrier, but also releases nutrients promoting growth of the microorganism once applied. There are a variety of application methods, for instance as liquid sprays or drenches, seed-dressings, and pelleted formulations which slowly release the BCA into the environment. As with chemicals the choice depends upon the target pathogen, and the mode of action of the BCA.

Examples of successful biocontrol agents for control of plant pathogens

Table 13.3 lists several BCAs which have been used, or are being used, on a commercial scale to control plant pathogens. There are many more examples which have shown promise in experimental trials, but have yet to make a significant impact in the market place.

Some case histories

Crown gall is a plant tumour caused by the soil-borne bacterium *Agrobacterium tumefaciens*; galls occur on the roots or often at the crown of the plant between root and shoot (see Fig. 8.11). The pathogen has a wide host range but the disease is especially important in stone fruits such as peaches, on grapevine, and in woody ornamentals such as roses. A highly effective biocontrol method has been developed using a related non-pathogenic bacterium, *Agrobacterium radiobacter*, to treat roots during transplanting. The BCA is commercially available (Table 13.3) and gives cheaper, more effective control than antibacterial chemicals.

Not all strains of *A. radiobacter* are able to protect plants from the disease. Effective strains possess two important properties. Firstly, they are able to colonize host roots to a higher population density than ineffective strains. Secondly, biologically active strains produce an antibiotic, agrocin, which is toxic to *A. tumefaciens*. This molecule is a 'rogue' nucleotide (Fig. 13.8) which interferes with DNA synthesis. Agrocin production is encoded by a plasmid which also carries genes for insensitivity to the toxin. As plasmids are often able to transfer between bacterial strains and even species, one concern with the use of this BCA is the possibility that the genes for agrocin insensitivity might 'jump' into the pathogen, rendering it immune to the toxin. Work in Australia has therefore concentrated on engineering *A. radiobacter* strains carrying modified plasmids which are unable to transfer between bacteria. This was done by deleting the region of the agrocin plasmid encoding transfer functions. Such *Tra⁻* strains are now registered for use in the field.

The soil-borne fungus *Heterobasidion annosum* is a serious problem in forestry plantations, where it causes root and butt rot of conifers (Fig. 13.9). The fungus is not an aggressive pathogen of intact trees but instead exploits cut tree stumps to establish a food base from which it can spread along the roots and infect adjacent hosts. As forestry practice involves frequent thinning and felling of trees, this provides the pathogen with ample opportunity to gain access. Treating stumps with chemicals only delays infection, as the concentration of fungicide eventually declines to an ineffective level. Instead, a biological solution was sought by looking for microbial antagonists able to colonize cut timber and

Table 13.3 Some commercial products for biocontrol of plant pathogens.

Agent	Target	Tradename(s)
Bacteria		
Agrobacterium radiobacter	Crown gall	Nogall, Galltrol
Pseudomonas fluorescens	Damping-off	Dagger
Streptomyces griseoviridis	Seedling diseases	Mycostop
Bacillus subtilis	Growth promotion Seedling diseases	Quantum 4000
Fungi		
Peniophora (*Phlebia*) gigantea	*Heterobasidion* rot	PG suspension Rotstop
Trichoderma harzianum	Mushroom dry bubble Silverleaf of fruit trees	Binab-T

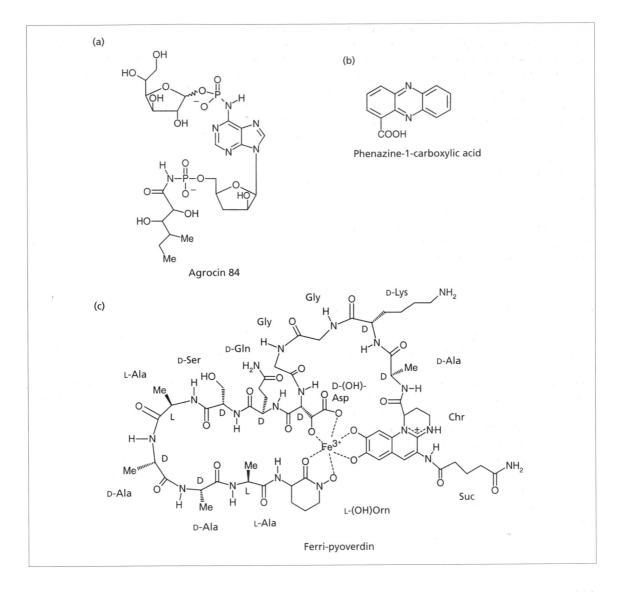

(a)

(b)

Phenazine-1-carboxylic acid

Agrocin 84

(c)

Ferri-pyoverdin

Fig. 13.8 Some compounds produced by bacterial antagonists which play a role in biological control. (a) The bacteriocin, agrocin. (b) The antibiotic, phenazine. (Structure courtesy of Mike Beale.) (c) The siderophore, pyoverdin. (After Mohn *et al.* 1994.)

prevent the pathogen from becoming established. The most effective agent proved to be another wood-rotting basidiomycete, *Peniophora (Phlebia) gigantea*. Spores of this fungus are painted or sprayed onto cut stumps or can now be applied automatically by timber-harvesting machines. Around 10^6 spores are used per square metre of stump surface, which ensures rapid colonization of the timber by the antagonist, thereby occupying the infection court and denying access to the pathogen. In the UK this treatment has provided cost-effective and environmentally safe control of the disease in pine plantations for more than 30 years, while in Scandinavia strains of the BCA have been developed for use on Norway spruce.

This example of protection of an infection court in a woody host has parallels in several other plant diseases. For instance, many pathogens, such as *Nectria galligena*, which causes silverleaf of fruit trees, gain

Fig. 13.9 Damage caused to conifer timber due to infection by butt rot, *Heterobasidion annosum*. Note rot of heartwood. (Courtesy of Steve Gregory, Forestry Commission.)

entry to the host via pruning wounds. These sites can be treated with a fast-growing antagonist, such as formulations of the saprophytic fungus *Trichoderma*, to prevent infection by the pathogen.

Soil-borne fungal pathogens which infect seeds and roots are a serious constraint to agricultural production as they affect crop establishment, leading to patchy growth and delayed development. Examples include damping-off diseases, caused by *Pythium* species and *Rhizoctonia solani*, and take-all of cereals, caused by *Gaeumannomyces graminis*. One feature common to the infection cycle of these diverse pathogens is the need to colonize the zone surrounding seeds or roots prior to penetration of host tissue. Opportunities exist therefore to interfere with this step by introducing aggressive microbial competitors, or to manipulate resident microbial populations to reduce infection. We have already noted (above) that the decline of take-all disease in cereal monocultures coincides with changes in the microbiology of the rhizosphere. Bacteria isolated from this zone (often described as 'rhizobacteria') have been intensively studied as potential biocontrol agents for take-all and several seedling diseases. The most promising candidate strains are almost invariably isolates of *Pseudomonas fluorescens*, a group of bacteria which are well-adapted to growth in the rhizosphere. Several strains of *P. fluorescens* have been tested in plot and field trials for the control of soil-borne fungal pathogens with varying degrees of success (Fig. 13.10), although to date only one commercial

product based on this bacterium has been launched (Table 13.3).

An interesting alternative to *P. fluorescens*, especially for use as a seed treatment, is *Bacillus*. Unlike most bacteria, species of *Bacillus* produce resistant endospores which can be kept as a dry formulation for long periods without losing viability. This property might also aid use in situations exposed to drying, such as on leaves and other aerial plant surfaces.

Mode of action of biocontrol agents against plant pathogens

The examples discussed above almost all involve protection of an infection court by prior treatment with a microbial antagonist. Effective suppression of the pathogen requires the establishment of a metabolically active threshold population of the BCA at the infection site. There are also examples of BCAs which act against pathogens in soil or on plant debris, affecting the survival of resting structures or propagules, rather than preventing infection. The exact mechanism of antagonism is often not known. There are, however, a number of possible modes of action (Fig. 13.11).

At one extreme, suppression might simply be due to occupation of a particular niche by the BCA, leading to physical exclusion of the pathogen, or to competition for essential nutrients. At one time it was believed that control of *Heterobasidion* by *Penio-*

Fig. 13.10 Protection of pea seedlings against pre-emergence damping-off caused by *Pythium ultimum*. The same number of seeds were sown in each tray. Treatments (left to right) are: compost + pathogen; compost + pathogen + *Pseudomonas fluorescens* drench; compost + pathogen + fungicide (metalaxyl) drench; compost without pathogen.

phora was due to this type of effect. Now it is suspected that more direct interactions between the hyphae of the two fungi may occur, in a type of territorial combat known as **hyphal interference**. A more clearly characterized mode of action is production of a toxic or inhibitory compound by the BCA, such as the example of agrocin mentioned above. Antibiotics have also been shown to be important in the biological control of root-infecting fungi including take-all. Strains of *P. fluorescens* active against *G. graminis* produce an antibiotic known as phenazine (Fig. 13.8); mutants which have lost the ability to produce this antibiotic give poor control. This correlation has been confirmed by molecular genetic analysis in which inactivation of antibiotic biosynthesis genes by transposons in the bacterium abolishes biocontrol activity, while introduction of a functional gene restores it. Such experiments suggest that more than 50% of the control effect is due to production of this compound. Different antibiotics have been implicated in the control of *Pythium* and other soil-borne fungal pathogens.

Siderophores are low molecular weight compounds which have a high affinity for iron and aid transport into cells. These chemicals are efficient scavengers of iron and may thus mop up all of the available supply in the immediate environment. Many pathogens require iron as an essential mineral nutrient for growth, and in some cases iron is required for virulence. Hence, production of a siderophore by a BCA may reduce the growth of a pathogen, or its ability to attack the host. Fluorescent pseudomonads produce several siderophores, such as the pigmented compound pyoverdin (Fig. 13.8), and the most convincing evidence that these contribute to disease control again comes from studies on non-producing mutants which are less effective than wild-type strains.

Direct parasitism or predation of the pathogen can also occur and may be particularly important in reducing the viability of spores or survival structures such as sclerotia. Fungal pathogens forming sclerotia are difficult to control as these propagules persist for long periods in soil. Several fungi which invade sclerotia and act as parasites have now been identified. One of the most interesting is *Sporidesmium sclerotivorum*, which is an obligate parasite of sclerotia of several important pathogens, including species of *Sclerotinia*, *Sclerotium* and *Botrytis cinerea*. Multicellular conidia of the parasite are stimulated to germinate by chemicals diffusing from nearby sclerotia,

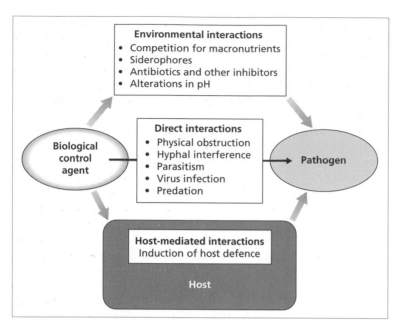

Fig. 13.11 Modes of action of microbial biological control agents.

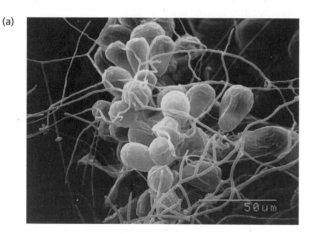

(a)

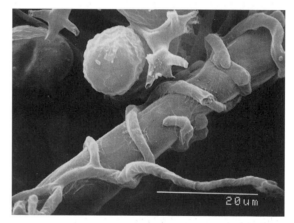

(b)

Fig. 13.12 An example of mycoparasitism. Infection of the grapevine downy mildew pathogen, *Plasmopara viticola*, by *Fusarium proliferatum* showing: (a) thin hyphae of the mycoparasite overgrowing sporangia of the downy mildew; and (b) parasite hyphae curling around and penetrating sporangiophores. (From Falk *et al*. 1996.)

and germ tubes infect the sclerotia, causing eventual lysis. Spores of this parasite added in sufficient quantities to soil have been shown to give good control of diseases such as lettuce drop, caused by *Sclerotinia minor*. Unfortunately the BCA is difficult to produce on a large scale in pure culture. Another sclerotial

parasite, *Coniothyrium minitans*, has also demonstrated biocontrol potential, but only when added to soil at high inoculum rates.

Fungi which infect other fungi are usually described as **mycoparasites** (Fig. 13.12). These agents themselves have diverse host–parasite relationships, ranging from necrotrophy to biotrophy, and may infect different stages in fungal life cycles. Some are close relatives of plant-pathogenic fungi; for instance, mycoparasitic species of *Pythium*, such as *P. nunn* and *P. oligandrum*, coil around and lyse hyphae of the damping-off fungus *P. ultimum*. The diversity of interactions occurring in natural environments means that it is usually possible to find an organism parasitic on the target pathogen or pest. Hence fungi attacking other fungi, nematodes and insects are all being studied as potential BCAs.

Predation is an important mode of action in the biocontrol of insect pests but has not been exploited to any significant extent in the suppression of pathogens. Fungal spores are subject to predation in soils, for instance by large, mobile amoebae which are able to penetrate even highly resistant spores and destroy them. These protozoa, known as vampyrellids, contact spores by means of thin pseudopodia, drill a hole in the spore wall, enter, and consume the spore contents. Such natural predation is no doubt important in reducing pathogen inoculum levels in soil, but so far these antagonists have proved difficult to produce in culture.

Hypovirulence

A unique form of natural biocontrol has been observed in the pandemic of the highly destructive chestnut blight disease, caused by *Cryphonectria parasitica*. This fungus infects via wounds causing aggressive lesions, or cankers, which girdle the stem, leading to the death of shoots above the infection site. The disease was first recorded in the USA in 1904, and despite attempts to eradicate it, spread to destroy most of the native chestnuts in the eastern states. The pathogen is believed to have originated from Asia, and the severity of the epidemic is consistent with a 'new encounter' disease (see p. 15), in which a pathogen infects a previously unexposed and highly susceptible host population. When, in 1938, the disease was recorded in Italy, it was feared that European chestnut trees would suffer a similar fate. The initial European outbreak was also severe, but subsequently many infected trees began to show signs of recovery, with spontaneous healing of cankers. Significantly, strains of the fungus isolated from such cankers were found to be less virulent than the original pathogen. Furthermore, if these **hypovirulent** isolates were co-inoculated with highly virulent strains, the resulting cankers also healed. The most intriguing observation, however, was that virulent strains could be converted to the hypovirulent phenotype by hyphal contact and fusion in culture. Some transmissible factor moved from the hypovirulent strain into the more aggressive one. The agent(s) responsible were shown to be cytoplasmic

and were subsequently identified as double-stranded (ds) RNA molecules. Several different-sized dsRNAs have been isolated from hypovirulent strains of the fungus, and the larger of these have sequence homology with certain plant virus genomes. It seems likely that hypovirulence is therefore a type of virus infection.

The natural transfer of hypovirulence between strains of *C. parasitica* suggests that the disease might be managed simply by introducing hypovirulent isolates into the pathogen population. In Europe this has to some extent occurred naturally, but the situation in the USA is more complex, due to the larger genetic variation between *Cryphonectria* strains in that region. For transfer to occur, hyphae must fuse, and this is often limited by natural compatibility barriers. To overcome this problem there is now interest in engineering replicating forms of the dsRNA into a range of *C. parasitica* strains representing the different compatibility types. In this way more efficient spread of hypovirulence should occur, and the epidemic should be restricted.

Induction of host resistance

An alternative, and quite different, mode of action of BCAs is disease suppression through the induction of host defence mechanisms. It has been known for many years that inoculation of plants with avirulent strains of pathogens, or agents causing necrosis, can trigger both local and systemic plant resistance (see p. 161). More recently, it has been shown that rhizobacteria applied to seeds or roots can induce a systemic resistance response expressed against pathogens infecting aerial tissues. Examples include treatment with *Pseudomonas* spp. increasing resistance to the vascular wilt fungus *Fusarium oxysporum* f.sp. *dianthi* in carnation, as well as to leaf pathogens such as *Colletotrichum orbiculare* and *Pseudomonas syringae* in cucumber (Fig. 13.13), and bacterial blight, *Pseudomonas syringae* pv. *phaseolicola*, in bean. This resistance appears to be associated with the induction of pathogenesis-related (PR) proteins, as seen in a typical systemic acquired resistance (SAR) response (see p. 162). Presumably the colonizing BCA produces a signal molecule(s) which activates the SAR pathway. However, the specific mechanism of induction is not known. One possibility is that toxic products released in the rhizosphere, such as hydrogen cyanide, might induce plant

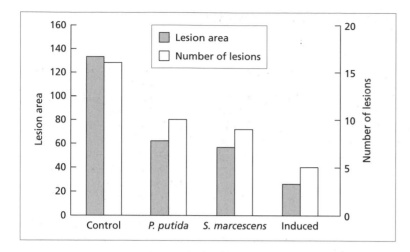

Fig. 13.13 Severity of angular leaf spot, caused by the bacterium *Pseudomonas syringae* pv. *lacrymans* on cucumber plants grown from seeds treated with the plant growth-promoting rhizobacteria *Pseudomonas putida* and *Serratia marcescens*. Control columns show values for unprotected plants, induced shows results for plants in which systemic acquired resistance (SAR) was induced by prior inoculation of the first leaf with the pathogen. (Data from Liu *et al.* 1995.)

defence. There is also recent evidence that some bio-control bacteria may act via a pathway distinct from a typical SAR response. When a bioactive strain of *P. fluorescens* was applied to the roots of *Arabidopsis* plants, resistance to both a vascular wilt fungus and a bacterial foliar pathogen was subsequently increased, but this change in disease reaction type was not accompanied by the accumulation of either PR proteins or salicylic acid. It seems likely therefore that different BCAs may operate through different pathways to influence host defence.

Engineering more effective biocontrol agents

The main obstacle to more widespread acceptance of biological control as a genuine alternative to chemicals has been the inconsistent performance of most BCAs in the field. Growers are reluctant to adopt methods which occasionally fail to benefit the crop. Generally, it is possible to find natural antagonists of pathogens which work well under certain conditions. The challenge therefore is to improve the performance of these agents over a range of conditions, either by devising more effective methods of formulation and delivery, or by improving the properties of the BCA itself. Experience in industrial microbiology has shown that microorganisms can be subjected to repeated rounds of mutation and selection to obtain higher yielding or better-adapted strains. Can we achieve similar results with BCAs?

The critical difference between an industrial process and biological disease control is that a BCA has to function in a complex and variable environ-ment which includes interactions with other organisms. In most cases we understand only some of the factors involved in effective antagonism of the pathogen. Using control of a root-infecting pathogen as an example, it is clear that any BCA must be able to colonize the rhizosphere and to interfere with infection by the pathogen. The BCA must enter the root zone, and spread and proliferate to maintain a threshold population in competition with the indigenous microflora. Properties aiding this process may include motility, chemotaxis towards chemical signals from the root, attachment to the surface, and ability to utilize nutrients present in root exudates. Interference with the pathogen might involve one or more of the mechanisms discussed above, such as nutrient competition, production of antibiotics or siderophores, direct parasitism, or induction of host defence. The genetic control of these different traits is likely to be complex, so that manipulation by mutation or, more precisely, recombinant DNA, may be difficult. Nevertheless there is some hope that genetic engineering might provide improved strains giving more consistent control performance. To date, examples are restricted to simple traits such as resistance to fungicides, aiding co-formulation of BCAs with antifungal chemicals, and hydrolytic enzymes such as chitinase which are active in the degradation of fungal cell walls and the insect exoskeleton. It may also be feasible to enhance antibiotic production by altering the regulation of biosynthetic pathways. For instance, recombinant strains of *P. fluorescens* expressing higher levels of oomycin, an antibiotic implicated in the suppression of *Pythium* root rot, have been engineered by changing a regulatory gene. Also, as noted above, genetic engineering was used to disarm strains of *A. radiobacter* for the control of crown gall disease.

A further obstacle to the utility of genetically engineered strains of BCAs is public acceptability. The issue of transgenic organisms is controversial, and their use is regulated by strict legislation. While transgenic crops are now permitted, subject to controls, in several countries, there is greater concern about engineered microorganisms, and this may limit their application in the field.

The future of biocontrol

In some respects the record of biocontrol of plant diseases has been a disappointment. Few commercial products have reached the market place and much of the early promise has not been fulfilled in practical terms. Is this a fair assessment?

To a large extent the problems of biocontrol are due to two factors. Firstly, we still have an inadequate understanding of microbial ecology and the factors leading to sustained performance of a BCA in a natural environment. This is a state-of-knowledge problem. Secondly, the expectation that natural control agents will substitute for chemicals in terms of instant results is unreasonable. There needs to be a more subtle approach in which these agents are seen as part of an integrated strategy for disease management. Already there are indications that mixtures of BCAs with different modes of action are more effective than application of a single agent. Such cocktails can combine strains which vary in environmental tolerance, thereby extending the range of conditions under which the formulation will perform.

Finally, it should be noted that the classical concept of biocontrol, involving introduction of a specific antagonist, is now being superseded by a wider vision. Biologically active molecules produced by BCAs are a starting point for natural product chemistry aimed at discovering new classes of chemical control agents. Similarly, it may be possible to engineer plants to express novel defence compounds originating from biological agents. This has already been achieved with the insecticidal toxin from *Bacillus thuringiensis*. Hence control methods based on chemical, biological and genetic approaches are converging, and the future will see increasing integration of these once separate areas of discovery.

Further reading

Books

Campbell, R. (1989) *Biological Control of Microbial Plant Pathogens*. Cambridge University Press, Cambridge.

Chet, I. (ed.) (1987) *Innovative Approaches to Plant Disease Control*. John Wiley & Sons, New York.

Cook, R.J. & Baker, K.F. (1983) *The Nature and Practice of Biological Control of Plant Pathogens*. The American Phytopathological Society, St Paul, Minn.

Hornby, D. (ed.) (1990) *Biological Control of Soil-Borne Plant Pathogens*. CAB International, Wallingford.

Palti, J. (1981) *Cultural Practices and Infectious Crop Diseases*. Springer-Verlag, Berlin.

Wood, R.K.S. & Way, M.J. (eds) (1988) *Biological Control of Pests, Pathogens and Weeds: Developments and Prospects*. Royal Society, London.

Reviews and papers

Adams, P.B. (1990) The potential of mycoparasites for biological control of plant diseases. *Annual Review of Phytopathology* **28**, 59–72.

Baker, C.A. & Henis, J.M.S. (1990) Commercial production and formulation of microbial biocontrol agents. In: *New Directions in Biological Control: Alternatives for Suppressing Agricultural Pests and Diseases* (eds R.A. Baker & P.E. Dunn), pp. 333–344. Alan Liss, New York.

Basallote-Ureba, M.J. & Melero-Vala, J.M. (1993) Control of garlic white rot by soil solarization. *Crop Protection* **12**, 219–223.

Bullock, D.G. (1992) Crop rotation. *Critical Reviews in Plant Sciences* **11**, 309–326.

Campbell, R. (1994) Biological control of soilborne diseases: some present problems and different approaches. *Crop Protection* **13**, 4–13.

Capper, A.L. & Campell, R. (1986) The effect of artificially inoculated antagonistic bacteria on the prevalence of take-all disease of wheat in field experiments. *Journal of Applied Bacteriology* **60**, 155–160.

Cooke, R.J., Thomashow, L.S., Weller, D.M. *et al.* (1995) Molecular mechanisms of defense by rhizobacteria against root disease. *Proceedings of the National Academy of Sciences of the USA* **92**, 4197–4201.

Grondeau, C. & Samson, R. (1994) A review of thermotherapy to free plant materials from pathogens, especially seeds from bacteria. *Critical Reviews in Plant Sciences* **13**, 57–75.

Heiniger, W. & Rigling, D. (1994) Biological control of chestnut blight in Europe. *Annual Review of Phytopathology* **32**, 581–599.

Keck, M., Chartier, R., Zislavsky, W., Lecomte, P. & Paulin, J.P. (1995) Heat treatment of plant propagation material for the control of fireblight. *Plant Pathology* **44**, 124–129.

Lewis, J.A. & Papavizas, G.C. (1991) Biocontrol of plant diseases: the approach for tomorrow. *Crop Protection* **10**, 95–105.

Liu, L., Kloepper, J.W. & Tuzun, S. (1995) Induction of systemic resistance in cucumber against bacterial angular leaf spot by plant growth-promoting rhizobacteria. *Phytopathology* **85**, 843–847.

Nuss, D.L. (1992) Biological control of chestnut blight: an example of virus-mediated attenuation of fungal pathogenesis. *Microbiological Reviews* **56**, 561–576.

Stapleton, J.J. & DeVay, J.E. (1986) Soil solarization: a non-chemical approach for the management of plant pathogens and pests. *Crop Protection* **5**, 190–198.

Weller, D.M. (1988) Biological control of soilborne plant pathogens in the rhizosphere with bacteria. *Annual Review of Phytopathology* **26**, 379–407.

14   An Integrated Approach to Disease Management

'A sustainable system of crop production is one that may be used continuously for many years, is soundly based on the potential and within the limitations of a particular region, does not unduly deplete its resources or degrade its environment, makes the best use of energy and materials, ensures good and reliable yields, and benefits the health and wealth of the local population.' [R.K.S. Wood, 1993]

From a historical perspective, many of the problems encountered in crop protection have arisen from a 'quick-fix' approach, in which a simple strategy has been used, often intensively, to control a disease or pest. The 'boom-and-bust' cycle seen with many *R* genes and agrochemicals is a direct consequence of this simplistic approach. What is required is a more lasting solution, using a combination of different control strategies. The overall aim is to develop **sustainable** systems of disease and pest management, based on a sound understanding of the whole crop ecosystem. Ideally, such an integrated approach will not only maintain the efficacy of host resistance and agrochemicals, but also bring other benefits such as reduced environmental impacts, and lower control costs.

Much of the early progress towards an integrated approach was made by entomologists attempting to manage insect pests. The concept arose largely from problems with insecticides, including the rapid development of resistance, and a growing awareness of the environmental hazards associated with their use. Also, it was recognized that chemicals used to control pests often affected non-target species, including natural enemies of the pest. Hence the emphasis shifted to a more ecological approach, in which insecticides are used sparingly, and the natural constraints on pest populations enhanced rather than suppressed. Integrated pest management (IPM) views the crop and its environs as a single system, and incorpo-

rates information on pest ecology and behaviour, as well as natural predators, parasites and other factors regulating pest populations. It also defines action thresholds at which pesticides might need to be strategically deployed. Finally, an IPM system must take account of economic and sociological factors, so that it is sustainable within the overall agricultural context.

Integrated control of plant disease

Until recently there was less emphasis on integrated systems to control plant pathogens than in pest management, for a number of reasons. Initially at least, there were fewer problems with fungicides than with insecticides. Host genetic resistance to pathogens was a relatively successful strategy, albeit short-lived in some cases. Also, there were fewer obvious biological options for controlling pathogens, such as directly acting predators. Finally, it should be noted that many of the traditional practices for limiting disease losses, such as crop rotation, already included an element of integrated control.

Current pressures to adopt a more carefully coordinated approach to disease management derive as much from the need to reduce input costs and environmental impacts, as from the shortcomings of any individual control strategy. Again this is part of the trend towards more sustainable crop production systems. In subsistence agriculture the need for low-cost, integrated approaches to disease control has always been pre-eminent. To date there are few examples of fully integrated systems which combine cultural, chemical, biological and host genetic factors to control plant pathogens. Nevertheless, progress is being made by putting together packages of measures which offer significant advantages to growers as well as reducing reliance on chemical agents.

Integration of fungicides with host resistance

One salutary fact in plant pathology is that some well-known historical problems are still with us today. The infamous potato late blight pathogen, *Phytophthora infestans* (Plate 1, facing p. 12), which caused destruction and famine more than 150 years ago, continues to damage potato crops throughout the world. Single-gene, vertical resistance has not provided a lasting solution, and the pathogen has rapidly developed resistance to some of the most effective fungicides used for control of the disease, such as metalaxyl. However, cost-effective control can be achieved by combining several different measures. For example, it has been demonstrated that polygenic, horizontal resistance to late blight can reduce fungicide use by about 0.5 kg ha^{-1} week^{-1}, compared with cultivars possessing no resistance. Further improvements in the efficiency of fungicide use may be achieved if spray treatments are applied according to a forecasting system such as BLITE-CAST (see p. 66). Consequent reductions in the number of spray applications (Table 14.1) not only reduce costs but also limit mechanical damage to the crop and reduce fungicide residues. To summarize, the integration of host horizontal resistance with mixtures of fungicides with different modes of action (e.g. a phenylamide and a dithiocarbamate) applied only when necessary according to a computer-based warning system, can manage late blight while reducing inputs into the crop. Further refinements to the system include improving the resolution of the computer models, so that disease risk can be assessed on a local scale, even down to individual fields. This requires on-farm monitoring of the main environmental parameters regulating disease epidemics.

Control of lettuce downy mildew, *Bremia lactucae*, is also based on the tactical use of host resistance combined with fungicides. In this case, combinations of six different host *R* genes have helped to limit losses, used according to the incidence of pathotypes with corresponding virulence to these genes. The fungicide metalaxyl gives good control of the disease, but as with potato late blight, fungicide-insensitive strains of the pathogen are now common. However, resistance to metalaxyl occurs in some, rather than all pathotypes of *B. lactucae*, and provided the genetic composition of the pathogen population is known, the fungicide can be used in combination with host resistance genes known to be effective against metalaxyl-resistant pathotypes. The success of this strategy depends on careful monitoring of the pathogen population from season to season, to detect any changes in virulence or fungicide sensitivity which might increase the vulnerability of the lettuce cultivars planted.

A further extension of this strategy has been recommended for the management of powdery mildew disease in cereals (see p. 227). Careful choice of crop genotypes for use in mixtures which are regularly changed should avoid major shifts to virulence in the pathogen, and hence prevent the disease reaching epidemic levels. Occasional use of fungicides to reduce the size or structure of the pathogen population may be necessary, but the total input of chemicals in such a system is much less than in a conventional monoculture.

Integration of cultural and biological measures

Many *Phytophthora* species are soil-borne pathogens which cause root rot and crown rot diseases of important crops such as soybean, peppers, citrus (Plate 2,

Year	Criterion for spray application	No. of fungicide applications	Final amount of foliar blight
1973	7-day intervals	10	0.3
	BLITECAST forecast	6	0.2
1974	7-day intervals	9	<0.1
	BLITECAST forecast	5	<0.1
	Unsprayed	0	89.0

Table 14.1 Control of late blight of potato in small field plots sprayed with a fungicide at either weekly intervals or according to a disease forecasting system. (Data from Fry & Thurston 1980.)

In both years differences between the final amount of blight with weekly and BLITECAST spray regimes were not significantly different.

facing p. 12) and avocado. *Phytophthora* root rot (PRR) caused by *P. cinnamomi* occurs in avocado (*Persea americana*) groves in many parts of the world. In plantations established on shallow desert soils in southern California, the disease can result in a severe decline, with many trees being killed outright. As PRR occurs in more than 60% of groves in this region, it poses a serious threat to avocado production.

The initial response to the problem of PRR was to expand new plantings onto virgin land free from the pathogen. However, this provided only a short-term solution, and current attempts to manage the disease are based on a series of different measures (Fig. 14.1). The avocado cultivars originally planted are highly susceptible to PRR, and hence there has been a systematic programme to find more resistant rootstocks, by screening germplasm from the centre of genetic diversity of *Persea* in Central America. Several new sources have been discovered, including wild species such as *P. schiedeana*. Rootstocks developed from such material do not completely prevent infection but reduce the severity of symptoms caused by PRR.

Several routine precautions and cultural measures can also limit the spread and effects of this disease. Rigorous hygiene in the nurseries supplying seedlings, combined with systematic checks and quarantine measures, can prevent distribution of the disease to new areas. It has also been shown that careful management of the soil and water regime in avocado groves can reduce infection. It is well known that many soil-borne *Phytophthora* species are favoured by irrigation (Fig. 14.2), which encourages the production and dispersal of zoospores. Planting trees on mounds, which aids drainage, together with precise regulation of irrigation to prevent waterlogging, is beneficial. Most of the soils used for avocado production are relatively low in organic matter, and mulching or manuring can improve soil structure and may also boost populations of potentially antagonistic soil microorganisms. Certain soils, for instance in Australia, have been shown to be naturally suppressive to *P. cinnamomi*, and such biological control might be supplemented by use of appropriate microbial antagonists capable of colonizing the rhizosphere. Finally, at least two classes of fungicides, the

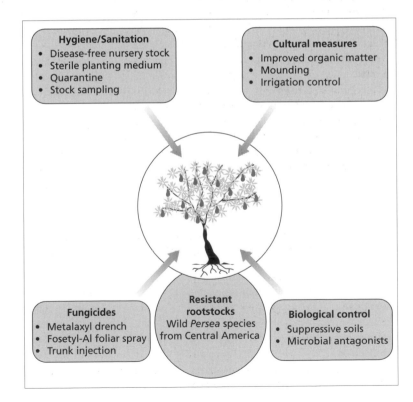

Fig. 14.1 Strategies for integrated control of *Phytophthora* root rot of avocado (*Persea americana*).

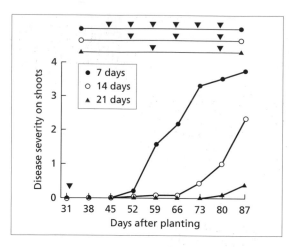

Fig. 14.2 Effect of irrigation on severity of root rot of squash (*Cucurbito pepo*) caused by *Phytophthora capsici*. Plots with infested soil were irrigated at 7-, 14- or 21-day intervals. The arrow on the *x*-axis indicates day when soil was inoculated, and the arrows above the graph indicate the times of different irrigations. (After Café-Filho *et al.* 1995.)

phenylamides and phosphonates, are active against *P. cinnamomi* and have been used to control PRR in the field. Metalaxyl is usually applied as a soil drench, but Fosetyl-Al, being phloem-mobile (see p. 199) can be used as a foliar spray or introduced into individual trees by trunk injection (see Fig. 11.6). The latter method is very efficient, with two injections a year giving good protection of large trees. The various strategies for integrated control of PRR in avocado are summarized in Fig. 14.1.

Good crop hygiene and cultural practices combined with host resistance and the strategic use of chemicals have also been employed to control the bacterial disease citrus canker, caused by *Xanthomonas campestris* pv. *citri*. The stringent quarantine measures applied to prevent introduction of this pathogen have been described earlier (see p. 85). In the south of Brazil the disease was introduced in 1957 and attempts to eradicate it proved unsuccessful. Integrated management is possible, however, as the pathogen does not survive for long in soil or on non-host plants. Current recommendations are to plant citrus cultivars which possess a degree of resistance, in farms which are free from the disease, or in which eradication measures have been applied at least one year previously. The health of nursery stock is carefully monitored to ensure that diseased plants are not

distributed. These hygiene measures are combined with the use of windbreaks to reduce dispersal of the bacterium, and applications of copper-based bactericides to trees each year during the new flush of growth.

Taking decisions

Practical crop protection requires the grower to take decisions such as which cultivar to plant, and when or whether to apply agrochemicals or some other control measure. In situations where an obviously destructive disease outbreak is occurring, such as late blight in a potato crop, the decision is usually simple. But for many diseases the critical factors may not be obvious, and the final outcome may be far from clear. There is a need for additional information on which to base a course of action. Improving the accuracy of decision-taking is therefore an important aspect of integrated disease control. Hence the current interest in the development of **decision support systems** to help growers make the correct choices for pest and disease management. Such systems derive from long-established principles of disease forecasting (see p. 64), but differ in the scale and complexity of data they can handle, and the active participation of growers in the decision process. An appropriate analogy is with financial markets, where investment decisions are taken based on share indices and predictions of future events. Furthermore, as field crops are exposed to simultaneous attack by several different pathogens or pests, any management system must take these different threats into account within an integrated scheme of control measures.

The basic components of a typical decision support system are shown in Fig. 14.3. While each system will vary in its design and specific purpose, most combine information from databases with simulation models and expert systems. The central resource is usually a database which holds very large sets of information on past patterns of disease incidence, climatic conditions, details of crop development and physiology, cultivar resistance, and responses to different chemical treatments. The simulation models carry out calculations of particular processes, such as epidemic development or the multiplication of disease vectors, based on existing data and any new information input by the user. As such estimates integrate many factors and handle large amounts of data they require the speed and power of modern computers to func-

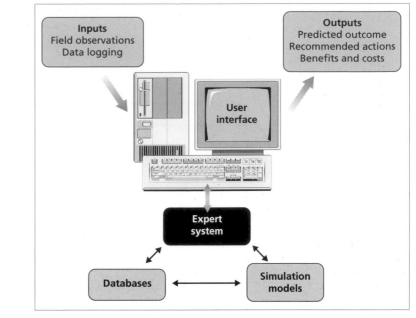

Fig. 14.3 Components of a typical decision support system.

tion effectively. The expert system acts as an interactive interface which requests information from the user, and then outputs estimates of disease risk, as well as the options for disease control. Unlike simple forecasting models, decision support systems also predict the likely consequences of different courses of action, and hence allow users to compare a series of different scenarios.

At present most decision support systems are at an experimental stage, based on incomplete datasets and relatively simple process models. Nevertheless, practical implementation of these systems is going ahead in Europe and the USA, with consequent reductions in pesticide use. Figure 14.4 shows the relative need for fungicide treatments to control three foliar diseases of wheat in Denmark during the 1991 and 1992 seasons, based on estimates of disease risk. In 1991 there was little need for control of yellow stripe rust, a fairly constant risk of mildew, and a high risk of *Septoria*. In the following season there was little risk of damage from any of these diseases, and any fungicide applications would have been a wasted investment. Such systems are under constant refinement to improve the accuracy of disease risk estimates, and also to provide more precise, site-specific recommendations, which take account of local variations in climate, topography, soil and so on. With improved access to personal computers (PCs) and computer

networks, growers will be able to access control advice simply by logging on to the appropriate system. Ultimately these systems may become fully interactive, so that local observations on, for instance, disease incidence or the occurrence of vectors, can be fed in, with continuous updating of the database.

The future of disease control

At the dawn of a new millennium, predicting future events has become a popular pastime. Developments in plant pathology — which remains an applied science, concerned with seeking practical solutions to disease problems — can only be realistically assessed within the broader context of future trends in crop production. The increasingly competitive global trade in plant products, with less market protection, is likely to affect the economics of production, with reduced subsidies and consequent pressure to reduce the level of inputs into crops. Hence the current surpluses of some commodities will disappear, and there will be an increasing emphasis on the efficient use of resources, and the development and use of sustainable production systems. More efficient management of weeds, pests and pathogens will be a vital part of this process. Routine use of pesticides is likely to be replaced by a more targeted approach, linked to

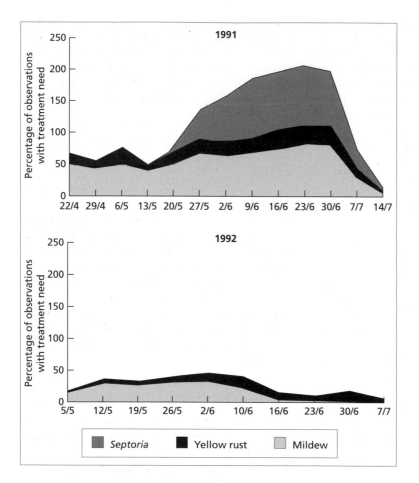

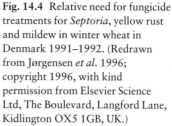

Fig. 14.4 Relative need for fungicide treatments for *Septoria*, yellow rust and mildew in winter wheat in Denmark 1991–1992. (Redrawn from Jørgensen *et al.* 1996; copyright 1996, with kind permission from Elsevier Science Ltd, The Boulevard, Langford Lane, Kidlington OX5 1GB, UK.)

improved assessments of disease risk. More accurate detection and diagnosis of disease problems, especially in the early stages of epidemic development, will be necessary to enable this change. The very rapid progress being made in the field of molecular diagnostics suggests that the main difficulty in practice will be in devising appropriate sampling procedures for field-grown crops, rather than any limitations in the technology itself. The shift to a more strategic use of agrochemicals linked to diagnosis of a developing problem will bring the chemical control of plant disease closer to clinical medicine, where the emphasis has always been on curative measures rather than routine prophylaxis. Consumer pressure is likely to contribute to this change, with the detection of pesticide residues being included in overall assessments of crop quality.

The impact of biotechnology

The rapid developments in plant biotechnology will have an increasing impact on agricultural practice. The first successful genetic engineering of a crop plant, tobacco, was only reported in 1983, but since then all of the major crop species have been stably transformed, including initially difficult groups such as the cereals and woody perennials. The ability to engineer precisely a series of plant characters, including resistance to pests and diseases, will have a major influence on the range of crop genotypes grown, provided consumer concerns about the acceptability of transgenic crops can be resolved. These developments will affect not only the production of crops for traditional uses, such as food and fibres, but also create novel applications, with 'designer' crops serving as renewable sources of high-value chemicals, pharmaceuticals and vaccines. Alternative crops for industry, or for use as renewable sources of energy to replace fossil fuels, are already being widely planted, and will no doubt bring new disease and pest problems. At the same time, improvements in techniques for the selection of agronomically important traits, based on molecular markers, along with an exponential increase in our understanding of plant genome structure and organization, will change the face of plant breeding. Within the next few years the entire sequence of the *Arabidopsis* genome will be known, and the information gained is already being applied to the genetic analysis of major crop species.

What are the likely implications of this revolution for practical plant pathology? Firstly, it will lead to novel crop genotypes being deployed on a large scale in the field. It can be argued that this is merely an extension of normal agricultural practice, whereby elite cultivars are regularly superseded by improved types. Transgenic crops will, however, contain novel combinations of genes, some of them transferred across wide taxonomic barriers. How pathogen populations will adapt to these genetic changes is difficult to predict, but past experience suggests that the capacity of microorganisms for rapid evolution to exploit new niches will ensure that disease remains a threat. The potentially widespread use of virus gene sequences to confer protection against virus infection in transgenic crops (see p. 229) has already raised concerns as to whether such plants might serve as sites for the evolution of novel virus strains through recombination events. Careful risk assessment will be a vital part of any programme to deploy engineered crops on a large scale. Secondly, fundamental breakthroughs in understanding the molecular basis of host–pathogen recognition, along with genetic dissection of the signal and response pathways, will provide new options for manipulating crop resistance. The ability to clone *R* genes from a wide range of sources, and to introduce them directly into commercial genotypes, will accelerate the breeding cycle, but doubts remain over the potential durability of such novel resistance. The challenge here is to devise genetic puzzles for the pathogen which will provide long-term protection of the crop in the field. Altering the specificity of the recognition step itself is now a possibility, through protein engineering to design receptors which bind alternative ligands. Hence one might envisage introducing *R* genes tailored to recognize some essential part of the pathogen, such as a cell wall or membrane component. Engineering combinations of genes with quite different modes of action is a further option, so that the pathogen would need to undergo several mutations to counter the novel resistance.

The accelerating pace of biotechnology will also lead to innovations in the discovery and development of agrochemicals, based on 'smart' screens designed to identify compounds active against known molecular targets. Vast numbers of compounds can already be processed, utilizing libraries of chemicals generated by new methods of combinatorial synthesis. This search for new activity should also explore the biological interactions occurring in natural ecosystems, to identify new signal molecules and analyse the chemical basis of biological control mechanisms. Such an approach, which is part of the wider imperative to catalogue and conserve global biodiversity, has already led to the discovery of promising new classes of compounds active against insect pests.

The impact of information technology

The second revolution affecting all aspects of our lives, including agriculture, is, of course, information technology. The continual introduction of more powerful microprocessors linked to computer networks is rapidly changing the way information can be handled and disseminated, with corresponding changes in the nature and organization of commerce. The implications for the farmer are wide-ranging, from improved access to information on disease identification and control, to the application of complex crop models which will aid decision-taking by integrating numerous factors to predict future events. The ability to automatically log climate and other environmental data within crops will lead to more accurate predictions of disease risk. At the same time the remarkable resolution now provided by networks of global positioning satellites, which can pinpoint locations on the Earth's surface to within a metre, can be used to monitor variations in crop performance, and to improve the targeting of inputs such as fertilizers or other agrochemicals. This is part of the trend towards 'precision agriculture', which aims to improve efficiency by placing crop treatments exactly where they are needed. Developments in imaging will also provide remote-sensing technology to track the progress of disease outbreaks, or to monitor the movement of pests. These advances will greatly increase the amount and accuracy of information which is instantly available to growers, and should lead to improvements in crop management, including more efficient control of weeds, pests and diseases.

The technology gap

In all the excitement generated by these unprecedented technological advances we must not lose sight of the fact that for many crops, in large regions of the world, productivity is poor and still limited by problems of knowledge and resources. This dilemma is exacerbated by the continual growth in human populations in such regions. Problems of deforestation and soil erosion also affect production. Here the real challenge is to boost the yield and reliability of staple crops, through improvements in agronomy and a better understanding of the main constraints on productivity. The scale of losses due to pests and pathogens remains unacceptably high, especially if post-harvest losses are included. For some tropical crops, the principal disease problems are poorly defined, and basic information on pathogen epidemiology is still lacking. In this situation, improved diagnosis and advice on low-cost control measures is required. But even if improved productivity can keep pace with population growth, there are other uncertainties, such as the possible impact of global climate change, which may lead to wide fluctuations in rainfall, and additional stresses on crops. Alterations in global climate patterns will affect the distribution of pathogens and their vectors, and hence the incidence and severity of disease. The predicted outcome of such change is largely a matter of speculation. But one thing is certain: new problems will continue to arise, and the plant pathologist will still have an important role to play in limiting the threat posed by plant disease.

Further reading

Book

Norton, G.A. & Mumford, J.D. (eds) (1993) *Decision Tools for Pest Management*. CAB International, Wallingford.

Reviews and papers

Café-Filho, A.C., Duniway, J.M., & Davis, R.M. (1995) Effects of the frequency of furrow irrigation on root and fruit rots of squash caused by *Phytophthora capsici*. *Plant Disease* **79**, 44–48.

Coffey, M.D. (1987) Phytophthora root rot of avocado. An integrated approach to control in California. *Plant Disease* **71**, 1046–1052.

Crute, I.R. (1992) The role of resistance breeding in the integrated control of downy mildew (*Bremia lactucae*) in protected lettuce. *Euphytica* **63**, 95–102.

Cu, R.M. & Line, R.F. (1994) An expert advisory system for wheat disease management. *Plant Disease* **78**, 209–215.

Fry, W.E. (1977) Integrated control of potato late blight—effects of polygenic resistance and techniques of timing fungicide applications. *Phytopathology* **67**, 415–420.

Gliessman, S.R. (1995) Sustainable agriculture: an agroecological perspective. *Advances in Plant Pathology* **11**, 45–57.

Jørgensen, L.N., Secher, B.J.M. & Nielsen, G.C. (1996) Monitoring disease of winter wheat on both a field and a national level in Denmark. *Crop Protection* **15**, 383–390.

Knight, J.D. & Mumford, J.D. (1994) Decision-support systems in crop protection. *Outlook on Agriculture* **23**, 281–285.

Leite, R.P. & Mohan, S.K. (1990) Integrated management of the citrus bacterial canker disease caused by *Xanthomonas campestris* pv. *citri* in the state of Parana, Brazil. *Crop Protection* **9**, 3–7.

Paveley, N.D., Lockley, K.D., Sylvester-Bradley, R. & Thomas, J. (1997) Determinants of fungicide spray decisions for wheat. *Pesticide Science* **49**, 379–388.

Stevenson, W.R. (1993) IPM for potatoes: A multifaceted approach to disease management and information delivery. *Plant Disease* **77**, 309–311.

Van Bruggen, A.H.C. (1995) Plant disease severity in high-input compared to reduced-input and organic farming systems. *Plant Disease* **79**, 976–983.

Appendix
Annotated List of Pathogens
and the Diseases they Cause

Main diseases considered in the text. These are arranged according to the type of pathogen, and within each group on an alphabetic basis. Footnote symbols in notes indicate the reference text(s) where further information can be found (see end of Appendix).

Pathogen	Host	Disease	Notes
Fungi			
Albugo candida	Brassicas	White blister	Obligate pathogen occurring on many crucifer species§
Alternaria solani	Potato and tomato	Early blight	Common in warmer countries‡§
Armillaria mellea	Trees	Butt rot and root rot	Common in plantations and gardens*†§
Bipolaris maydis (Helminthosporium maydis, Cochliobolus heterostrophus)	Maize	Leaf blight	Cause of southern corn leaf blight, now controlled by resistant varieties*§
Botrytis cinerea	Vegetables and vines	Grey mould	Opportunistic pathogen of a wide range of fruits and vegetables. Control by fungicides but resistance is common*‡§
Botrytis fabae	Broad beans (Vicia)	Chocolate spot	Aggressive pathogen in wet conditions‡§
Bremia lactucae	Lettuce	Downy mildew	Important disease controlled by host resistance and fungicides†§
Ceratocystis fagacearum	Oak	Wilt	Spreads by root grafts and beetle vectors§
Cercospora nicotianae	Tobacco	Frog-eye	Favoured by warm weather§
Cladosporium fulvum	Tomato	Leaf mould	Occurs as many races. Well-defined gene-for-gene system§
Claviceps purpurea	Cereals and grasses	Ergot	Replaces grain with black sclerotia containing alkaloids†‡§
Cochliobolus (Helminthosporium) carbonum	Maize	Leaf spot	Worldwide occurrence, rarely serious
Cochliobolus (Helminthosporium) victoriae	Oats	Victoria blight	Susceptible cultivars sensitive to host-specific toxin§
Colletotrichum circinans	Onions	Smudge disease	Attacks white-skinned varieties*§

Pathogen	Host	Disease	Notes
Colletotrichum gloeosporioides	Fruits and plantation crops	Post-harvest rot; anthracnose of leaves	Common in tropics*§
Colletotrichum lindemuthianum	Bean	Anthracnose	Occurs as races differentiated on bean cultivars§
Crinipellis perniciosa	Cocoa	Witch's broom	Damaging disease in South America
Cronartium ribicola	Pines	White blister rust	Alternate host is *Ribes* spp.*
Cryphonectria parasitica	Chestnut	Chestnut blight	Causes cankers and death of trees. Serious in North America and Europe*§
Dothidella ulei	Rubber	Leaf blight	Endemic in South America, but not in Malaysia, from which it is excluded by quarantine
Epichloe typhina	Grasses	Choke disease	Systemic infection which inhibits flowering§
Erysiphe graminis	Cereals and grasses	Powdery mildew	Prevalent in Europe, control by host resistance and systemic fungicides‡§
Fomes lignosus	Trees, especially rubber	White root disease	Pentachloronitrobenzene used to arrest growth of rhizomorphs on roots
Fusarium culmorum	Cereals	Foot rot	One of a complex of pathogens which infect the stem base‡§
Fusarium oxysporum	Many hosts	Vascular wilt	Soil-borne pathogen occurring as distinct form species infecting different hosts*§
For example, *F.o.* f.sp. *cubense*	Banana	Panama disease	Destructive disease now controlled by host resistance†
Fusarium solani f.sp. *pisi* (*Nectria haematococca*)	Pea	Foot rot and root rot	Other form species attack beans and other legumes§
Fusicoccum amygdali	Almond and peach	Canker	Produces a toxin causing excessive transpiration§
Gaeumannomyces graminis	Cereals	Take-all	Soil-borne pathogen infecting roots‡§
Gibberella fujikuroi	Rice	Foolish seedling or Bakanae	Seed- or soil-borne disease*§
Helminthosporium sacchari	Sugarcane	Eyespot	Produces host-specific toxin
Hemileia vastatrix	Coffee	Leaf rust	Most destructive disease of , coffee, now worldwide*†
Heterobasidion annosum	Conifers	Root rot and butt rot	Common in plantations, controlled using antagonistic fungus§
Magnaporthe grisea	Rice	Rice blast	Major disease of rice, control by resistant varieties and fungicides*†
Melampsora lini	Flax	Rust	Widespread disease, best known as basis of gene-for-gene theory§
Microsphaera alphitoides	Oak	Powdery mildew	Can be serious on young trees in nurseries§
Monilinia fructicola	Apple and plum	Brown rot	Post-harvest disease of fruit*§
Mycosphaerella fijiensis	Bananas	Black Sigatoka	Variant of Sigatoka (see below)

Pathogen	Host	Disease	Notes
Mycosphaerella musicola	Bananas	Sigatoka disease	Infects leaves, causing fruit loss*
Nectria galligena	Apples and pears	Canker	Sanitation important, plus fungicides and biocontrol*§
Nectria haematococca	Peas	Foot rot and root rot	See *Fusarium solani* f.sp. *pisi*
Ophiostoma novo-ulmi	Elm	Dutch elm	New aggressive strain responsible for pandemic in northern hemisphere*§
Penicillium digitatum	Citrus fruit	Green mould	Post-harvest rot, infects via wounds*§
Penicillium expansum	Apples and pears	Blue mould	Post-harvest soft rot*§
Penicillium italicum	Citrus fruit	Blue mould	Post-harvest rot occurs with green mould (above)*§
Periconia circinata	Sorghum	Milo disease	Soil-borne pathogen producing toxin
Peronospora parasitica	Brassica	Downy mildew	Occurs on many crucifers, including *Arabidopsis*‡§
Peronospora pisi (P. viciae)	Pea and other legumes	Downy mildew	Transmitted by seed- and soil-borne oospores§
Peronospora tabacina	Tobacco	Blue mould	Destructive disease controlled by host resistance and fungicides†§
Phytophthora cinnamomi	Trees and shrubs	Root rot	Many woody hosts attacked, including avocado and ornamental species. Also causes Jarrah die-back, a devastating disease of native Australian forests†§
Phytophthora erythroseptica	Potato	Pink rot	Occurs in waterlogged soils‡
Phytophthora infestans	Potato	Late blight	Affects haulm and tubers, control mainly by fungicide sprays based on disease forecasts*†‡§
Phytophthora megasperma	Vegetable crops and trees	Root rot	Wide host range, but host-adapted forms occur, e.g. *P.m.* var. *sojae* on soybean*†§
Phytophthora palmivora	Cocoa	Black pod	Spreads by zoospores*
Phytophthora parasitica	Tobacco	Black shank	Infects roots and stem base
Plasmodiophora brassicae	Brassica	Club root	Causes galls on roots. Persists in soil for long periods*‡§
Plasmopara viticola	Vines	Downy mildew	European grape varieties are most susceptible, control by fungicides*†§
Polymyxa graminis	Cereals		Chytrid fungi which infect host roots. Unimportant as pathogens but vectors of serious virus diseases, e.g. rhizomania§
P. betae	Sugarbeet		
Pseudocercosporella (Tapesia) herpotrichoides	Wheat and other cereals	Eyespot	Infects stem base, causing lodging. Control with fungicides†‡§
Pseudoperonospora cubensis	Cucurbits	Downy mildew	Particularly important in glasshouse crops. Control by fungicides§

Pathogen	Host	Disease	Notes
Pseudoperonospora humuli	Hops	Downy mildew	Outbreaks in wet conditions, control by resistant cultivars and fungicides§
Puccinia graminis	Cereals	Black stem rust	Form species on different hosts, e.g. f.sp. *tritici* (wheat) and f.sp. *hordei* (barley)*†‡§
Puccinia recondita	Wheat and other cereals	Brown rust/leaf rust	Mainly controlled by host resistance‡§
Puccinia striiformis	Cereals	Yellow stripe rust	Sporadic occurrence, usually in cool conditions. Has recently spread to South America and Australasia†‡§
Pyrenophora avenae	Oats	Seedling blight, leaf stripe	Seed-borne disease effectively controlled by seed-dressing with fungicides§
Pyrenophora teres	Barley	Netblotch	Seed-borne disease, occurs as net and spot forms*‡§
Pyrenophora tritici-repentis	Wheat	Tan spot	Survives on stubble, infection from airborne ascospores*§
Pythium species	Numerous	Damping-off	Seedling diseases prevalent in wet soils*‡§
Rhizoctonia cerealis	Cereals	Sharp eyespot	Survives on infected stubble, part of stem-base disease complex‡§
Rhizoctonia solani	Potato	Stem canker	Also affects tubers, causing black scurf‡§
Rhynchosporium secalis	Barley	Leaf blotch/scald	Splash-dispersed pathogen prevalent in wet conditions‡§
Sclerotium cepivorum	Onion and garlic	White rot	Survives in soil as resistant sclerotia*§
Sclerotium rolfsii	Numerous	Root rot and collar rot	Very wide host range, especially on vegetable crops in warm climates*§
Septoria lycopersici	Tomato and potato	Leaf spot	Potato form restricted to South America§
Septoria (Stagonospora) nodorum	Wheat	Leaf blotch and glume blotch	Late-season infections attack ears, causing severe lesions‡§
Septoria tritici	Wheat	Leat blotch	Occurs worldwide. Has increased in importance in recent years‡§
Sphaerotheca fuliginea	Cucurbits	Powdery mildew	Important in glasshouses§
Taphrina deformans	Peach	Leaf curl	Can also infect fruits; control by fungicide spray in autumn or spring before bud break*§
Thielaviopsis basicola	Cotton and tobacco	Black root rot	Soil-borne pathogen with wide host range, including vegetables and ornamentals§
Tilletia indica	Wheat	Karnal bunt	Originally in Asia, has now spread to Mexico*§
Uromyces appendiculatus	Bean	Rust	One of a number of *Uromyces* species causing rust on legumes§

Pathogen	Host	Disease	Notes
Ustilago maydis	Maize	Corn smut	Forms galls on aerial parts of plant, especially on ears*§
Ustilago nuda	Barley	Loose smut	Controlled by treating seeds with fungicides*‡§
Venturia inaequalis	Apple	Scab	Almost all commercial cultivars are susceptible, hence control relies on fungicides, often applied according to a disease forecast*†§
Verticillium albo-atrum	Hop, alfalfa, tomato and others	Wilt	Soil-borne pathogen, difficult to control§
Verticillium dahliae	Numerous including cotton, potatoes and tree species	Wilt	Produces microsclerotia. More common in warmer regions§
Bacteria			
Agrobacterium tumefaciens	Numerous dicotyledons	Crown gall	Plant cancer, worldwide distribution, can be controlled by bacterial antagonist*§
Agrobacterium rhizogenes	Numerous dicotyledons	Hairy root	Closely related to crown gall§
Clavibacter insidiosum	Alfalfa	Bacterial wilt	Important in North America; control by resistant cultivars*§
Erwinia amylovora	Pears and apples	Fireblight	Control by sanitation and copper sprays*†§
Erwinia carotovora pv. *atroseptica*	Potato	Blackleg	Attacks stem base and tubers causing necrosis*§
Erwinia carotovora pv. *carotovora*	Numerous vegetables	Soft rot	Often a post-harvest pathogen; control by sanitation and reducing humidity*§
Erwinia tracheiphila	Cucurbits	Bacterial wilt	Spread by insect vectors; may by controlled by insecticides*
Pseudomonas syringae	Numerous	Diverse	Important pathogen existing as numerous pathovars adapted to different hosts
P. syringae pv. *phaseolicola*	Bean	Halo blight	Seed-borne disease spread by rain-splash§
P. syringae pv. *savastanoi*	Olives, oleander and privet	Knot and canker	Enters by pruning wounds, leaf scars and wounds. Control by sanitation and copper sprays§
P. syringae pv. *tabaci*	Tobacco	Wildfire	Control using resistant cultivars, seed disinfection, sanitation and copper or streptomycin sprays*§
Ralstonia solanacearum	Banana, tomato, potato, tobacco	Bacterial wilt Moko disease (banana)	Mainly important in the tropics; control by resistant varieties*§
Streptomyces scabies	Potato	Common scab	Favoured by low soil moisture during tuber development and high pH*‡§

Pathogen	Host	Disease	Notes
Xanthomonas campestris	Numerous	Diverse	Occurs as pathovars adapted on different hosts
Xanthomonas campestris pv. *citri*	Citrus	Canker	Destructive disease usually controlled by quarantine and eradication measures*†
Xanthomonas campestris pv. *manihotis*	Cassava	Bacterial blight	Occurs in Asia, Africa and Latin America. Often transmitted by cuttings. Control by cultural measures, sanitation and host resistance
Xanthomonas campestris pv. *oryzae*	Rice	Leaf blight	Important in Far East†
Xanthomonas campestris pv. *vesicatoria*	Tomato and pepper	Bacterial spot	Control by sanitation, seed certification, cultural measures and copper sprays*§
Xylella fastidiosa	Citrus	Citrus variegated chlorosis	Limited to xylem tissues, spread by vegetative propagation and probably insect vectors*

Mycoplasma-like organisms (MLOs)

Pathogen	Host	Disease	Notes
Phytoplasma sp.	Coconut	Lethal yellowing	High mortality; control by resistant varieties and hybrids*
Phytoplasma sp.	Maize	Corn stunt	Transmitted by leaf-hopper vectors, controlled by resistant cultivars*
Spiroplasma citri	Citrus	Stubborn disease	Occurs in phloem tissues, spread by leaf-hopper vectors*§

Viruses

Pathogen	Host	Disease	Notes
African cassava mosaic (ACMV)	Cassava	Mosaic	Widespread in Africa, transmitted by the whitefly *Bemisia tabaci**
Barley yellow dwarf (BYDV)	Cereals	Yellow dwarf	Spread by aphid vectors†‡§¶
Barley yellow mosaic (BaYMV)	Barley	Yellow mosaic	Spread by the soil-borne fungus *Polymyxa graminis*‡§
Barley mild mosaic (BaMMV)	Barley	Mild mosaic	Spread by the fungus *Polymyxa graminis*
Beet necrotic yellow vein (BNYVV)	Sugarbeet	Rhizomania	Spread by another root-invading fungus *Polymyxa betae*‡§
Cocoa swollen shoot (CSSV)	Cocoa	Swollen shoot	Serious in West Africa. Spread by scale-insect vectors; control by eradication*¶
Cucumber mosaic (CMV)	Numerous	Mosaic and stunting	Worldwide distribution. Has the widest host range of any plant virus*§¶
Grapevine fanleaf (GFLV)	Vine	Fanleaf and mosaic	Transmitted by persistent nematode vectors and vegetative propagation; control by cloning virus-free stocks§¶

Pathogen	Host	Disease	Notes
Lettuce big vein (LBV)	Lettuce	Big vein	Transmitted by zoospores of *Olpidium brassicae*, a root-infecting fungus§¶
Potato leaf roll virus (PLRV)	Potato	Leaf roll	Transmitted by aphids in a persistent manner‡§¶
Potato virus X (PVX)	Potato	Mild mosaic, mottle	Often asymptomatic, but may cause severe disease when another virus is present, e.g. potato virus Y§¶
Prunus necrotic ringspot (PNRV)	Stone-fruit trees	Diverse	Transmitted by budding, grafting, seed and pollen. Control by using virus-free nursery stock*§¶
Raspberry bushy dwarf (RBDV)	*Rubus* spp.	Bushy dwarf	Transmitted by seed and pollen; control by using resistant cultivars§¶
Raspberry ringspot (RRV)	Raspberry, tomato, gooseberry and many perennial plants		Nematode and seed transmitted; control involves eradicating nematodes
Tobacco mosaic virus (TMV)	Numerous	Mosaic	Wide host range but important only in Solanaceae. Often spread by contact*§¶
Tobacco stunt virus (TSV)	Tobacco	Stunt	Transmitted by *Olpidium brassicae*, a root-infecting fungus

Viroids

Pathogen	Host	Disease	Notes
Coconut cadang cadang (CCCVd)	Coconut	Trans.: 'dying, dying'	Occurs in Philippines, transmitted mechanically and probably by insects*
Potato spindle tuber viroid (PSTVd)	Potato and tomato	Spindle tuber	Spread by contact, insects and seed. Control by planting viroid-free stock*§

Angiosperms

Pathogen	Host	Disease	Notes
Orobanche spp.	Sunflower, tobacco, tomato and faba bean	Broomrape	Root parasite, seeds survive in infested soil for long periods*
Striga hermonthica	Maize, rice, sorghum, millet and sugarcane	Witchweed	Root parasite, control difficult. Quarantine, sanitation, catch crops and resistant cultivars may be used*

References:
* Agrios, G.N. (1997) *Plant Pathology*, 4th edn. Academic Press, San Diego, Calif.
† Mukhopadhyay, A.N., Singh, U.S., Kumar, J. & Chaube, H.S. (1992) *Plant Diseases of International Importance*. Prentice Hall, Des Moines, Iowa.
‡ Parry, D. (1990) *Plant Pathology in Agriculture*. Cambridge University Press, Cambridge.
§ Smith, I.M., Dunez, J., Lelliott, R.A., Phillips, D.H. & Archer, S.A. (1988) *European Handbook of Plant Diseases*. Blackwell Scientific Publications, Oxford.
¶ Smith, K.M. (1972) *A Textbook of Plant Virus Diseases*. Longman, London.

Index